Astroparticle Physics

Claus Grupen

Astroparticle Physics

With contributions from Glen Cowan, Simon Eidelman, and Tilo Stroh

 Springer

Prof. Dr. Claus Grupen

University of Siegen
Department of Physics
Walter-Flex-Strasse 3
57068 Siegen
Germany
e-mail: grupen@hep.physik.uni-siegen.de

With contributions from:

Dr. Glen Cowan

Royal Holloway, University of London
Physics Department
England
e-mail: G.Cowan@rhul.ac.uk

Dipl. Phys. Tilo Stroh

University of Siegen
Department of Physics
Germany
e-mail: stroh@sirs02.physik.uni-siegen.de

Prof. Dr. Simon Eidelman

Budker Institute of Nuclear Physics
Novosibirsk
Russia
e-mail: Simon.Eidelman@cern.ch

ISBN-13 978-3-642-06455-5   e-ISBN-13 978-3-540-27670-8

Springer is a part of Springer Science+Business Media
springeronline.com
© Springer-Verlag Berlin Heidelberg 2010
Printed in Germany

Cover design: *design & production* GmbH, Heidelberg based on an idea of the author
Images: ArmbrustDesign, Siegen, Germany

Printed on acid-free paper

Preface

"The preface is the most important part of a book. Even reviewers read a preface."

Philip Guedalla

Preface to the English Translation

This book on astroparticle physics is the translation of the book on 'Astroteilchenphysik' published in German by Vieweg, Wiesbaden, in the year 2000. It is not only a translation, however, but also an update. The young field of astroparticle physics is developing so rapidly, in particular with respect to 'new astronomies' such as neutrino astronomy and the detailed measurements of cosmic background radiation, that these new experimental results and also new theoretical insights need to be included.

The details of the creation of the universe are not fully understood yet and it is still not completely clear how the world will end, but recent results from supernovae observations and precise measurement of the primordial blackbody radiation seem to indicate with increasing reliability that we are living in a flat Euclidean universe which expands in an accelerated fashion.

In the last couple of years cosmology has matured from a speculative science to a field of textbook knowledge with precision measurements at the percent level.

The updating process has been advanced mainly by my colleague Dr. Glen Cowan who is lecturing on astroparticle physics at Royal Holloway College, London, and by myself. The chapter on 'Cosmology' has been rewritten, and chapters on 'The Early Universe', 'Big Bang Nucleosynthesis', 'The Cosmic Microwave Background', and 'Inflation' as well as a section on gravitational astronomy have been added. The old chapter on 'Unsolved Problems' was moved into a new chapter on 'Dark Matter', and part of it went into chapters on primary and secondary cosmic rays.

The book has been extended by a large number of problems related to astroparticle physics. Full solutions to all problems are given. To ease the understanding of theoretical aspects and the interpretation of physics data, a mathematical appendix is offered where most of the formulae used are presented and/or derived. In addition, details on the thermodynamics of the early universe have been treated in a separate appendix.

Prof. Dr. Simon Eidelman from the Budker Institute of Nuclear Physics in Novosibirsk and Dipl.Phys. Tilo Stroh have carefully checked the problems and proposed new ones. Dr. Ralph Kretschmer contributed some interesting and very intricate problems. I have also received many comments from my colleagues and students in Siegen.

The technical aspects of producing the English version lay in the hands of Ms. Ute Smolik, Lisa Hoppe, and Ms. Angelika Wied (text), Dipl.Phys. Stefan Armbrust (updated the figures), Dr. Glen Cowan and Ross Richardson (polished my own English translation), and M.Sc. Mehmet T. Kurt (helped with the editing). The final appearance of the book including many comments on the text, the figures, and the layout was accomplished by Dipl.Phys. Tilo Stroh and M.Sc. Nadir Omar Hashim.

Without the help of these people, it would have been impossible for me to complete the translation in any reasonable time, if at all. In particular, I would like to thank my colleague Prof. Dr. Torsten Fließbach, an expert on Einstein's theory of general relativity, for his critical assessment of the chapter on cosmology and for proposing significant improvements. Also the contributions by Dr. Glen Cowan on the new insights into the evolution of the early universe and related subjects are highly appreciated. Dr. Cowan has really added essential ingredients with the last chapters of the book. Finally, Prof. Dr. Simon Eidelman, Dr. Armin Böhrer, and Dipl.Phys. Tilo Stroh read the manuscript with great care and made invaluable comments. I thank all my friends for their help in creating this English version of my book.

Siegen, February 2005   *Claus Grupen*

Preface to the German Edition

The field of astroparticle physics is not really a new one. Up until 1960, the physics of cosmic rays essentially represented this domain. Elementary particle physics in accelerators has evolved from the study of elementary-particle processes in cosmic radiation. Among others, the first antiparticles (positrons) and the members of the second lepton generation (muons) were discovered in cosmic-ray experiments.

The close relationship between cosmology and particle physics was, however, recognized only relatively recently. Hubble's discovery of the expanding universe indicates that the cosmos originally must have had a very small size. At such primeval times, the universe was a microworld that can only be described by quantum-theoretical methods of elementary particle physics. Today, particle physicists try to recreate the conditions that existed in the early universe by using electron–positron and proton–antiproton collisions at high energies to simulate 'mini Big Bangs'.

The popular theories of elementary particle physics attempt to unify the various types of interactions in the Standard Model. The experimental confirmation of the existence of heavy vector bosons that mediate weak interactions (W^+, W^-, Z^0), and progress in the theoretical understanding of strong interactions seem to indicate that one may be able to understand the development of the universe just after the Big Bang. The high temperatures or energies that existed at the time of the Big Bang will, however, never be reached in earthbound laboratories. This is why a symbiosis of particle physics, astronomy, and cosmology is only too natural. Whether this new field is named astroparticle physics or particle astrophysics is more or less a matter of taste or the background of the author. This book will deal both with astrophysics and elementary particle physics aspects. We will equally discuss the concepts of astrophysics focusing on particles and particle physics using astrophysical methods. The guiding line is physics with astroparticles. This is why I preferred the term astroparticle physics over particle astrophysics.

After a relatively detailed historical introduction (Chap. 1) in which the milestones of astroparticle physics are mentioned, the basics of elementary particle physics (Chap. 2), particle interactions (Chap. 3), and measurement techniques (Chap. 4) are presented. Astronomical aspects prevail in the discussion of acceleration mechanisms (Chap. 5) and primary cosmic rays (Chap. 6). In these fields, new disciplines such as neutrino and gamma-ray astronomy represent a close link to particle physics. This aspect is even more pronounced in the presentation of secondary cosmic rays (Chap. 7). On the one hand, secondary cosmic rays have been a gold mine for discoveries in elementary particle physics. On the other hand, however, they sometimes represent an annoying background in astroparticle observations.

The highlight of astroparticle physics is surely cosmology (Chap. 8) in which the theory of general relativity, which describes the macrocosm, is united with the successes of elementary particle physics. Naturally, not all questions have been answered; therefore a final chapter is devoted to open and unsolved problems in astroparticle physics (Chap. 9).

The book tries to bridge the gap between popular presentations of astroparticle physics and textbooks written for advanced students. The necessary basics from elementary particle physics, quantum physics, and special relativity are carefully introduced and applied, without rigorous derivation from appropriate mathematical treatments. It should be possible to understand the calculations presented with the knowledge of basic A-level mathematics.

On top of that, the basic ideas discussed in this book can be followed without referring to special mathematical derivations.

I owe thanks to many people for their help during the writing of this book. Dr. Armin Böhrer read the manuscript with great care. Ms. Ute Bender and Ms. Angelika Wied wrote the text, and Ms. Claudia Hauke prepared the figures that were finalized by Dipl.Phys. Stefan Armbrust. I owe special thanks to Dr. Klaus Affholderbach and Dipl.Phys. Olaf Krasel who created the computer layout of the whole book in the LATEX style. I am especially indebted to Dipl.Phys. Tilo Stroh for his constant help, not only as far as physics questions are concerned, but in particular for applying the final touch to the manuscript with his inimitable, masterful eye for finding the remaining flaws in the text and the figures. Finally, I owe many thanks to the Vieweg editors, Ms. Christine Haite and Dipl.Math. Wolfgang Schwarz.

Geneva, July 2000

Table of Contents

"The most technologically efficient machine that man has invented is the book."

Northrop Frye

1 Historical Introduction

"Look into the past as guidance for the future."

Robert Jacob Goodkin

The field of astroparticle physics, or particle astrophysics is relatively new. It is therefore not easy to describe the history of this branch of research. The selection of milestones in this book is necessarily subject to a certain arbitrariness and personal taste.

Historically, astroparticle physics is based on optical astronomy. As detector techniques improved, this observational science matured into astrophysics. This research topic involves many subfields of physics, like mechanics and electrodynamics, thermodynamics, plasma physics, nuclear physics, and elementary particle physics, as well as special and general relativity. Precise knowledge of particle physics is necessary to understand many astrophysical contexts, particularly since comparable experimental conditions cannot be prepared in the laboratory. The astrophysical environment therefore constitutes an important laboratory for high energy physicists.

astrophysics as laboratory for high energy physics

The use of the term astroparticle physics is certainly justified, since astronomical objects have been observed in the 'light' of elementary particles. Of course, one could argue that X-ray or gamma-ray astronomy is more closely related to astronomy rather than to astroparticle physics. To be on the safe side, the new term astroparticle physics, should be restricted to 'real' elementary particles. The observations of our Sun in the light of neutrinos in the Homestake Mine (Davis experiment) in 1967, constitutes the birth of astroparticle physics, even though the first measurements of solar neutrinos by this radiochemical experiment were performed without directional correlation. It is only since the Kamiokande[1] experiment of 1987, that one has been able to 'see' the Sun in real time, whilst additionally being able to measure the direction of the emitted neutrinos. Nature was also kind enough to explode a supernova in the Large Mag-

justification of the nomenclature

Davis experiment

Kamiokande experiment

SN 1987A

---

[1] Kamiokande – Kamioka Nucleon Decay Experiment

Fig. 1.1
Crab Nebula {1}

ellanic Cloud in 1987 (SN 1987A), whose neutrino burst could be recorded in the large water Cherenkov detectors of Kamiokande and IMB[2] and in the scintillator experiment at Baksan.

Presently, the fields of gamma and neutrino astronomy are expanding rapidly. Astronomy with charged particles, however, is a different matter. Irregular interstellar and intergalactic magnetic fields randomize the directions of charged cosmic rays. Only particles at very high energies travel along approximately straight lines through magnetic fields. This makes astronomy with charged particles possible, if the intensity of energetic primaries is sufficiently high.

Actually, there are hints that the highest-energy cosmic rays ($> 10^{19}$ eV) have a non-uniform distribution and possibly originate from the supergalactic plane. This plane is an accumulation of galaxies in a disk-like fashion, in a similar way that stars form the Milky Way. Other possible sources, however, are individual galactic nuclei (M87?) at cosmological distances.

The milestones which have contributed to the new discipline of astroparticle physics shall be presented in chronological order. For that purpose, the relevant discoveries in astronomy, cosmic rays, and elementary particle physics will be considered in a well-balanced way. It is, of course, true that this selection is subject to personal bias.

Vela supernova
It is interesting to point out the observations of the Vela supernova by the Sumerians 6000 years ago. This supernova exploded in the constellation Vela at a distance of 1500 light-years. Today the remnant of this explosion is visible, e.g., in **Vela X1** the X-ray and gamma range. Vela X1 is a binary, one component of which is the Vela pulsar. With a rotational period of 89 ms the Vela pulsar is one of the 'slowest' pulsars so far observed in binaries. The naming scheme of X-ray sources is such that Vela X1 denotes the strongest ($\widehat{=}$ 'the first') X-ray source in the constellation Vela.

The second spectacular supernova explosion was observed in China in 1054. The relic of this outburst is the Crab **Crab Nebula** Nebula, whose remnant also emits X rays and gamma rays like Vela X1. Because of its time-independent brightness the Crab is often used as a 'standard candle' in gamma-ray astronomy (Fig. 1.1).

northern lights The observation of the northern lights (Gassendi 1621 **(aurora borealis)** and Halley 1716) as the aurora borealis ('northern dawn')

---

[2] IMB – Irvine Michigan Brookhaven collaboration

lead Mairan, in 1733, to the idea that this phenomenon might be of solar origin. Northern and southern lights are caused by solar electrons and protons incident in the polar regions traveling on helical trajectories along the Earth's magnetic field lines. At high latitudes, the charged particles essentially follow the magnetic field lines. This allows them to penetrate much deeper into the atmosphere, compared to equatorial latitudes where they have to cross the field lines perpendicularly (Fig. 1.2).

It is also worth mentioning that the first correct interpretation of nebulae, as accumulations of stars which form galaxies, was given by a philosopher (Kant 1775) rather than by an astronomer.

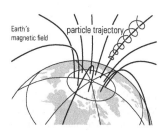

Fig. 1.2
Helical trajectory of an electron in the Earth's magnetic field

1.1 Discoveries in the 20th Century

> *"Astronomy is perhaps the science whose discoveries owe least to chance, in which human understanding appears in its whole magnitude, and through which man can best learn how small he is."*
>
> *Georg Christoph Lichtenberg*

The discovery of X rays (Röntgen 1895, Nobel Prize 1901), radioactivity (Becquerel 1896, Nobel Prize 1903), and the electron (Thomson 1897, Nobel Prize 1906) already indicated a particle physics aspect of astronomy. At the turn of the century Wilson (1900) and Elster & Geitel (1900) were concerned with measuring the remnant conductivity of air. Rutherford realized in 1903 that shielding an electroscope reduced the remnant conductivity (Nobel Prize 1908 for investigations on radioactive elements). It was only natural to assume that the radioactivity of certain ores present in the Earth's crust, as discovered by Becquerel, was responsible for this effect.

astronomy and particle physics

In 1910, Wulf measured a reduced intensity in an electrometer at the top of the Eiffel tower, apparently confirming the terrestrial origin of the ionizing radiation. Measurements by Hess (1911/1912, Nobel Prize 1936) with balloons at altitudes of up to 5 km showed that, in addition to the terrestrial component, there must also be a source of ionizing radiation which becomes stronger with increasing altitude (Figs. 1.3 and 1.4).

cosmic rays

Fig. 1.3
Victor Hess at a balloon ascent for
measuring cosmic radiation {2}

Fig. 1.4
Robert Millikan at a take-off of
balloon experiments in Bismarck,
North Dakota (1938) {3}

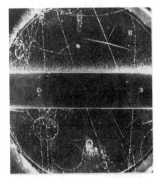

This extraterrestrial component was confirmed by Kohl-
hörster two years later (1914). By developing the cloud
chamber in 1912, Wilson made it possible to detect and follow the tracks left by ionizing particles (Nobel Prize 1927,
Fig. 1.5).

The extraterrestrial cosmic radiation that increases with
altitude ('Höhenstrahlung') has numerous experimental possibilities (Fig. 1.6) and is of special importance to the development of astroparticle physics.

Fig. 1.5
Tracks of cosmic particles in a
cloud chamber {4}

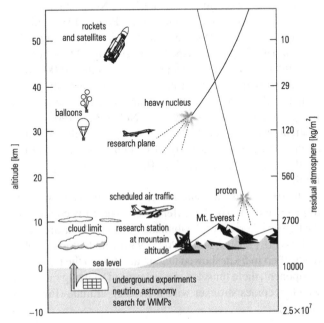

Fig. 1.6
Possibilities for experiments in the
field of cosmic rays

In parallel to these experimental observations, Einstein developed his theories of special and general relativity (1905 and 1916). The theory of *special relativity* is of paramount importance for particle physics, while the prevailing domain of *general relativity* is cosmology. Einstein received the Nobel Prize in 1921 not, however, for his fundamental theories on relativity and gravitation, but for the correct quantum-mechanical interpretation of the photoelectric effect and the explanation of Brownian motion. Obviously the Nobel committee in Stockholm was not aware of the outstanding importance of the theories of relativity or possibly not even sure about the correctness of their predictions. This occurred even though Schwarzschild had already drawn correct conclusions for the existence of black holes as early as 1916, and Eddington had verified the predicted gravitational bending of light passing near the Sun during the solar eclipse in 1919. The experimental observation of the deflection of light in gravitational fields also constituted the discovery of *gravitational lensing*. This is when the image of a star appears to be displaced due to the gravitational lensing of light that passes near a massive object. This effect can also lead to double, multiple, or ring-shaped images of a distant star or galaxy if there is a massive object in the line of sight between the observer on Earth and the star (Fig. 1.7). It was only in 1979 that multiple images of a quasar (double quasar) could be observed. This was followed in 1988 by an Einstein ring in a radio galaxy, as predicted by Einstein in 1936.

theories of relativity

black holes

gravitational lensing

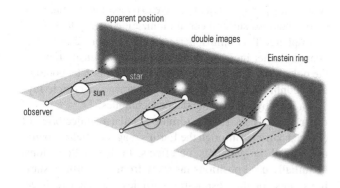

apparent position

double images

Einstein ring

star

sun

observer

Fig. 1.7
Gravitational lensing by a massive object:
a) deflection of light,
b) double images,
c) Einstein ring

In the field of astronomy, stars are classified according to their brightness and colour of the spectrum (Hertzsprung–Russell diagram 1911). This scheme allowed a better understanding of the stellar evolution of main-sequence stars to

stellar evolution

Hertzsprung–Russell diagram

red giants and white dwarves. In 1924 Hubble was able to confirm Kant's speculation that 'nebulae' are accumulations of stars in galaxies, by resolving individual stars in the Andromeda Nebula. Only a few years later (1929), he observed the redshift of the spectral lines of distant galaxies, thereby demonstrating experimentally that the universe is expanding.

expansion of the universe

In the meantime, a clearer picture about the nature of cosmic rays had emerged. Using new detector techniques in 1926, Hoffmann observed particle multiplication under absorbing layers ('Hoffmann's collisions'). In 1927, Clay demonstrated the dependence of the cosmic-ray intensity on the geomagnetic latitude. This was a clear indication of the charged-particle nature of cosmic rays, since photons would not have been influenced by the Earth's magnetic field.

penetration of cosmic rays into the atmosphere

Primary cosmic rays can penetrate deep into the atmosphere at the Earth's poles, by traveling parallel to the magnetic field lines. At the equator they would feel the full component of the Lorentz force ($F = e(v \times B)$; F – Lorentz force, v – velocity of the cosmic-ray particle, B – Earth's magnetic field, e – elementary charge: at the poles $v \parallel B$ holds with the consequence of $F = 0$, while at the equator one has $v \perp B$ which leads to $|F| = e\, v\, B$). This *latitude effect* was controversial at the time, because expeditions starting from medium latitudes ($\approx 50°$ north) to the equator definitely showed this effect, whereas expeditions to the north pole observed no further increase in cosmic-ray intensity. This result could be explained by the fact that charged cosmic-ray particles not only have to overcome the magnetic cutoff, but also suffer a certain ionization energy loss in the atmosphere. This atmospheric cutoff of about 2 GeV prevents a further increase in the cosmic-ray intensity towards the poles (Fig. 1.8). In 1929 Bothe and Kohlhörster could finally confirm the charged-particle character of cosmic rays at sea level by using coincidence techniques.

latitude effect

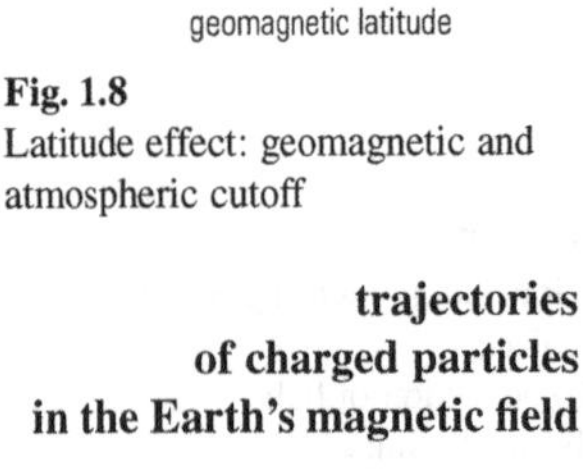

Fig. 1.8
Latitude effect: geomagnetic and atmospheric cutoff

trajectories of charged particles in the Earth's magnetic field

In as early as 1930, Störmer calculated trajectories of charged particles through the Earth's magnetic field to better understand the geomagnetic effects. In these calculations, he initially used positions far away from the Earth as starting points for the cosmic-ray particles. He soon realized, however, that most particles failed to reach sea level due to the action of the magnetic field. The low efficiency of this approach led him to the idea of releasing antiparticles from sea level to discover where the Earth's magnetic field would guide them. In these studies, he observed that particles with certain momenta could be trapped by the magnetic field, which caused them to propagate back and forth from one magnetic pole to the other in a process called 'magnetic mirroring'. The accumulated particles form radiation belts, which were discovered in 1958 by Van Allen with experiments on board the Explorer I satellite (Fig. 1.9, see also Fig. 7.2).

Van Allen belts

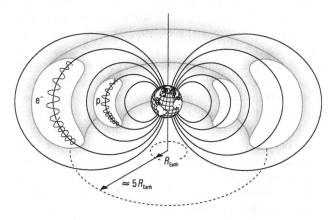

Fig. 1.9
Van Allen belts

The final proof that primary cosmic rays consist predominantly of positively charged particles was established by the observation of the *east–west effect* (Johnson and Alvarez & Compton, Nobel Prize Alvarez 1968, Nobel Prize Compton 1927). Considering the direction of incidence of cosmic-ray particles at the north pole, one finds a higher intensity from the west compared to the east. The origin of this asymmetry relates to the fact that some possible trajectories of positively charged particles from easterly directions do not reach out into space (dashed tracks in Fig. 1.10). Therefore, the intensity from these directions is reduced.

In 1933, Rossi showed in a coincidence experiment that secondary cosmic rays at sea level initiate cascades in a lead absorber of variable thickness ('Rossi curve'). The absorption measurements in his apparatus also indicated that cosmic rays at sea level consist of a soft and a penetrating component.

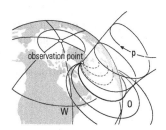

Fig. 1.10
East–west effect

1.2 Discoveries of New Elementary Particles

*"If I would remember the names of all
these particles, I'd be a botanist."*

Enrico Fermi

discovery of the positron

Up to the thirties, only electrons, protons (as part of the nucleus), and photons were known as elementary particles. The positron was discovered in a cloud chamber by Anderson in 1932 (Nobel Prize 1936). This was the antiparticle of the electron, which was predicted by Dirac in 1928 (Nobel Prize 1933). This, and the discovery of the neutron by Chadwick in 1932 (Nobel Prize 1935), started a new chapter in elementary particle and astroparticle physics. Additionally in 1930, Pauli postulated the existence of a neutral, massless spin-$\frac{1}{2}$ particle to restore the validity of the energy, momentum, and angular-momentum conservation laws that appeared to be violated in nuclear beta decay (Nobel Prize 1945). This hypothetical enigmatic particle, the neutrino, could only be shown to exist in a reactor experiment in 1956 (Cowan & Reines, Nobel Prize 1995). It eventually lead to a completely new branch of astronomy; *neutrino astronomy* is a classic example of a perfect interplay between elementary particle physics and astronomy.

neutrino postulate

neutrino astronomy

It was reported that Landau (Nobel Prize 1962), within several hours of hearing about the discovery of the neutron, predicted the existence of cold, dense stars which consisted mainly of neutrons. In 1967, the existence of rotating neutron stars (pulsars) was confirmed by observing radio signals (Hewish and Bell, Nobel Prize for Hewish 1975).

neutron stars

pulsars

stability of a neutron star

Neutrons in a neutron star do not decay. This is because the Pauli exclusion principle (1925) forbids neutrons to decay into occupied electron states. The Fermi energy of remnant electrons in a neutron star is at around several 100 MeV, while the maximum energy transferred to electrons in neutron decay is 0.77 MeV. There are therefore no vacant electron levels available.

After discovering the neutron, the second building block of the nucleus, the question of how atomic nuclei could stick together arose. Although neutrons are electrically neutral, the protons would electrostatically repel each other. Based on the range of the nuclear force and Heisenberg's uncertainty principle (1927, Nobel Prize 1932), Yukawa conjectured in 1935 that unstable mesons of 200-fold electron mass could possibly mediate nuclear forces (Nobel Prize 1949).

Yukawa particle

Initially it appeared that the muon discovered by Anderson and Neddermeyer in a cloud chamber in 1937, had the required properties of the hypothetical Yukawa particle. The muon, however, has no strong interactions with matter, and it soon became clear that the muon was a heavy counterpart of the electron. The fact that another electron-like particle existed caused Rabi (Nobel Prize 1944) to remark: "Who ordered this?" Rabi's question remains unanswered to this day. The situation became even more critical when Perl (Nobel Prize 1995) discovered another, even heavier lepton, the tau, in 1975.

discovery of muons

discovery of the tau

The discovery of the strongly interacting charged pions ($\pi^{\pm}$) in 1947 by Lattes, Occhialini, Powell, and Muirhead, using nuclear emulsions exposed to cosmic rays at mountain altitudes, solved the puzzle about the Yukawa particles (Nobel Prize 1950 to Cecil Powell for his development of the photographic method of studying nuclear processes and his discoveries regarding mesons made with this method). The pion family was supplemented in 1950 by the discovery of the neutral pion (π^0). Since 1949, pions can also be produced in particle accelerators.

discovery of pions

Up to this time, elementary particles were predominantly discovered in cosmic rays. In addition to the muon ($\mu^{\pm}$) and the pions (π^+, π^-, π^0), tracks of charged and neutral kaons were observed in cloud-chamber events. Neutral kaons revealed themselves through their decay into two charged particles. This made the K^0 appear as an upside down 'V', because only the ionization tracks of the charged decay products of the K^0 were visible in the cloud chamber (Rochester & Butler 1947, Fig. 1.11).

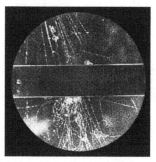

Fig. 1.11
Decays of neutral kaons in a cloud chamber {4}

V particles

In 1951, part of the Vs were recognized as Lambda baryons, which also decayed relatively quickly into two charged secondaries ($\Lambda^0 \rightarrow p + \pi^-$). In addition, the Ξ and Σ hyperons were discovered in cosmic rays (Ξ: Armenteros et al., 1952; Σ: Tomasini et al., 1953).

baryons and hyperons

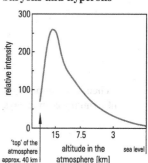

Apart from studying local interactions of cosmic-ray particles, their global properties were also investigated. The showers observed under lead plates by Rossi were also found in the atmosphere (Pfotzer, 1936). The interactions of primary cosmic rays in the atmosphere initiate *extensive air showers*, see Sect. 7.4, (Auger, 1938). These showers lead to a maximum intensity of cosmic rays at altitudes of 15 km above sea level ('Pfotzer maximum', Fig. 1.12).

Fig. 1.12
Intensity profile of cosmic particles in the atmosphere

One year earlier (1937), Bethe and Heitler, and at the same time Carlson and Oppenheimer, developed the theory of electromagnetic cascades, which was successfully used to describe the extensive air showers.

energy source of stars

In 1938, Bethe together with Weizsäcker, solved the long-standing mystery of the energy generation in stars. The

nuclear fusion

fusion of protons leads to the production of helium nuclei, in which the binding energy of 6.6 MeV per nucleon is released, making the stars shine (Nobel Prize 1967).

solar wind

In 1937, Forbush realised that a significant decrease of the cosmic-ray intensity correlated with an increased solar activity. The active Sun appears to create some sort of *solar wind* which consists of charged particles whose flux generates a magnetic field in addition to the geomagnetic field. The solar activity thereby modulates the galactic component of cosmic rays (Fig. 1.13).

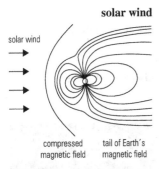

Fig. 1.13
Influence of the solar wind on the Earth's magnetic field

The observation that the tails of comets always point away from the Sun led Biermann to conclude in 1951, that some kind of solar wind must exist. This more or less continuous particle flux was first directly observed by the Mariner 2 space probe in 1962. The solar wind consists predominantly of electrons and protons, with a small admixture of α particles. The particle intensities at a distance of one astronomical unit (the distance from Sun to Earth) are 2×10^8 ions/(cm^2 s). This propagating solar plasma carries part of the solar magnetic field with it, thereby preventing some primary cosmic-ray particles to reach the Earth.

In 1949 it became clear that primary cosmic rays consisted mainly of protons. Schein, Jesse, and Wollan used balloon experiments to identify protons as the carriers of cosmic radiation.

Fermi
acceleration mechanism

Fermi (Nobel Prize in 1938 for experiments on radioactivity and the theory of nuclear beta decay) investigated the interactions of cosmic-ray particles with atmospheric atomic nuclei and with the solar and terrestrial magnetic fields. By as early as 1949, he also had considered possible mechanisms that accelerated cosmic-ray particles to very high energies.

chemical composition
of primary cosmic rays

Meanwhile, it had been discovered that in addition to electrons, protons, and α particles, the whole spectrum of heavy nuclei existed in cosmic radiation (Freier, Bradt, Peters, 1948). In 1950, ter Haar discussed supernova explosions as the possible origin of cosmic rays, an idea that was later confirmed by simulations and measurements.

discovery of antiparticles

After discovering the positron in 1932, the antiproton, the second known antiparticle, was found in an accelerator experiment by Chamberlain and Segrè in 1955 (Nobel Prize in 1959). Positrons (Meyer & Vogt, Earl, 1961) and antiprotons (Golden, 1979) were later observed in primary cosmic rays. It is, however, assumed that these cosmic-ray antiparticles do not originate from sources consisting of antimatter, but are produced in secondary interactions between primary cosmic rays and the interstellar gas or in the upper layers of the atmosphere.

1.3 Start of the Satellite Era

> *"Having probes in space was like having a cataract removed."*
>
> *Hannes Alfvén*

The launch of the first artificial satellite (Sputnik, October 4th, 1957) paved the way for developments that provided completely new opportunities in astroparticle physics. The atmosphere represents an absorber with a thickness of ≈ 25 radiation lengths. The observation of primary X rays and gamma radiation was previously impossible due to their absorption in the upper layers of the atmosphere. This electromagnetic radiation can only be investigated – undisturbed by atmospheric absorption – at very high altitudes near the 'top' of the atmosphere. It still took some time until the first **X-ray satellites** X-ray satellites (e.g., 1970 Uhuru, 1978 Einstein Observatory, 1983 Exosat; Nobel Prize for R. Giacconi 2002) and gamma satellites (e.g., 1967 Vela, 1969 OSO-3, 1972 SAS-2, 1975 COS-B)[3] were launched. They provided a wealth **γ satellite** of new data in a hitherto unaccessible spectral range. The galactic center was found to be bright in X rays and gamma rays, and the first point sources of high-energy astroparticles could also be detected (Crab Nebula, Vela X1, Cygnus X3, . . .).

With the discovery of *quasistellar radio sources* (quasars, **quasars** 1960), mankind advanced as far as to the edge of the universe. Quasars appear to outshine whole galaxies if they are really located at cosmological distances. Their distance is determined from the redshift of their spectral lines. The most distant quasar currently known, was discovered in 2001 and has a redshift of $z = \frac{\lambda - \lambda_0}{\lambda_0} = 6.28$. An object even farther away is the galaxy Abell 1835 IR 1916 with a redshift

---

[3] OSO – Orbiting Solar Observatory
SAS – Small Astronomy Satellite

of $z = 10$. Its discovery was made possible through light amplification by a factor of about 50 resulting from strong gravitational lensing by a very massive galactic cluster in the line of sight to the distant galaxy [1]. As will be discussed in Chap. 8, this implies that the quasar is seen in a state when the universe was less than 5% of its present age. Consequently, this quasar resides at a distance of 13 billion light-years.[4] Initially, there was some controversy about whether the observed quasar redshifts were of gravitational or cosmological origin. Today, there is no doubt that the observed large redshifts are a consequence of the Hubble expansion of the universe.

Big Bang theory   The expansion of the universe implies that it began in a giant explosion, some time in the past. Based on this Big Bang hypothesis, one arrives at the conclusion that this must have occurred about 15 billion years ago. The *Big Bang model* was in competition with the idea of a steady-state **steady-state universe**   universe for quite some time. The steady-state model was based on the assumption that the universe as a whole was time independent with new stars being continuously created while old stars died out. On the other hand, Gamow had been speculating since the forties that there should be a residual radiation from the Big Bang. According to his estimate, the temperature of this radiation should be in the **cosmic background radiation**   range of a few Kelvin. Penzias and Wilson (Nobel Prize 1978) detected this echo of the Big Bang by chance in 1965, while they were trying to develop low-noise radio antennae (Fig. 1.14).[5] With this discovery, the Big Bang model finally gained general acceptance. The exact temperature of this blackbody radiation was measured by the COBE[6] satellite in 1992 as 2.726 ± 0.005 Kelvin.[7]

---

[4] It has become common practice in the scientific literature that the number 10^9 is called a billion, while in other countries the billion is 10^{12}. Throughout this book the notation that a billion is equal to a thousand millions is used.

[5] The excrements of pigeons presented a severe problem during an attempt to reduce the noise of their horn antenna. When, after a thorough cleaning of the whole system, a residual noise still remained, Arno Penzias was reported to have said: "Either we have seen the birth of the universe, or we have seen a pile of pigeon shit."

[6] COBE – COsmic ray Background Explorer

[7] The presently (2004) most accurate value of the blackbody temperature is 2.725 ± 0.001 K.

COBE also found spatial asymmetries of the 2.7 Kelvin blackbody radiation at a level of $\Delta T/T \approx 10^{-5}$. This implies that the early universe had a lumpy structure, which can be considered as a seed for galaxy formation.

In parallel with the advance of cosmology, the famous two-neutrino experiment of Lederman, Schwartz, and Steinberger in 1962 (Nobel Prize 1988) represented an important step for the advancement of astroparticle physics. This experiment demonstrated that the neutrino emitted in nuclear beta decay is not identical with the neutrino occurring in pion decay ($\nu_\mu \neq \nu_e$). At present, three generations of neutrinos are known (ν_e, ν_μ, and ν_τ). The direct observation of the tau neutrino was established only relatively recently (July 2000) by the DONUT[8] experiment.

The observation of solar neutrinos by the Davis experiment in 1967 marked the beginning of the discipline of neutrino astronomy (Nobel Prize for R. Davis 2002). In fact, Davis measured a deficit in the flux of solar neutrinos, which was confirmed by subsequent experiments, GALLEX[9], SAGE[10], and Kamiokande (Nobel Prize for M. Koshiba 2002). It is considered unlikely that a lack of understanding of solar physics is responsible for the solar neutrino problem. In 1958 Pontecorvo highlighted the possibility of *neutrino oscillations*. Such oscillations ($\nu_e \rightarrow \nu_\mu$) are presently generally accepted as explanation of the solar neutrino deficit. This would imply that neutrinos have a very small non-vanishing mass. In the framework of the electroweak theory (Glashow, Salam, Weinberg 1967; Nobel Prize 1979) that unifies electromagnetic and weak interactions, a non-zero neutrino mass was not foreseen. The introduction of quarks as fundamental constituents of matter (Gell-Mann and Zweig 1964, Nobel Prize for Gell-Mann 1969), and their description by the theory of quantum chromodynamics extended the electroweak theory to the *Standard Model of elementary particles* (Veltman, t'Hooft; Nobel Prize 1999).

In this model, the masses of elementary particles cannot be calculated a priori. Therefore, small non-zero neutrino masses should not represent a real problem for the standard model, especially since it contains 18 free parameters that

Fig. 1.14
Penzias and Wilson in front of their horn antenna used for measuring of the blackbody radiation {5}

neutrino oscillation

electroweak theory

quarks

Standard Model

---

[8] DONUT – Direct Observation of NU Tau (ν_τ)

[9] GALLEX – German–Italian GALLium EXperiment

[10] SAGE – Soviet American Gallium Experiment

have to be determined by experimental information. However, three neutrino generations with non-zero mass would add another 7 parameters (three for the masses and four mixing parameters). It is generally believed that the standard model will not be the final word of the theoreticians.

charmed mesons

fourth quark

The discovery of charmed mesons in cosmic rays (Niu et al. 1971) and the confirmation, by accelerator experiments, for the existence of a fourth quark (Richter & Ting 1974, Nobel Prize 1976, Fig. 1.15) extended the standard model of Gell-Mann and Zweig (up, down, strange, and charm).

The theory of general relativity and Schwarzschild's ideas on the formation of gravitational singularities were supported in 1970 by precise investigations of the strong X-ray source Cygnus X1. Optical observations of Cygnus X1 indicated that this compact X-ray source is ten times more massive than our Sun. The rapid variation in the intensity of X rays from this object leads to the conclusion that this source only has a diameter of about 10 km. A typical neutron star has a similar diameter to this, but is only three times as heavy as the Sun. An object that was as massive as Cygnus X1 would experience such a large gravitational contraction, which would overcome the Fermi pressure of degenerate neutrons. This leads to the conclusion that a black hole must reside at the center of Cygnus X1.

Cygnus X1 and gravitational singularities

black hole in Cygnus X1?

Hawking radiation

evaporation of black holes

By 1974, Hawking had already managed to unify some aspects of the theory of general relativity and quantum physics. He was able to show that black holes could evaporate by producing fermion pairs from the gravitational energy outside the event horizon. If one of the fermions escaped from the black hole, its total energy and thereby its mass would be decreased (*Hawking radiation*). The time constants for the evaporation process of massive black holes, however, exceed the age of the universe by many orders of magnitude.

There were some hopes that gravitational waves, which would be measured on Earth, could resolve questions on the formation of black holes and other cosmic catastrophes. These hopes were boosted by gravitational-wave experiments by Weber in 1969. The positive signals of these early experiments have, so far, not been confirmed. It is generally believed that the findings of Weber were due to mundane experimental backgrounds.

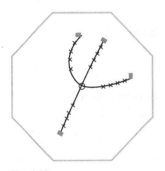

Fig. 1.15
Decay of an excited charm particle ($\psi' \to \psi + \pi^+ + \pi^-$, with the subsequent decay $\psi \to \mu^+ + \mu^-$)

In contrast, Taylor and Hulse succeeded in providing indirect evidence for the emission of gravitational waves in

1974, by observing a binary star system that consisted of a pulsar and a neutron star (Nobel Prize 1993). They were able to precisely test the predictions of general relativity using this binary star system. The rotation of the orbital ellipse (periastron rotation) of this system is ten thousand times larger than the perihelion rotation of the planet Mercury. The decreasing orbital period of the binary is directly related to the energy loss by the emission of gravitational radiation. The observed speeding-up rate of the orbital velocities of the partners of the binary system and the slowing-down rate of the orbital period agree with the prediction based on the theory of general relativity to better than 1‰.

It is to be expected that there are processes occurring in the universe, which lack an immediate explanation. This was underlined by the discovery of *gamma-ray bursters* (GRB) in 1967. It came as a surprise when gamma-ray detectors on board military reconnaissance satellites, which were in orbit to check possible violations of the test-ban treaty on thermonuclear explosions, observed γ bursts. This discovery was withheld for a while due to military secrecy. However, when it became clear that the γ bursts did not originate from Earth but rather from outer space, the results were published. Gamma-ray bursters light up only once and are very short-lived, with burst durations lasting from 10 ms to a few seconds. It is conceivable that γ bursts are caused by supernova explosions or by collisions between neutron stars.

γ burster

It might appear that the elementary-particle aspect of astroparticle physics has been completed by the discovery of the b quark (Lederman 1977) and t quark (CDF collaboration 1995). There are now six known leptons (v_e, e^-; v_μ, μ^-; v_τ, τ^-) along with their antiparticles ($\bar{v}_e$, e^+; $\bar{v}_\mu$, μ^+; $\bar{v}_\tau$, τ^+). These are accompanied by six quarks (up, down; charm, strange; top, bottom) and their corresponding six antiquarks. These matter particles can be arranged in three families or 'generations'. Measurements of the primordial deuterium, helium, and lithium abundance in astrophysics had already given some indication that there may be only three families with light neutrinos. This astrophysical result was later confirmed beyond any doubt by experiments at the electron–positron collider LEP[11] in 1989 (see also Fig. 2.1). The standard model of elementary particles, with its three fermion generations, was also verified by the discovery of gluons, the carriers of the strong force (DESY[12], 1979), and

top and bottom quarks

gluons

[11] LEP – Large Electron–Positron collider at CERN in Geneva

W^+, W^-, Z

Fig. 1.16
Supernova explosion SN 1987A in
the Tarantula Nebula {6}

the bosons of the weak interaction (W^+, W^-, Z; CERN[13] 1983; Nobel Prize for Rubbia and van der Meer 1984). The discovery of asymptotic freedom of quarks in the theory of the strong interaction by Gross, Politzer, and Wilczek was honored by the Nobel Prize in 2004.

The observation of the supernova explosion 1987A, along with the burst of extragalactic neutrinos, represented the birth of real astroparticle physics. The measurement of only 20 neutrinos out of a possible 10^{58} emitted, allowed elementary particle physics investigations that were hitherto inaccessible in laboratory experiments. The dispersion of arrival times enabled physicists to derive an upper limit of the neutrino mass ($m_{\nu_e} < 10\,\mathrm{eV}$). The mere fact that the neutrino source was 170 000 light-years away in the Large Magellanic Cloud, allowed a lower limit on the neutrino lifetime to be estimated. The gamma line emission from SN 1987A gave confirmation that heavy elements up to iron, cobalt, and nickel were synthesized in the explosion, in agreement with predictions of supernova models. As the first optically visible supernova since the discovery of the tele-scope, SN 1987A marked an ideal symbiosis of astronomy, astrophysics, and elementary particle physics (Fig. 1.16).

ROSAT

Hubble telescope

The successful launch of the high-resolution X-ray satel-lite ROSAT[14] in 1990, paved the way for the discovery of numerous X-ray sources. The Hubble telescope, which was started in the same year, provided optical images of stars and galaxies in hitherto unprecedented quality, once the slightly defocusing mirror had been adjusted by a spectacular re-pair in space. The successful mission of ROSAT was fol-lowed by the X-ray satellites Chandra (named after Sub-rahmanyan Chandrasekhar, Nobel Prize 1983) and XMM[15] both launched in 1999.

CGRO

The Compton Gamma-Ray Observatory (CGRO, launched in 1991) opened the door for GeV gamma astronomy. Ground-based atmospheric air Cherenkov tele-scopes and extensive air-shower experiments were able to identify TeV point sources in our Milky Way (Crab Nebula,

[12] DESY – Deutsches Elektronen Synchrotron in Hamburg

[13] CERN – Conseil Européen pour la Recherche Nucléaire

[14] ROSAT – ROentgen SATellite of the Max-Planck Institute for Extraterrestrial Physics, Munich

[15] XMM – X-ray Multi-Mirror mission, renamed Newton Obser-vatory in 2002

1989) and at extragalactical distances (1992, Markarian 421, Markarian 501). The active galactic nuclei of the Markarian galaxies are also considered excellent candidate sources of high-energy hadronic charged cosmic rays.

1.4 Open Questions

"We will first understand how simple the universe is, when we realize, how strange it is."

Anonymous

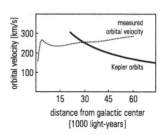

Fig. 1.17
Orbital velocities of stars in the Milky Way in comparison with Keplerian trajectories

A still unsolved question of astroparticle physics is the problem of *dark matter* and *dark energy*. From the observation of orbital velocities of stars in our Milky Way and the velocities of galaxies in galactic clusters, it is clear that the energy density of the visible matter in the universe is insufficient to correctly describe the dynamics (Fig. 1.17).

Since the early nineties, the MACHO[16] and EROS[17] experiments have searched for compact, non-luminous, Jupiter-like objects in the halo of our Milky Way, using the technique of microlensing. Some candidates have been found, but their number is nowhere near sufficient to explain the missing dark matter in the universe. One can conjecture that exotic, currently unknown particles (supersymmetric particles, WIMPs[18], . . .), or massive neutrinos may contribute to solve the problem of the missing dark matter. A non-vanishing vacuum energy density of the universe is also known to play a decisive rôle in the dynamics and evolution of the universe.

In 1998 the Super-Kamiokande experiment found evidence for a non-zero neutrino mass by studying the relative abundances of atmospheric electron and muon neutrinos. The observed deficit of atmospheric muon neutrinos is most readily and elegantly explained by the assumption that neutrinos oscillate from one lepton flavour to another ($\nu_\mu \rightarrow \nu_\tau$). This is only possible if neutrinos have mass. The presently favoured mass of 0.05 eV for ν_τ, however, is insufficient to explain the dynamics of the universe alone.

MACHO, EROS

supersymmetric particles
WIMPs

non-zero neutrino masses

---

[16] MACHO – search for MAssive Compact Halo Objects
[17] EROS – Expérience pour la Recherche d'Objets Sombres
[18] WIMP – Weakly Interacting Massive Particles

The oscillation scenario for solar neutrinos was confirmed in 2001 by the SNO[19] experiment by showing that the total flavour-independent neutrino flux from the Sun arriving at Earth (ν_e, ν_μ, ν_τ) was consistent with solar-model expectations, demonstrating that some of the solar electron neutrinos had oscillated into a different neutrino flavour.

Higgs particle

The generation of the masses for elementary particles is still an open question. In the standard model of electroweak and strong interactions, the mass generation is believed to come about by a *spontaneous symmetry breaking*, the so-called Higgs mechanism. This process favours the existence of at least one additional massive neutral boson. Whether the LEP experiments at CERN have seen this enigmatic particle at the kinematic limit of LEP, with a mass of about 115 GeV, needs to be confirmed by future hadron colliders.

A very recent, equally exciting discovery, is the measurement of the acceleration parameter of the universe. Based on the ideas of the classical Big Bang, one would assume that the initial thrust of the explosion would be slowed

accelerating universe?

down by gravitation. Observations on distant supernova explosions (1998) however, appeared to indicate that in early cosmological epochs, the rate of expansion was smaller than today. The finding of an *accelerating universe* – which is now generally accepted – has important implications for cosmology. It suggests that the largest part of the missing dark matter is stored as dark energy in a dynamical vacuum ('quintessence').

extrasolar planets

Finally, it should be highlighted that the discovery of *extrasolar planets* (Mayor and Queloz 1995) has led to the resumption of discussions on the existence of extraterrestrial intelligence. So far, the nearest extrasolar planet ('Millennium') has been observed in the Tau Boötis solar system, by the Herschel telescope on the Canary Islands. The planet is twice as large as Jupiter and eight times as heavy, and is situated at a distance of 55 light-years. Until now about 100 extrasolar planets have been discovered. Possibly we are not the only intelligent beings in the universe pursuing astroparticle physics.

---

[19] SNO – Sudbury Neutrino Observatory

1.5 Problems

1. Work out the
   a) velocity of an Earth satellite in a low-altitude orbit,
   b) the escape velocity from Earth,
   c) the altitude of a geostationary satellite above ground level. Where can such a geostationary satellite be positioned?
2. What is the bending radius of a solar particle (proton, 1 MeV kinetic energy) in the Earth's magnetic field (0.5 Gauss) for vertical incidence with respect to the field? Use the relation between the centrifugal force and the Lorentz force (6.1), and argue whether a classical calculation or a relativistic calculation is appropriate.
3. Estimate the average energy loss of a muon in the atmosphere (production altitude 20 km, muon energy $\approx$ 10 GeV; check with Fig. 4.2).
4. What is the ratio of intensities of two stars which differ by one unit in magnitude only (for the definition of the magnitude see the Glossary)?
5. Small astronomical objects like meteorites and asteroids are bound by solid-state effects while planets are bound by gravitation. Estimate the minimum mass from where on gravitational binding starts to dominate as binding force. Gravitational binding dominates if the potential gravitational energy exceeds the total binding energy of the solid material, where the latter is taken to be proportional to the number of atoms in the object. The average atomic number is A, from which together with the Bohr radius r_B the average density can be estimated.
6. There is a statement in this chapter that a quasar at a redshift of $z = 6.68$ gives us information on the universe when it was only about 3% of its present age. Can you convert the redshift into the age of the universe?

2 The Standard Model of Elementary Particles

"Most basic ideas of science are essentially simple and can usually be expressed in a language that everyone understands."

Albert Einstein

Over the last years a coherent picture of elementary particles has emerged. Since the development of the atomic model, improvements in experimental resolution have allowed scientists to investigate smaller and smaller structures. Even the atomic nucleus, which contains practically the total mass of the atom, is a composite object. Protons and neutrons, the building blocks of the nucleus, have a granular structure that became obvious in electron–nucleon scattering experiments. In the naïve quark parton model, a nucleon consists of three quarks. The onion-type phenomenon of ever smaller constituents of particles that were initially considered to be fundamental and elementary, may have come to an end with the discovery of quarks and their dynamics. While atoms, atomic nuclei, protons, and neutrons can be observed as free particles in experiments, quarks can never escape from their hadronic prison. In spite of an intensive search by numerous experiments, nobody has ever been able to find free quarks. Quantum chromodynamics, which describes the interaction of quarks, only allows the *asymptotic freedom* of quarks at high momenta. Bound quarks that are inside nucleons typically have low momenta and are subject to '*infrared slavery*'. This confinement does not allow the quarks to separate from each other.

composition of atomic nuclei

composition of nucleons

quark confinement

Quarks are constituents of strongly interacting hadronic matter. The size of quarks is below 10^{-17} m. In addition to quarks, there are leptons that interact weakly and electromagnetically. With the resolution of the strongest microscopes (accelerators and storage rings), quarks and leptons appear to be pointlike particles, having no internal structure. Three different types of leptons are known: electrons, muons, and taus. Each charged lepton has a separate neutrino: ν_e, ν_μ, ν_τ. Due to the precise investigations of the Z particle, which is the neutral carrier of weak interactions, it is known that there are exactly three particle families with

quarks and leptons

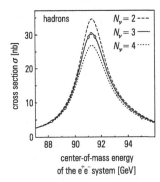

center-of-mass energy
of the e⁺e⁻ system [GeV]

Fig. 2.1
Determination of the number of
neutrino generations from Z decay

light neutrinos (Fig. 2.1). This result was obtained from
the measurement of the total Z decay width. According to
Heisenberg's uncertainty principle, the resolution of complementary quantities is intrinsically limited by Planck's
constant ($h = 6.626\,0693 \times 10^{-34}$ J s). The relation between
the complementary quantities of energy and time is

$$\Delta E \, \Delta t \geq \hbar/2 \quad (\hbar = h/2\pi) \, . \tag{2.1}$$

If $\Delta t = \tau$ is the lifetime of the particle, relation (2.1) implies
that the decay width $\Delta E = \Gamma$ is larger when τ is shorter. If
there are many generations of light neutrinos, the Z particle
can decay into all these neutrinos,

$$Z \to \nu_x + \bar{\nu}_x \, . \tag{2.2}$$

neutrino generations

These decays can occur even if the charged leptons ℓ_x associated with the respective generation are too heavy to be
produced in Z decay. A large number of different light neutrinos will consequently reduce the Z lifetime, thereby increasing its decay width. The exact measurement of the Z
decay width took place at the LEP storage ring (Large Electron–Positron collider) in 1989, enabling the total number
of neutrino generations to be determined: there are exactly
three lepton generations with light neutrinos.

primordial
helium abundance

The measurement of the primordial helium abundance
had already allowed physicists to derive a limit for the number of neutrino generations. The nucleosynthesis in the early
universe was essentially determined by the number of relativistic particles, which were able to cool down the universe
after the Big Bang. At temperatures of $\approx 10^{10}$ K, which
correspond to energies where nucleons start to bind in nuclei (≈ 1 MeV), these relativistic particles would have consisted of protons, neutrons, electrons, and neutrinos. If many

nucleosynthesis

different neutrino flavours exist, a large amount of energy
would have escaped from the original fireball, owing to the
low interaction probability of neutrinos. This has the consequence that the temperature would have decreased quickly.
A rapidly falling temperature means that the time taken for
neutrons to reach nuclear binding energies would have been
very short, and consequently they would have had very little
time to decay (lifetime $\tau_n = 885.7$ s). If there were many
neutrons that did not decay, they would have been able to
form helium together with stable protons. The primordial
helium abundance is therefore an indicator of the number

Table 2.1: Periodic table of elementary particles: matter particles (fermions) [2]

LEPTONS ℓ, spin $\frac{1}{2}\hbar$ (antileptons $\bar{\ell}$)						
electr. charge [e]	1. generation		2. generation		3. generation	
	flavour	mass [GeV/c^2]	flavour	mass [GeV/c^2]	flavour	mass [GeV/c^2]
0	ν_e electron neutrino	$< 2.5 \times 10^{-9}$ at 95% CL	ν_μ muon neutrino	$< 1.9 \times 10^{-4}$ at 90% CL	ν_τ tau neutrino	< 0.018 at 95% CL
-1	e electron	5.11×10^{-4}	μ muon	0.106	τ tau	1.777
QUARKS q, spin $\frac{1}{2}\hbar$ (antiquarks $\bar{q}$)						
electr. charge [e]	flavour	$\simeq$ mass [GeV/c^2]	flavour	$\simeq$ mass [GeV/c^2]	flavour	$\simeq$ mass [GeV/c^2]
$+2/3$	u up	1.5×10^{-3} to 4×10^{-3}	c charm	1.15 to 1.35	t top	174.3
$-1/3$	d down	4×10^{-3} to 8×10^{-3}	s strange	0.08 to 0.13	b bottom	4.1 to 4.4

of neutrino generations. In 1990, the experimentally determined primordial helium abundance allowed physicists to conclude that the maximum number of different light neutrinos is four.

In addition, there are also three quark generations, which **properties of quarks** have a one-to-one correspondence with the three lepton generations:

$$\begin{pmatrix} \nu_e \\ e^- \end{pmatrix} \begin{pmatrix} \nu_\mu \\ \mu^- \end{pmatrix} \begin{pmatrix} \nu_\tau \\ \tau^- \end{pmatrix}$$
$$\begin{pmatrix} u \\ d \end{pmatrix} \begin{pmatrix} c \\ s \end{pmatrix} \begin{pmatrix} t \\ b \end{pmatrix} \quad . \tag{2.3}$$

The properties of these fundamental matter particles are listed in Table 2.1. Quarks have fractional electric charges (in units of the elementary charge). The different kinds of **flavour** quarks (u, d; c, s; t, b) in the three respective generations (families) are characterized by a different flavour. The masses of neutrinos from direct measurements are compati- **masses of neutrinos** ble with being zero, therefore only upper limits can be found experimentally. It must be emphasized, however, that neu- **neutrino oscillations** trinos do have a small mass, as indicated by the Super-

Kamiokande and SNO experiments which are interpreted in terms of neutrino oscillations. Actually, in grand unified theories (GUTs) unifying electroweak and strong interactions, neutrinos are predicted to have small but non-zero masses. Only approximate values of masses for quarks can be given, because free quarks do not exist and the binding energies of quarks in hadrons can only be estimated roughly. For each particle listed in Table 2.1 there exists an antiparticle, which is in all cases different from the original particle. This means that there are actually 12 fundamental leptons and an equal number of quarks.

interactions between elementary particles

The interactions between elementary particles are governed by different forces. There are four forces in total, distinguished by strong, electromagnetic, weak, and gravitational interactions. In the 1960s, it was possible to unite the electromagnetic and weak interactions into the electroweak theory. The carriers of all the interactions are particles with integer spin (*bosons*), in contrast to the matter particles that all have half-integer spin (*fermions*). The properties of these bosons are compiled in Table 2.2.

electroweak theory

bosons and fermions

Table 2.2
Periodic table of elementary particles: carriers of the forces (bosons) [2]

electroweak interaction	γ	W^-	W^+	Z
spin [$\hbar$]	1	1	1	1
electric charge [e]	0	-1	$+1$	0
mass [GeV/c^2]	0	80.4	80.4	91.2

strong interaction	gluon g
spin [$\hbar$]	1
electric charge [e]	0
mass [GeV/c^2]	0

gravitational interaction	graviton G
spin [$\hbar$]	2
electric charge [e]	0
mass [GeV/c^2]	0

While the existence of the gauge bosons of electroweak interactions, and the gluon of strong interactions are well established, the *graviton*, the carrier of the gravitational force, has not yet been discovered. The properties of interactions

are compared in Table 2.3. It is apparent that gravitation can **weakness of gravitation**
be completely neglected in the microscopic domain, because
its strength in relation to strong interactions is only 10^{-40}.

Table 2.3
Properties of interactions

inter- $\rightarrow$ action	gravitation	electroweak		strong
property $\downarrow$		weak	electro-magnetic	
acts on	mass–energy	flavour	electric charge	colour charge
affected particles	all	quarks, leptons	all charged particles	quarks, gluons
exchange particle	graviton G	W^+, W^-, Z	γ	gluons g
relative strength	10^{-40}	10^{-5}	10^{-2}	1
range	∞	$\approx 10^{-3}$ fm	∞	≈ 1 fm
example	system Earth–Moon	β decay	atomic binding	nuclear binding

In the primitive quark model, all strongly interacting
particles (hadrons) are composed of valence quarks. A **valence quarks**
baryon is a three-quark system, whereas a meson consists
of a quark and an antiquark. Examples of baryons include
the proton, which is a uud system, and the neutron is a udd
composite. Correspondingly, an example of a meson is the
positively charged pion, which is a $u\bar{d}$ system. The existence
of baryons consisting of three identical quarks with parallel
spin ($\Omega^- = (sss)$, spin $\frac{3}{2}\hbar$) indicates that quarks must have
a hidden quantum number, otherwise the Pauli exclusion **hidden quantum numbers**
principle would be violated. This hidden quantum number is
called *colour*. Electron–positron interactions show that there **colour of quarks**
are exactly three different colours. Each quark therefore
comes in three colours, however all observed hadrons have
neutral colour. If the three degrees of freedom in colour
are denoted by red (r), green (g), and blue (b), the proton
is a composite object made up from $u_{red}u_{green}d_{blue}$. In **sea quarks**
addition to valence quarks, there is also a sea of virtual
quark–antiquark pairs in hadrons.

binding of quarks in hadrons
by gluon exchange

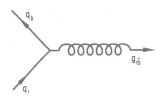

Fig. 2.2
Creation of coloured gluons by
quarks

binding of nucleons in nuclei

The quarks that form hadrons are held together by the exchange of gluons. Since gluons mediate the interactions between quarks, they must possess two colours: they carry a colour and an anticolour. Since there are three colours and anticolours each, one would expect that $3 \times 3 = 9$ gluons exist. The strong interaction, however, is only mediated by eight gluons. Gluons are no pure colour–anticolour systems like, for example, $r\bar{g}$, but rather mixed states. In quantum chromodynamics the possible 9 gluons form an octet of coloured gluons and a singlet consisting of a colour-neutral mixed state of *all* colours and anticolours, $(r\bar{r} + g\bar{g} + b\bar{b})$. In a very simplified picture, the gluon radiation of a quark can be illustrated by the diagram shown in Fig. 2.2.

Nucleons in a nucleus are bound together by the residual interaction of gluons, in very much the same way as molecular binding is a result of the residual interactions of electric forces.

2.1 Examples of Interaction Processes

"It is possible in quantum mechanics to sneak quickly across a region which is illegal energetically."

Richard P. Feynman

Interactions of elementary particles can be graphically represented by Feynman diagrams[1], which present a short-hand for the determination of cross sections. In the following, the underlying quark–lepton structure will be characterized for some interaction processes.

Rutherford scattering of electrons on protons is mediated by photons (Fig. 2.3).

Fig. 2.3
Rutherford scattering of electrons
on protons

Fig. 2.4
Rutherford scattering as
photon–quark subprocess

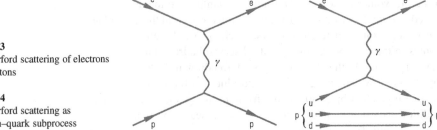

[1] see the Glossary

At high energies however, the photon does not interact with the proton as a whole, but rather only with one of its constituent quarks (Fig. 2.4). The other quarks of the nucleon participate in the interaction only as spectators. As photons are electrically neutral particles, they cannot change the nature of a target particle in an interaction. In weak interactions however, there are charged bosons which can cause an interchange between particles within a family. As an example, Fig. 2.5 shows the scattering of an electron neutrino on a neutron via a charged-current (W^+, W^- exchange) reaction.

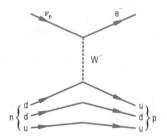

Fig. 2.5
Neutrino–neutron scattering by charged currents

In a neutral-current interaction (Z exchange), the neutrino would not alter its nature when scattered off the neutron. If electron neutrinos are scattered on electrons, charged and neutral currents can contribute. This is also true for scattering of muon or tau neutrinos on electrons (Fig. 2.6).

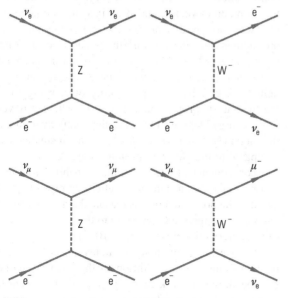

Fig. 2.6
Different Feynman diagrams contributing to the scattering of neutrinos on electrons

Decays of elementary particles can be described in a similar way. Nuclear beta decay of the neutron $n \rightarrow p + e^- + \bar{\nu}_e$ is mediated by a weak charged current (Fig. 2.7), where a d quark in the neutron is transformed into a u quark by the emission of a virtual W^-. The W^- immediately decays into members of the first lepton family ($W^- \rightarrow e^-\bar{\nu}_e$). In principle, the W^- can also decay according to $W^- \rightarrow \mu^-\bar{\nu}_\mu$ or $W^- \rightarrow \bar{u}d$, but this is not kinematically allowed.

nuclear beta decay

muon decay

Muon decay can be described in a similar fashion (Fig. 2.8). The muon transfers its charge to a W^-, thereby transforming itself into the neutral lepton of the second family, the ν_μ. The W^- in turn decays again into in $e^- \bar{\nu}_e$.

Fig. 2.7
Neutron decay

Fig. 2.8
Muon decay

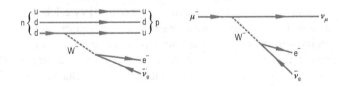

pion decay

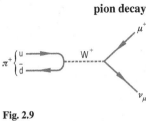

Fig. 2.9
Pion decay

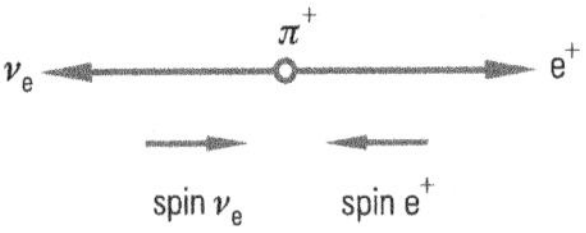

Fig. 2.10
Helicity conservation in π^+ decay

Finally, pion decay will be discussed (Fig. 2.9). In principle, the W^+ can also decay in this case, into an $e^+ \nu_e$ state. Helicity reasons, however, strongly suppress this decay: as a spin-0 particle, the pion decays into two leptons that must have antiparallel spins due to angular-momentum conservation. The *helicity* is the projection of the spin onto the momentum vector, and it is fixed for the neutrino (for massless particles the spin is either parallel or antiparallel to the momentum). Particles normally carry negative helicity (spin $\parallel -\boldsymbol{p}$, left-handed) so that the positron, as an antiparticle (spin $\parallel \boldsymbol{p}$, right-handed), must take on an unnatural helicity (Fig. 2.10). The probability of carrying an abnormal helicity is proportional to $1 - \frac{v}{c}$ (where v is velocity of the charged lepton). Owing to the relatively high mass of the muon ($m_\mu \gg m_e$), it takes on a much smaller velocity compared to the electron in pion decay, i.e., $v(\mu) \ll v(e)$. The consequence of this is that the probability for the decay muon to take on an unnatural helicity is much larger compared to the positron. For this reason, the $\pi^+ \to e^+ \nu_e$ decay is strongly suppressed compared to the $\pi^+ \to \mu^+ \nu_\mu$ decay (the suppression factor is 1.23×10^{-4}).

quantum numbers

lepton number

The various elementary particles are characterized by quantum numbers. In addition to the electric charge, the membership of a quark generation (quark flavour) or lepton generation (lepton number) is introduced as a quantum number. Leptons are assigned the lepton number $+1$ in their respective generation, whereas antileptons are given the lepton number -1. Lepton numbers for the different lepton families (L_e, L_μ, L_τ) are separately conserved, as is shown in the example of the muon decay:

$$
\begin{array}{ccccc}
\mu^- & \to & \nu_\mu & + e^- & + \bar{\nu}_e \\
L_\mu \quad 1 & & 1 & 0 & 0 \\
L_e \quad 0 & & 0 & 1 & -1
\end{array}
\tag{2.4}
$$

The parity transformation P is the space inversion of a physical state. Parity is conserved in strong and electromagnetic interactions, however, in weak interactions it is maximally violated. This means that the mirror state of a weak process does not correspond to a physical reality. Nature distinguishes between the right and left in weak interactions.

parity

parity violation

The operation of charge conjugation C applied to a physical state changes all the charges, meaning that particles and antiparticles are interchanged, whilst leaving quantities like momentum or spin untouched. Charge conjugation is also violated in weak interactions. In β decay, for example, left-handed electrons (negative helicity) and right-handed positrons (positive helicity) are favoured. Even though the symmetry operations P and C are not conserved individually, their combination CP, which is the application of space inversion (parity operation P) with subsequent interchange of particles and antiparticles (charge conjugation C) is a well-respected symmetry. This symmetry, however, is still broken in certain decays (K^0 and B^0 decays), but it is a common belief that the CPT symmetry (CP symmetry with additional time inversion) is conserved under all circumstances.

charge conjugation

CP conservation in weak interactions?

CP violation
CPT symmetry

Some particles, like kaons, exhibit very strange behaviour. They are produced copiously, but decay relatively slowly. These particles are produced in strong interactions, but they decay via weak interactions. This property is accounted for by introducing the quantum number *strangeness*, which is conserved in strong interactions, but violated in weak decays. Owing to the conservation of strangeness in strong interactions, only the associate production of strange particles, i.e., the combined production of hadrons one of which contains a strange and the other an anti-strange quark, is possible, such as

strange particles

strangeness

$$\pi^- + p \rightarrow K^+ + \Sigma^- . \tag{2.5}$$

In this process, the $\bar{s}$ quark in the K^+ ($= u\bar{s}$) receives the strangeness $+1$, whilst the s quark in the Σ^- ($= dds$) is assigned the strangeness -1. In the weak decay of the $K^+ \rightarrow \pi^+\pi^0$, the strangeness is violated, since pions do not contain strange quarks (s).

Certain particles that behave in an identical way under strong interactions, but differ in their charge state, are integrated into isospin multiplets. Protons and neutrons are nucleons that form an isospin doublet of $I = 1/2$. When the nucleon isospin is projected onto the z axis, the state with $I_z = +1/2$ corresponds to a proton whereas the $I_z = -1/2$

isospin multiplet

isospin doublet of nucleons

isospin triplet of pions

state relates to the neutron. The three pions (π^+, π^-, π^0) combine to form an isospin triplet with $I = 1$. In this case, $I_z = -1$ corresponds to the π^-, $I_z = +1$ is the π^+, whilst $I_z = 0$ relates to the π^0. The particle multiplicity m in an isospin multiplet is related to the isospin via the equation

$$m = 2I + 1 . \tag{2.6}$$

baryon number

Finally, the *baryon number* should be mentioned. Quarks are assigned the baryon number $1/3$, and antiquarks are given $-1/3$. All baryons consisting of three quarks are therefore assigned the baryon number 1, whereas all other particles get the baryon number 0.

conservation laws of particle physics

The properties of the conservation laws for the different interaction types in elementary particle physics are compiled in Table 2.4.

Table 2.4
Conservation laws of particle physics (conserved: +; violated: –)

physical	interaction		
quantity	strong	electromagnetic	weak
momentum	+	+	+
energy (incl. mass)	+	+	+
ang. momentum	+	+	+
electric charge	+	+	+
quark flavour	+	+	−
lepton number*	./.	+	+
parity	+	+	−
charge conjugation	+	+	−
strangeness	+	+	−
isospin	+	−	−
baryon number	+	+	+

*the lepton number is not relevant for strong interactions

Unfortunately, there is a small but important complication in the quark sector. As can be seen from Table 2.1, there is a complete symmetry between leptons and quarks. Leptons, however, participate in interactions as free particles, whereas quarks do not. Due to quark confinement, spectator quarks always participate in the interactions in some way. For charged leptons, there is a strict law of lepton-number conservation: The members of different generations do not mix with each other. For the quarks, it was seen that weak processes can change the strangeness. In Λ decay, the s quark belonging to the second generation can transform into a u quark of the first generation. This would otherwise only be allowed to happen to a d quark (Fig. 2.11).

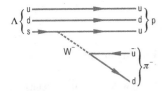

Fig. 2.11
Lambda decay: $\Lambda \to p + \pi^-$

It appears as if the s quark can sometimes behave like the d quark. It is, in fact, the d' and s' quarks that couple to weak interactions, rather than the d and s quarks. The d' and s' quarks can be described as a rotation with respect to the d and s quarks. This rotation is expressed by

$$d' = d \cos \theta_C + s \sin \theta_C ,$$
$$s' = -d \sin \theta_C + s \cos \theta_C ,$$

(2.7)

where θ_C is the mixing angle (Cabibbo angle).

Cabibbo angle

The reason that angles are used for weighting is based on the fact that the sum of the squares of the weighting factors, $\cos^2 \theta + \sin^2 \theta = 1$, automatically guarantees the correct normalization. θ_C has been experimentally obtained to be approximately 13 degrees ($\sin \theta_C \approx 0.2235$). Since $\cos \theta_C \approx 0.9747$, the d' quark predominantly behaves like the d quark, albeit with a small admixture of the s quark.

The quark mixing originally introduced by Cabibbo was extended by Kobayashi and Maskawa to all three quark families, such that d', s', and b' are obtained from d, s, b by a rotation matrix. This matrix is called the Cabibbo–Kobayashi–Maskawa matrix (CKM matrix),

CKM matrix

$$\begin{pmatrix} d' \\ s' \\ b' \end{pmatrix} = U \begin{pmatrix} d \\ s \\ b \end{pmatrix} .$$

(2.8)

The elements on the main diagonal of the (3×3) matrix U are very close to unity. The off-diagonal elements indicate the strength of the quark-flavour violation. A similar complication in the neutrino sector will be discussed later, where the eigenstates of the mass are not identical with the eigenstates of the weak interaction (see Sect. 6.2.1).

The Standard Model of electroweak and strong interactions cannot be the final theory. The model contains too many free parameters, which have to be adjusted by hand. In addition, the masses of all fundamental fermions are initially zero. They only get their masses by a mechanism of spontaneous symmetry breaking (*Higgs mechanism*). Another very important point to note is that gravitation is not considered in this model at all, whereas it is the dominant force in the universe as a whole. There have been many attempts to formulate a Theory of Everything (TOE) that unites all interactions. A very promising candidate for such a global description is the *string theory*. String theory is based on the assumption that elementary particles are not point-like, but

limits of the Standard Model

Higgs mechanism

Theory of Everything

string theory

supersymmetry

are one-dimensional strings. Different string excitations or oscillations correspond to different particles. In addition, certain string theories are supersymmetric. They establish a symmetry between fermions and bosons. String theories, and in particular superstring theories, are constructed in a higher-dimensional space. Out of the original 11 dimensions in the so-called M superstring theory, 7 must be compacted to a very small size, because they are not observed in nature.

weakness of gravity

String theories are presently considered as best candidates to unite quantum field theories and general relativity. They might even solve the problem of the three generations of elementary particles. In the framework of string theories in eleven dimensions the weakness of gravity might be related to the fact that part of the gravitational force is leaking into extra dimensions, while, e.g., electromagnetism, in contrast, is confined to the familiar four dimensions.

If gravity were really leaking into extra dimensions, the energy sitting there could give rise to dark energy influencing the structure of the universe (see Chap.13 on Dark Matter). Gravitational matter in extra dimensions would only be visible by its gravitational interactions.

holographic universe

It is also conceivable that we live in a *holographic universe* in the sense that all informations from a higher-dimensional space could be coded into a lower-dimensional space, just like a three-dimensional body can be represented by a two-dimensional hologram.

In Fig. 2.12, an overview of the historical successes of the unification of different theories is displayed with a projection into the future. One assumes that with increasing temperature ($\widehat{=}$ energy), nature gets more and more sym-

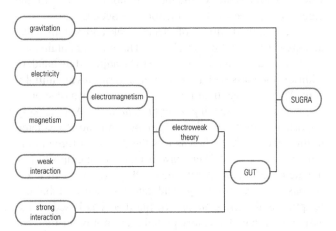

Fig. 2.12
Unification of all different
interactions into a Theory of
Everything
(GUT – Grand Unified Theory,
SUGRA – Super Gravitation)

metric. At very high temperatures, as existed at the time of the Big Bang, the symmetry was so perfect that all interactions could be described by one universal force. The reason that increasingly large accelerators with higher energies are being constructed is to track down this universal description of all forces.

universal force

According to present beliefs, the all-embracing theory of supergravity (SUGRA) is embedded in the M theory, an 11-dimensional superstring theory. The smallest constituents of this superstring theory are p-dimensional objects ('*branes*') of the size of the Planck length $L_P = \sqrt{\hbar G/c^3}$ (where G is the gravitational constant, $\hbar$ is Planck's constant, and c is the velocity of light). Seven of the ten spatial dimensions are compacted into a Calabi–Yau space. According to taste, the 'M' in the M theory stands for 'membrane', 'matrix', 'mystery', or 'mother (of all theories)'.

M theory

2.2 Problems

1. Which of the following reactions or decays are allowed?
   a) $\mu^- \to e^- + \gamma$,
   b) $\mu^+ \to e^+ + \nu_e + \bar{\nu}_\mu + e^+ + e^-$,
   c) $\pi^0 \to \gamma + e^+ + e^-$,
   d) $\pi^+ \to \mu^+ + e^-$,
   e) $\Lambda \to p + K^-$,
   f) $\Sigma^+ \to n + \pi^+$,
   g) $K^+ \to \pi^+ + \pi^- + \pi^+$,
   h) $K^+ \to \pi^0 + \pi^0 + e^+ + \nu_e$.

2. What is the minimum kinetic energy of a cosmic-ray muon to survive to sea level from a production altitude of $20\,\text{km}$ ($\tau_\mu = 2.197\,03\,\mu\text{s}$, $m_\mu = 105.658\,37\,\text{MeV}$)? For this problem one should assume that all muons have the given lifetime in their rest frame.

3. Work out the Coulomb force and the gravitational force between two singly charged particles of the Planck mass at a distance of $r = 1\,\text{fm}$!

4. In a fixed-target experiment positrons are fired at a target of electrons at rest. What positron energy is required to produce a Z ($m_Z = 91.188\,\text{GeV}$)?

3 Kinematics and Cross Sections

> *"The best way to escape a problem is to solve it."*
>
> *Alan Saporta*

In astroparticle physics the energies of participating parti- **relativistic kinematics**
cles are generally that high, that *relativistic kinematics* must
be used. In this field of science it becomes obvious that mass
and energy are only different facets of the same thing. Mass
is a particularly compact form of energy, which is related to
the total energy of a particle by the famous Einstein relation

$$E = mc^2 . \tag{3.1}$$

In this equation m is the mass of a particle, which moves
with the velocity v, and c is the velocity of light in vacuum.

The experimental result that the velocity of light in vac- **relativistic mass increase**
uum is the maximum velocity in all inertial systems leads to
the fact that particles with velocity near the velocity of light
do not get much faster when accelerated, but mainly only
become heavier,

$$m = \frac{m_0}{\sqrt{1 - \beta^2}} = \gamma m_0 . \tag{3.2}$$

In this equation m_0 is the rest mass, $\beta = v/c$ is the particle
velocity, normalized to the velocity of light, and

$$\gamma = \frac{1}{\sqrt{1 - \beta^2}} \tag{3.3}$$

is the Lorentz factor. Using this result, (3.1) can also be writ- **Lorentz factor**
ten as

$$E = \gamma m_0 c^2 , \tag{3.4}$$

where $m_0 c^2$ is the rest energy of a particle. The momentum
of a particle can be expressed as

$$p = mv = \gamma m_0 \beta c . \tag{3.5}$$

Using (3.3), the difference

$$E^2 - p^2 c^2 = \gamma^2 m_0^2 c^4 - \gamma^2 m_0^2 \beta^2 c^4$$

can be written as

$$E^2 - p^2 c^2 = \frac{m_0^2 c^4}{1 - \beta^2}(1 - \beta^2) = m_0^2 c^4 \ . \tag{3.6}$$

invariant mass This result shows that $E^2 - p^2 c^2$ is a Lorentz-invariant quantity. This quantity is the same in all systems and it equals the square of the rest energy. Consequently, the total energy of a relativistic particle can be expressed by

$$E = c\sqrt{p^2 + m_0^2 c^2} \ . \tag{3.7}$$

This equation holds for all particles. For massless particles or, more precisely, particles with rest mass zero, one obtains

$$E = cp \ . \tag{3.8}$$

mass equivalent Particles of total energy E without rest mass are also subject to gravitation, because they acquire a mass according to

$$m = E/c^2 \ . \tag{3.9}$$

classical approximation The transition from relativistic kinematics to classical (Newtonian) mechanics ($p \ll m_0 c$) can also be derived from (3.7) by series expansion. The kinetic energy of a particle is obtained to

$$E^{\text{kin}} = E - m_0 c^2 = c\sqrt{p^2 + m_0^2 c^2} - m_0 c^2$$

$$= m_0 c^2 \sqrt{1 + \left(\frac{p}{m_0 c}\right)^2} - m_0 c^2$$

$$\approx m_0 c^2 \left(1 + \frac{1}{2}\left(\frac{p}{m_0 c}\right)^2\right) - m_0 c^2$$

$$= \frac{p^2}{2m_0} = \frac{1}{2}m_0 v^2 \ , \tag{3.10}$$

in accordance with classical mechanics. Using (3.4) and (3.5), the velocity can be expressed by

$$v = \frac{p}{\gamma m_0} = \frac{c^2 p}{E}$$

or

$$\beta = \frac{cp}{E} \ . \tag{3.11}$$

In relativistic kinematics it is usual to set $c = 1$. This simplifies all formulae. If, however, numerical quantities have to be calculated, the actual value of the velocity of light has to be considered.

3.1 Threshold Energies

> *"Energy has mass and mass represents its energy."*
>
> *Albert Einstein*

In astroparticle physics frequently the problem occurs to determine the *threshold energy* for a certain process of particle production. This requires that in the center-of-mass system of the collision at least the masses of all particles in the final state of the reaction have to be provided. In storage rings the center-of-mass system is frequently identical with the laboratory system so that, for example, the creation of a particle of mass M in an electron–positron head-on collision (e^+ and e^- have the same total energy E) requires

threshold energy

$$2E \geq M . \tag{3.12}$$

If, on the other hand, a particle of energy E interacts with a target at rest as it is characteristic for processes in cosmic rays, the center-of-mass energy for such a process must first be calculated.

determination of the center-of-mass energy

For the general case of a collision of two particles with total energy E_1 and E_2 and momenta p_1 and p_2 the Lorentz-invariant center-of-mass energy E_{CMS} can be determined using (3.7) and (3.11) in the following way:

$$
\begin{aligned}
E_{CMS} &= \sqrt{s} \\
&= \left\{ (E_1 + E_2)^2 - (p_1 + p_2)^2 \right\}^{1/2} \\
&= \left\{ E_1^2 - p_1^2 + E_2^2 - p_2^2 + 2E_1 E_2 - 2p_1 \cdot p_2 \right\}^{1/2} \\
&= \left\{ m_1^2 + m_2^2 + 2E_1 E_2 (1 - \beta_1 \beta_2 \cos\theta) \right\}^{1/2} . \tag{3.13}
\end{aligned}
$$

In this equation θ is the angle between p_1 and p_2. For high energies ($\beta_1, \beta_2 \to 1$ and $m_1, m_2 \ll E_1, E_2$) and not too small angles θ (3.13) simplifies to

$$E_{CMS} = \sqrt{s} \approx \{ 2E_1 E_2 (1 - \cos\theta) \}^{1/2} . \tag{3.14}$$

If one particle (for example, the particle of the mass m_2) is at rest (laboratory system $E_2 = m_2$, $p_2 = 0$), (3.13) leads to

$$\sqrt{s} = \{ m_1^2 + m_2^2 + 2E_1 m_2 \}^{1/2} . \tag{3.15}$$

Using the relativistic approximation ($m_1^2, m_2^2 \ll 2E_1 m_2$) one gets

$$\sqrt{s} \approx \sqrt{2E_1 m_2} \ . \tag{3.16}$$

In such a reaction only particles with total masses $M \leq \sqrt{s}$ can be produced.

$p\bar{p}$ production **Example 1:** Let us assume that a high-energy cosmic-ray proton (energy E_p, momentum $\boldsymbol{p}$, rest mass m_p) produces a proton–antiproton pair on a target proton at rest:

$$p + p \to p + p + p + \bar{p} \ . \tag{3.17}$$

According to (3.13) the center-of-mass energy can be calculated as follows:

$$\begin{aligned}
\sqrt{s} &= \left\{ (E_p + m_p)^2 - (\boldsymbol{p} - 0)^2 \right\}^{1/2} \\
&= \left\{ E_p^2 + 2m_p E_p + m_p^2 - p^2 \right\}^{1/2} \\
&= \left\{ 2m_p E_p + 2m_p^2 \right\}^{1/2} \ .
\end{aligned} \tag{3.18}$$

For the final state, consisting of three protons and one antiproton (the mass of the antiproton is equal to the mass of the proton), one has

$$\sqrt{s} \geq 4m_p \ . \tag{3.19}$$

From this the threshold energy of the incident proton can be derived to be

$$\begin{aligned}
2m_p E_p + 2m_p^2 &\geq 16\,m_p^2 \ , \\
E_p &\geq 7m_p \ (= 6.568\,\text{GeV}) \ , \quad (3.20) \\
E_p^{\text{kin}} = E_p - m_p &\geq 6m_p \ .
\end{aligned}$$

e^+e^- production For the equivalent process of e^+e^- pair production by an energetic electron on an electron target at rest,

$$e^- + e^- \to e^- + e^- + e^+ + e^- \ , \tag{3.21}$$

one would get the corresponding result $E_e^{\text{kin}} \geq 6m_e$.

photo pair production **Example 2:** Let us consider the photoproduction of an electron–positron pair on a target electron at rest,

$$\gamma + e^- \to e^- + e^+ + e^- \ ; \tag{3.22}$$

$$\begin{aligned}
\sqrt{s} = \{ m_e^2 + 2E_\gamma m_e \}^{1/2} &\geq 3m_e \ , \\
E_\gamma &\geq 4m_e \ , \\
E_\gamma &\geq 2.04\,\text{MeV} \ . \quad (3.23)
\end{aligned}$$

Example 3: Consider the photoproduction of a neutral pion $\quad$ **π^0 production**
(mass $m_{\pi^0} \approx 135$ MeV) on a target proton at rest (mass
m_p):

$$\gamma + p \rightarrow p + \pi^0 \; ; \tag{3.24}$$

$$\sqrt{s} = \{m_p^2 + 2E_\gamma m_p\}^{1/2} \geq m_p + m_{\pi^0} \; ,$$
$$m_p^2 + 2E_\gamma m_p \geq m_p^2 + m_{\pi^0}^2 + 2m_p m_{\pi^0} \; ,$$

$$E_\gamma \geq \frac{2m_p m_{\pi^0} + m_{\pi^0}^2}{2m_p} = m_{\pi^0} + \frac{m_{\pi^0}^2}{2m_p} \tag{3.25}$$
$$\geq m_{\pi^0} + 9.7 \, \text{MeV} \approx 145 \, \text{MeV} \; .$$

3.2 Four-Vectors

> *"The physicist in preparing for his work
> needs three things, mathematics, mathe-
> matics, and mathematics."*
>   *Wilhelm Conrad Röntgen*

For calculations of this kind it is practical to introduce $\quad$ **Lorentz-invariant**
Lorentz-invariant *four-vectors*. In the same way as time t $\quad$ **four-vectors**
and the position vector $s = (x, y, z)$ can be combined to
form a four-vector, also a four-momentum vector

$$q = \begin{pmatrix} E \\ p \end{pmatrix} \text{ with } p = (p_x, p_y, p_z) \tag{3.26}$$

can be introduced. Because of

$$q^2 = \begin{pmatrix} E \\ p \end{pmatrix}^2 = E^2 - p^2 = m_0^2 \tag{3.27}$$

the square of the four-momentum is equal to the square of
the rest mass. For photons one has

$$q^2 = E^2 - p^2 = 0 \; . \tag{3.28}$$

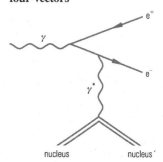

Fig. 3.1
The process
$\gamma + \text{nucleus} \rightarrow e^+ + e^- + \text{nucleus}'$

Those particles, which fulfill (3.27) are said to lie on the $\quad$ **mass shell**
mass shell. On-shell particles are also called real. Apart from
that, particles can also borrow energy for a short time from
the vacuum within the framework of Heisenberg's uncertainty principle. Such particles are called virtual. They are
not on the mass shell. In interaction processes virtual particles can only occur as exchange particles.

e^+e^- pair production in the Coulomb field of a nucleus

Example 4: Photoproduction of an electron–positron pair in the Coulomb field of a nucleus

In this example the incoming photon γ is real, while the photon γ^* exchanged between the electron and the nucleus is virtual (Fig. 3.1).

e^-p scattering

Example 5: Electron–proton scattering (Fig. 3.2)

The virtuality of the exchanged photon γ^* can easily be determined from the kinematics based on the four-momentum vectors of the electron and proton. The four-momentum vectors are defined in the following way: incoming electron $q_e = \binom{E_e}{\boldsymbol{p}_e}$, final-state electron $q'_e = \binom{E'_e}{\boldsymbol{p}'_e}$, incoming proton $q_p = \binom{E_p}{\boldsymbol{p}_p}$, final-state proton $q'_p = \binom{E'_p}{\boldsymbol{p}'_p}$. Since energy and momentum are conserved, also four-momentum conservation holds:

$$q_e + q_p = q'_e + q'_p . \tag{3.29}$$

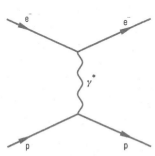

Fig. 3.2
The process $e^- + p \rightarrow e^- + p$

The four-momentum squared of the exchanged virtual photon $q^2_{\gamma^*}$ is determined to be

$$
\begin{aligned}
q^2_{\gamma^*} &= (q_e - q'_e)^2 \\
&= \left(\frac{E_e - E'_e}{\boldsymbol{p}_e - \boldsymbol{p}'_e}\right)^2 = (E_e - E'_e)^2 - (\boldsymbol{p}_e - \boldsymbol{p}'_e)^2 \\
&= E^2_e - \boldsymbol{p}^2_e + E'^2_e - \boldsymbol{p}'^2_e - 2E_e E'_e + 2\boldsymbol{p}_e \cdot \boldsymbol{p}'_e \\
&= 2m^2_e - 2E_e E'_e(1 - \beta_e \beta'_e \cos\theta) , \tag{3.30}
\end{aligned}
$$

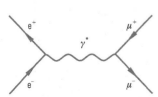

Fig. 3.3
The process $e^+e^- \rightarrow \mu^+\mu^-$

where β_e and β'_e are the velocities of the incoming and outgoing electron and θ is the angle between $\boldsymbol{p}_e$ and $\boldsymbol{p}'_e$. For high energies and not too small scattering angles (3.30) is simplified to

$$
\begin{aligned}
q^2_{\gamma^*} &= -2E_e E'_e(1 - \cos\theta) \\
&= -4E_e E'_e \sin^2\frac{\theta}{2} . \tag{3.31}
\end{aligned}
$$

If $\sin\frac{\theta}{2}$ can be approximated by $\frac{\theta}{2}$, one gets for not too small angles

$$q^2_{\gamma^*} = -E_e E'_e \theta^2 . \tag{3.32}$$

space-like photons

The mass squared of the exchanged photon in this case is negative! This means that the mass of γ^* is purely imaginary. Such photons are called *space-like*.

Example 6: Muon pair production in e^+e^- interactions   **μ pair production**
(Fig. 3.3)
Assuming that electrons and positrons have the same
total energy E and opposite momentum ($p_{e^+} = -p_{e^-}$),
one has

$$q_{\gamma^*}^2 = (q_{e^+} + q_{e^-})^2 = \left(\frac{E + E}{p_{e^+} + (-p_{e^+})} \right)^2$$
$$= 4E^2 . \qquad (3.33)$$

In this case the mass of the exchanged photon is $2E$,   **time-like photons**
which is positive. Such a photon is called *time-like*. The
muon pair in the final state can be created if $2E \geq 2m_\mu$.

The elegant formalism of four-momentum vectors for
the calculation of kinematical relations can be also extended
to decays of elementary particles. In a two-body decay of
an elementary particle at rest the two decay particles get
well-defined discrete energies because of momentum conservation.

Example 7: The decay $\pi^+ \rightarrow \mu^+ + \nu_\mu$ **two-body decay**
Four-momentum conservation yields

$$q_\pi^2 = (q_\mu + q_\nu)^2 = m_\pi^2 . \qquad (3.34)$$

In the rest frame of the pion the muon and neutrino are
emitted in opposite directions, $p_\mu = -p_{\nu_\mu}$,

$$\left(\frac{E_\mu + E_\nu}{p_\mu + p_{\nu_\mu}} \right)^2 = (E_\mu + E_\nu)^2 = m_\pi^2 . \qquad (3.35)$$

Neglecting a possible non-zero neutrino mass for this
consideration, one has

$$E_\nu = p_{\nu_\mu}$$

with the result

$$E_\mu + p_\mu = m_\pi .$$

Rearranging this equation and squaring it gives

$$E_\mu^2 + m_\pi^2 - 2E_\mu m_\pi = p_\mu^2 ,$$
$$2E_\mu m_\pi = m_\pi^2 + m_\mu^2 ,$$
$$E_\mu = \frac{m_\pi^2 + m_\mu^2}{2m_\pi} . \qquad (3.36)$$

decay kinematics

For $m_\mu = 105.658\,369\,\text{MeV}$ and $m_{\pi^\pm} = 139.570\,18$ MeV one gets $E_\mu^{\text{kin}} = E_\mu - m_\mu = 4.09\,\text{MeV}$. For the two-body decay of the kaon, $K^+ \to \mu^+ + \nu_\mu$, (3.36) gives $E_\mu^{\text{kin}} = E_\mu - m_\mu = 152.49\,\text{MeV}$ ($m_{K^\pm} = 493.677\,\text{MeV}$).

Due to helicity conservation the decay $\pi^+ \to e^+ + \nu_e$ is strongly suppressed (see Fig. 2.10). Using (3.36) the positron would get in this decay a kinetic energy of

$$E_{e^+}^{\text{kin}} = E_{e^+} - m_e = \frac{m_\pi}{2} + \frac{m_e^2}{2m_\pi} - m_e = \frac{m_\pi}{2}\left(1 - \frac{m_e}{m_\pi}\right)^2 \approx$$

69.3 MeV, which is approximately half the pion mass. This is not a surprise, since the 'heavy' pion decays into two nearly massless particles.

π^0 decay

Example 8: The decay $\pi^0 \to \gamma + \gamma$

The kinematics of the π^0 decay at rest is extremely simple. Each decay photon gets as energy one half of the pion rest mass. In this example also the decay of a π^0 in flight will be considered. If the photon is emitted in the direction of flight of the π^0, it will get a higher energy compared with the emission opposite to the flight direction. The decay of a π^0 in flight (Lorentz factor $\gamma = E_{\pi^0}/m_{\pi^0}$) yields a flat spectrum of photons between a maximum and minimum energy. Four-momentum conservation

$$q_{\pi^0} = q_{\gamma_1} + q_{\gamma_2}$$

leads to

$$q_{\pi^0}^2 = m_{\pi^0}^2 = q_{\gamma_1}^2 + q_{\gamma_2}^2 + 2q_{\gamma_1}q_{\gamma_2} \,. \tag{3.37}$$

Since the masses of real photons are zero, the kinematic limits are obtained from the relation

$$2q_{\gamma_1}q_{\gamma_2} = m_{\pi^0}^2 \,. \tag{3.38}$$

In the limit of maximum or minimum energy transfer to the photons they are emitted parallel or antiparallel to the direction of flight of the π^0. This leads to

$$\boldsymbol{p}_{\gamma_1} \parallel -\boldsymbol{p}_{\gamma_2} \,. \tag{3.39}$$

Using this, (3.38) can be expressed as

$$2(E_{\gamma_1}E_{\gamma_2} - \boldsymbol{p}_{\gamma_1} \cdot \boldsymbol{p}_{\gamma_2}) = 4E_{\gamma_1}E_{\gamma_2} = m_{\pi^0}^2 \,. \tag{3.40}$$

Because of $E_{\gamma_2} = E_{\pi^0} - E_{\gamma_1}$ (3.40) leads to the quadratic equation

$$E_{\gamma_1}^2 - E_{\gamma_1} E_{\pi^0} + \frac{m_{\pi^0}^2}{4} = 0 \tag{3.41}$$

with the symmetric solutions

**photon spectrum
from π^0 decay**

$$E_{\gamma_1}^{\max} = \frac{1}{2}(E_{\pi^0} + p_{\pi^0}) ,$$
$$E_{\gamma_1}^{\min} = \frac{1}{2}(E_{\pi^0} - p_{\pi^0}) . \tag{3.42}$$

Because of $E_{\pi^0} = \gamma m_{\pi^0}$ and $p_{\pi^0} = \gamma m_{\pi^0} \beta$ (3.42) can also be expressed as

$$E_{\gamma_1}^{\max} = \frac{1}{2}\gamma m_{\pi^0}(1 + \beta) = \frac{1}{2}m_{\pi^0}\sqrt{\frac{1 + \beta}{1 - \beta}} ,$$
$$E_{\gamma_1}^{\min} = \frac{1}{2}\gamma m_{\pi^0}(1 - \beta) = \frac{1}{2}m_{\pi^0}\sqrt{\frac{1 - \beta}{1 + \beta}} . \tag{3.43}$$

In the relativistic limit ($\gamma \gg 1$, $\beta \approx 1$) a photon emitted in the direction of flight of the π^0 gets the energy $E_\gamma^{\max} = E_{\pi^0} = \gamma m_{\pi^0}$ and the energy of the backward-emitted photon is zero.

From (3.43) it is clear that for any energy of a neutral pion, a range of possible photon energies contains $m_{\pi^0}/2$. If one has a spectrum of neutral pions, the energy spectra of the decay photons are superimposed in such a way that the resulting spectrum has a maximum at half the π^0 mass.

Much more difficult is the treatment of a three-body decay. Such a process is going to be explained for the example of the muon decay:

three-body decay

$$\mu^- \rightarrow e^- + \bar{\nu}_e + \nu_\mu . \tag{3.44}$$

Let us assume that the muon is originally at rest ($E_\mu = m_\mu$). Four-momentum conservation

$$q_\mu = q_e + q_{\bar{\nu}_e} + q_{\nu_\mu} \tag{3.45}$$

can be rephrased as

$$(q_\mu - q_e)^2 = (q_{\bar{\nu}_e} + q_{\nu_\mu})^2 ,$$

$$q_\mu^2 + q_e^2 - 2q_\mu q_e = m_\mu^2 + m_e^2 - 2\binom{m_\mu}{0}\binom{E_e}{\boldsymbol{p}_e}$$

$$= (q_{\bar{\nu}_e} + q_{\nu_\mu})^2 ,$$

$$E_e = \frac{m_\mu^2 + m_e^2 - (q_{\bar{\nu}_e} + q_{\nu_\mu})^2}{2m_\mu} . \qquad (3.46)$$

The electron energy is largest, if $(q_{\bar{\nu}_e} + q_{\nu_\mu})^2$ takes on a minimum value. For vanishing neutrino masses this means that the electron gets a maximum energy, if

$$q_{\bar{\nu}_e} q_{\nu_\mu} = E_{\bar{\nu}_e} E_{\nu_\mu} - \boldsymbol{p}_{\bar{\nu}_e} \cdot \boldsymbol{p}_{\nu_\mu} = 0 . \qquad (3.47)$$

Equation (3.47) is satisfied for $\boldsymbol{p}_{\bar{\nu}_e} \parallel \boldsymbol{p}_{\nu_\mu}$. This yields

$$E_e^{\max} = \frac{m_\mu^2 + m_e^2}{2m_\mu} \approx \frac{m_\mu}{2} = 52.83\,\text{MeV} . \qquad (3.48)$$

In this configuration the electron momentum $\boldsymbol{p}_e$ is antiparallel to both neutrino momenta which in turn are parallel to each other.

If the spins of all participating particles and the structure of weak interactions are taken into consideration, one obtains for the electron spectrum, using the shorthand $x = 2E_e/m_\mu \approx E_e/E_e^{\max}$,

$$N(x) = \text{const}\, x^2 (1.5 - x) . \qquad (3.49)$$

Just as in nuclear beta decay ($n \rightarrow p + e^- + \bar{\nu}_e$) the available decay energy in a three-body decay is distributed continuously among the final-state particles (Fig. 3.4).

electron spectrum in muon decay

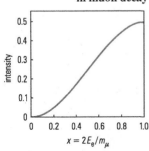

Fig. 3.4
Energy spectrum of electrons from muon decay

3.3 Lorentz Transformation

"We have learned something about the laws of nature, their invariance with respect to the Lorentz transformation, and their validity for all inertial systems moving uniformly, relative to each other. We have the laws but do not know the frame to which to refer them."

Albert Einstein

transformation between laboratory and center-of-mass system

If interaction or decay processes are treated, it is fully sufficient to consider the process in the center-of-mass system. In a different system (for example, the laboratory system) the energies and momenta are obtained by a *Lorentz transformation*. If E and p are energy and momentum in the center-of-mass system and if the laboratory system moves with the velocity β relative to $p_\parallel$, the transformed quantities E^* and $p_\parallel^*$ in this system are calculated to be (compare Fig. 3.5)

$$\begin{pmatrix} E^* \\ p_\parallel^* \end{pmatrix} = \begin{pmatrix} \gamma & -\gamma\beta \\ -\gamma\beta & \gamma \end{pmatrix}\begin{pmatrix} E \\ p_\parallel \end{pmatrix}, \quad p_\perp^* = p_\perp. \quad (3.50)$$

The transverse momentum component is not affected by this transformation. Instead of using the matrix notation, (3.50) can be written as

$$\begin{aligned} E^* &= \gamma E - \gamma\beta p_\parallel, \\ p_\parallel^* &= -\gamma\beta E + \gamma p_\parallel. \end{aligned} \quad (3.51)$$

For $\beta = 0$ and correspondingly $\gamma = 1$ one trivially obtains $E^* = E$ and $p_\parallel^* = p_\parallel$.

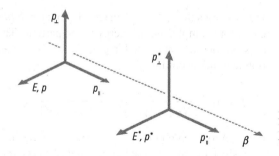

Fig. 3.5
Illustration of a Lorentz transformation

A particle of energy $E = \gamma_2 m_0$, seen from a system which moves with β_1 relative to the particle parallel to the momentum p, gets in this system the energy

$$\begin{aligned} E^* &= \gamma_1 E - \gamma_1\beta_1 p_\parallel \\ &= \gamma_1\gamma_2 m_0 - \gamma_1 \frac{\sqrt{\gamma_1^2-1}}{\gamma_1}\sqrt{(\gamma_2 m_0)^2 - m_0^2} \\ &= \gamma_1\gamma_2 m_0 - m_0\sqrt{\gamma_1^2-1}\sqrt{\gamma_2^2-1}. \end{aligned} \quad (3.52)$$

If $\gamma_1 = \gamma_2 = \gamma$ (for a system that moves along with a particle) one naturally obtains

$$E^* = \gamma^2 m_0 - m_0(\gamma^2 - 1) = m_0.$$

3.4 Cross Sections

"Physicists are, as a general rule, high-brows. They think and talk in long, Latin words, and when they write anything down they usually include at least one partial differential and three Greek letters."

Stephen White

cross section   Apart from the kinematics of interaction processes the *cross section* for a reaction is of particular importance. In the most simple case the cross section can be considered as an effective area which the target particle represents for the collision with a projectile. If the target has an area of πr_T^2 and the projectile size corresponds to πr_P^2, the geometrical cross section for a collision is obtained to be

$$\sigma = \pi (r_T + r_P)^2 . \tag{3.53}$$

In most cases the cross section also depends on other parameters, for example, on the energy of the particle. The atomic cross section σ_A, measured in cm^2, is related to the interac-
interaction length   tion length λ according to

$$\lambda \{cm\} = \frac{A}{N_A \{g^{-1}\} \varrho \{g/cm^3\} \sigma_A \{cm^2\}} \tag{3.54}$$

(N_A – Avogadro number; A – atomic mass of the target, ϱ – density). Frequently, the interaction length is expressed by
absorption coefficient $(\lambda \varrho) \{g/cm^2\}$. Correspondingly, the absorption coefficient is defined to be

$$\mu \{cm^{-1}\} = \frac{N_A \varrho \sigma_A}{A} = \frac{1}{\lambda} ; \tag{3.55}$$

equivalently, the absorption coefficient can also be expressed by $(\mu/\varrho) \{(g/cm^2)^{-1}\}$.
interaction rates   The absorption coefficient also provides a useful relation for the determination of interaction probabilities or rates,

$$\phi \{(g/cm^2)^{-1}\} = \frac{\mu}{\varrho} = \frac{N_A}{A} \sigma_A . \tag{3.56}$$

If σ_N is a cross section per nucleon, one has

$$\phi \{(g/cm^2)^{-1}\} = \sigma_N N_A . \tag{3.57}$$

If j is the particle flux per cm^2 and s, the number of particles dN scattered through an angle θ into the solid angle $d\Omega$ per unit time is

$$dN(\theta) = j\,\sigma(\theta)\,d\Omega\,,\tag{3.58}$$

where

$$\sigma(\theta) = \frac{d\sigma}{d\Omega}$$

is the differential scattering cross section, describing the probability of scattering into the solid-angle element $d\Omega$, where

differential scattering cross section

$$d\Omega = \sin\theta\,d\theta\,d\varphi\tag{3.59}$$

(φ – azimuthal angle, θ – polar angle).
    For azimuthal symmetry one has

$$d\Omega = 2\pi\,\sin\theta\,d\theta = -2\pi\,d(\cos\theta)\,.\tag{3.60}$$

Apart from the angular dependence the cross section can also depend on other quantities, so that a large number of differential cross sections are known, for example,

$$\frac{d\sigma}{dE}\,,\ \frac{d\sigma}{dp}\,,$$

or even double differential cross sections such as

double differential cross section

$$\frac{d^2\sigma}{dE\,d\theta}\,.\tag{3.61}$$

Apart from the mentioned characteristic quantities there is quite a large number of other kinematical variables which are used for the treatment of special processes and decays.

3.5 Problems

1. What is the threshold energy for a photon, E_{γ_1}, to produce a $\mu^+\mu^-$ pair in a collision with a blackbody photon of energy 1 meV?
2. The mean free path λ (in g/cm^2) is related to the nuclear cross section σ_N (in cm^2) by

$$\lambda = \frac{1}{N_A\sigma_N}\,,$$

where N_A is the Avogadro number, i.e., the number of nucleons per g, and σ_N is the cross section per nucleon. The number of particles penetrating a target x unaffected by interactions is

$$N = N_0 \, e^{-x/\lambda} \, .$$

How many collisions happen in a thin target of thickness x ($N_A = 6.022 \times 10^{23} \, g^{-1}$, $\sigma_N = 1 \, b$, $N_0 = 10^8$, $x = 0.1 \, g/cm^2$)?

3. The neutrino was discovered in the reaction

$$\bar{\nu}_e + p \rightarrow n + e^+ \, ,$$

where the target proton was at rest. What is the minimum neutrino energy to induce this reaction?

4. The scattering of a particle of charge z on a target of nuclear charge Z is mediated by the electromagnetic interaction. Work out the momentum transfer p_b, perpendicular to the momentum of the incoming particle for an impact parameter b (distance of closest approach)! For the calculation assume the particle track to be undisturbed, i.e., the scattering angle to be small.

5. The scattering of an electron of momentum p on a target nucleus of charge Z was treated in Problems 4 under the assumption that the scattering angle is small. Work out the general expression for the transverse momentum using the Rutherford scattering formula

$$\tan \frac{\vartheta}{2} = \frac{Z r_e}{b \beta^2} \, , \tag{3.62}$$

where ϑ is the scattering angle.

4 Physics of Particle and Radiation Detection

> *"Every physical effect can be used as a basis for a detector."*
>
> *Anonymous*

The measurement techniques relevant to astroparticle physics are rather diverse. The detection of astroparticles is usually a multistep process. In this field of research, particle detection is mostly indirect. It is important to identify the nature of the astroparticle in a suitable interaction process. The target for interactions is, in many cases, not identical with the detector that measures the interaction products. Cosmic-ray muon neutrinos, for example, interact via neutrino–nucleon interactions in the antarctic ice or in the ocean, subsequently producing charged muons. These muons suffer energy losses from electromagnetic interactions with the ice (water), which produces, among others, Cherenkov radiation. The Cherenkov light is recorded, via the photoelectric effect, by photomultipliers. This is then used to reconstruct the energy and the direction of incidence of the muon, which is approximately identical to the direction of incidence of the primary neutrino.

indirect particle detection

In this chapter, the primary interaction processes will first be described. The processes which are responsible for the detection of the interaction products in the detector will then be presented.

The cross sections for the various processes depend on the particle nature, the particle energy, and the target material. A useful relation to determine the interaction probability ϕ and the event rate is obtained from the atomic- (σ_A) or nuclear-interaction cross section (σ_N) according to

cross section

$$\phi\,\{(\mathrm{g/cm^2})^{-1}\} = \frac{N_A}{A}\sigma_A = N_A\,\{\mathrm{g}^{-1}\}\,\sigma_N\,\{\mathrm{cm^2}\}\,, \quad (4.1)$$

where N_A is Avogadro's number, A is the atomic mass of the target, and σ_A is the atomic cross section in $\mathrm{cm^2/atom}$ (σ_N in $\mathrm{cm^2/nucleon}$), see also (3.56) and (3.57). If the target represents an area density $d\,\{\mathrm{g/cm^2}\}$ and if the flux of primary particles is $F\,\{\mathrm{s}^{-1}\}$, the event rate R is obtained as

event rate

$$R = \phi\,\{(\mathrm{g/cm^2})^{-1}\}\,d\,\{(\mathrm{g/cm^2})\}\,F\,\{(\mathrm{s}^{-1})\}\,. \quad (4.2)$$

4.1 Interactions of Astroparticles

*"Observations are meaningless without
a theory to interpret them."*

Raymond A. Lyttleton

The primary particles carrying astrophysical information are nuclei (protons, helium nuclei, iron nuclei, ...), photons, or neutrinos. These three categories of particles are characterized by completely different interactions. Protons and other **measurement** nuclei will undergo strong interactions. They are also sub- **of primary nuclei** ject to electromagnetic and weak interactions, however, the corresponding cross sections are much smaller than those of strong interactions. Primary nuclei will therefore interact predominantly via processes of strong interactions. A typical interaction cross section for inelastic proton–proton scattering at energies of around $100\,\text{GeV}$ is $\sigma_N \approx 40\,\text{mb}$ ($1\,\text{mb} = 10^{-27}\,\text{cm}^2$). Since high-energy primary protons interact in the atmosphere via proton–air interactions, the cross section for proton–air collisions is of great interest. The dependence of this cross section on the proton energy is shown in Fig. 4.1.

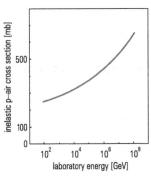

Fig. 4.1
Cross section for proton–air interactions

For a typical interaction cross section of $250\,\text{mb}$, the *mean free path* of protons in the atmosphere (for nitrogen: $A = 14$) is, see Chap. 3, (3.54),

$$\lambda = \frac{A}{N_A\,\sigma_A} \approx 93\,\text{g/cm}^2 . \tag{4.3}$$

This means that the first interaction of protons occurs in the **mean free path** upper part of the atmosphere. If the primary particles are not protons but rather iron nuclei (atomic number $A_{\text{Fe}} = 56$), the first interaction will occur at even higher altitudes because the cross section for iron–air interactions is correspondingly larger.

detection of primary photons   Primary high-energy photons (energy $\gg 10\,\text{MeV}$) interact via the electromagnetic process of electron–positron pair production. The characteristic interaction length[1] ('radiation

---

[1] The radiation length for electrons is defined in (4.7). It describes the degrading of the electron energy by bremsstrahlung according to $E = E_0\,\text{e}^{-x/X_0}$. This 'interaction length' X_0 is also characteristic for pair production by photons. The interaction length for hadrons (protons, pions, ...) is defined through (4.3), where σ_A is the total cross section. This length is sometimes also called collision length. If the total cross section in (4.3) is replaced by its inelastic part only, the resulting length is called absorption length.

length') for electrons in air is $X_0 \approx 36\,\text{g/cm}^2$. For high-energy photons (energy $\geq 10\,\text{GeV}$), where pair production dominates, the cross section is 7/9 of the cross section for electrons ([3], Chap. 1), so the radiation length for photons is 9/7 of that for electrons, i.e., $47\,\text{g/cm}^2$. The first interaction of photon-induced electromagnetic cascades therefore also occurs in the uppermost layers of the atmosphere.

The detection of cosmic-ray neutrinos is completely different. They are only subject to weak interactions (apart from gravitational interactions). The cross section for neutrino–nucleon interactions is given by

**detection
of cosmic-ray neutrinos**

$$\sigma_{\nu N} = 0.7 \times 10^{-38}\, E_\nu\,[\text{GeV}]\,\text{cm}^2/\text{nucleon} . \qquad (4.4)$$

Neutrinos of $100\,\text{GeV}$ possess a tremendously large interaction length in the atmosphere:

$$\lambda \approx 2.4 \times 10^{12}\,\text{g/cm}^2 . \qquad (4.5)$$

The vertex for possible neutrino–air interactions in the atmosphere should consequently be uniformly distributed.

Charged and/or neutral particles are created in the interactions, independent of the identity of the primary particle. These secondary particles will, in general, be recorded by the experiments or telescopes. To achieve this, a large variety of secondary processes can be used.

4.2 Interaction Processes
Used for Particle Detection

> *"I often say when you can measure what you are speaking about, and express it in numbers, you know something about it; but when you cannot measure it, when you cannot express it in numbers, your knowledge is of a meagre and unsatisfying kind."*
>
> *Lord Kelvin (William Thomson)*

Tables 4.1 and 4.2 show the main interaction processes of charged particles and photons, as they are typically used in experiments in astroparticle physics. In this overview, not only the interaction processes are listed, but also the typical detectors that utilize the corresponding interaction processes. The mechanism that dominates charged-particle interactions is the energy loss by ionization and excitation. This energy-loss process is described by the Bethe–Bloch formula:

interaction mechanisms

**energy loss
of charged particles**

Bethe–Bloch formula

Table 4.1
Overview of interaction processes
of charged particles

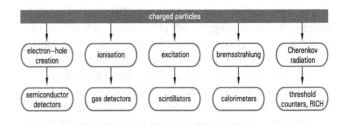

Table 4.2
Overview of interaction processes
of photons

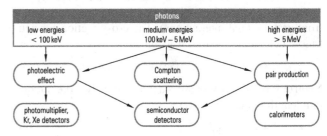

$$-\left.\frac{dE}{dx}\right|_{\text{ion}} = K\,z^2\,\frac{Z}{A}$$

$$\times \frac{1}{\beta^2}\left\{\frac{1}{2}\ln\frac{2m_e c^2\beta^2\gamma^2 T_{\max}}{I^2} - \beta^2 - \frac{\delta}{2}\right\}, \qquad (4.6)$$

where

K – $4\pi N_A r_e^2 m_e c^2 \approx 0.307\,\text{MeV}/(\text{g/cm}^2)$,

N_A – Avogadro's number,

r_e – classical electron radius ($\approx 2.82\,\text{fm}$),

$m_e c^2$ – electron rest energy ($\approx 511\,\text{keV}$),

z – charge number of the incident particle,

Z, A – target charge number and target mass number,

β – velocity ($= v/c$) of the incident particle,

γ – $1/\sqrt{1-\beta^2}$,

$T_{\max}$ – $\dfrac{2m_e p^2}{m_0^2 + m_e^2 + 2m_e E/c^2}$

         maximum energy transfer to an electron,

         m_0 – mass of the incident particle,

         p, E – momentum and total energy

                of the projectile,

I – average ionization energy of the target,

δ – density correction.

energy-loss dependence

The energy loss of charged particles, according to the Bethe–Bloch relation, is illustrated in Fig. 4.2. It exhibits a $1/\beta^2$ increase at low energies. The minimum ionization rate occurs at around $\beta\gamma \approx 3.5$. This feature is called the

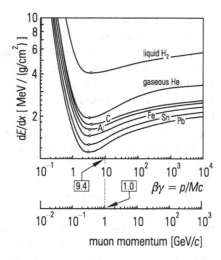

Fig. 4.2
Energy loss of charged particles in various targets [2]

minimum of ionization, and particles with such $\beta\gamma$ values are said to be *minimum ionizing*. For high energies, the energy loss increases logarithmically ('relativistic rise') and reaches a plateau ('Fermi plateau') owing to the density effect. The energy loss of gases in the plateau region is typically 60% higher compared to the ionization minimum. The energy loss of singly charged minimum-ionizing particles by ionization and excitation in air is $1.8\,\mathrm{MeV}/(\mathrm{g}/\mathrm{cm}^2)$ and $2.0\,\mathrm{MeV}/(\mathrm{g}/\mathrm{cm}^2)$ in water (ice).

Equation (4.6) only describes the average energy loss of charged particles. The energy loss is distributed around the most probable value by an asymmetric Landau distribution. The average energy loss is about twice as large as the most probable energy loss. The ionization energy loss is the basis of a large number of particle detectors.

Landau distribution

In particle astronomy, the Fly's Eye technique takes advantage of the scintillation mechanism in air to detect particles with energies $\geq$ EeV ($\geq 10^{18}\,\mathrm{eV}$). In this experiment, the atmosphere represents the target for the primary particle. The interaction products create scintillation light in the air, which is recorded by photomultipliers mounted in the focal plane of mirrors on the surface of the Earth.

Fly's Eye

For high energies, the bremsstrahlung process becomes significant. The energy loss of electrons due to this process can be described by

bremsstrahlung

$$-\left.\frac{\mathrm{d}E}{\mathrm{d}x}\right|_{\mathrm{brems}} = 4\alpha N_{\mathrm{A}}\frac{Z^2}{A}r_e^2 E \ln\frac{183}{Z^{1/3}} = \frac{E}{X_0}\,, \qquad (4.7)$$

radiation length

direct pair production

nuclear interactions

energy loss of muons

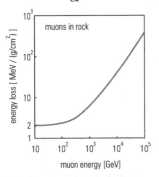

Fig. 4.3
Energy loss of muons in standard
rock

muon calorimetry

where α is the fine-structure constant ($\alpha^{-1} \approx 137$). The definition of the *radiation length* X_0 is evident from (4.7). The other quantities in (4.7) have the same meanings as in (4.6).

Energy loss due to bremsstrahlung is of particular importance for electrons. For heavy particles, the bremsstrahlung energy loss is suppressed by the factor $1/m^2$. The energy loss, however, increases linearly with energy, and is therefore important for all particles at high energies.

In addition to bremsstrahlung, charged particles can also lose some of their energy by direct electron–positron pair production, or by nuclear interactions. The energy loss due to these two interaction processes also varies linearly with energy. Muons as secondary particles in astroparticle physics play a dominant rôle in particle-detection techniques, e.g., in neutrino astronomy. Muons are not subject to strong interactions and they can consequently travel relatively large distances. This makes them important for particle detection in astroparticle physics. The total energy loss of muons can be described by:

$$-\frac{dE}{dx}\bigg|_{\text{muon}} = a(E) + b(E)\, E \ , \qquad (4.8)$$

where $a(E)$ describes the ionization energy loss, and $b(E)\,E$ summarizes the processes of muon bremsstrahlung, direct electron pair creation, and nuclear interaction. The energy loss of muons in standard rock depends on their energy. It is displayed in Fig. 4.3.

For particles with high energies, the total energy loss is dominated by bremsstrahlung and the processes that depend linearly on the particles' energies. These energy-loss mechanisms are therefore used as a basis for particle calorimetry. In calorimetric techniques, the total energy of a particle is dissipated in an active detector medium. The output signal of such a calorimeter is proportional to the absorbed energy. In this context, electrons and photons with energies exceeding 100 MeV can already be considered as high-energy particles because they initiate electromagnetic cascades. The mass of the muon is much larger than that of the electron, making *muon calorimetry* via energy-loss measurements only possible for energies beyond ≈ 1 TeV. This calorimetric technique is of particular importance in the field of TeV neutrino astronomy.

4.3 Principles of the Atmospheric Air Cherenkov Technique

> *"A great pleasure in life is doing what people say you cannot do."*
>
> Walter Bagehot

The atmospheric Cherenkov technique is becoming increasingly popular for TeV γ astronomy since it allows to identify photon-induced electromagnetic showers, which develop in the atmosphere. A charged particle that moves in a medium with refractive index n, and has a velocity v that exceeds the velocity of light $c_n = c/n$, emits electromagnetic radiation known as *Cherenkov radiation*. There is a threshold effect for this kind of energy loss; Cherenkov radiation only occurs if

Cherenkov effect

$$v \geq \frac{c}{n} \quad \text{or, equivalently,} \quad \beta = \frac{v}{c} \geq \frac{1}{n}. \tag{4.9}$$

Cherenkov radiation is emitted at an angle of

$$\theta_C = \arccos \frac{1}{n\beta} \tag{4.10}$$

relative to the direction of the particle velocity. Due to this process, a particle of charge number z creates a certain number of photons in the visible spectral range ($\lambda_1 = 400$ nm up to $\lambda_2 = 700$ nm). The number of photons is calculated from the following equation:

$$\frac{dN}{dx} = 2\pi\alpha z^2 \frac{\lambda_2 - \lambda_1}{\lambda_1 \lambda_2} \sin^2\theta_C$$
$$\approx 490\, z^2 \sin^2\theta_C\, \text{cm}^{-1}. \tag{4.11}$$

These photons are emitted isotropically (about the axis) in azimuth. For relativistic particles ($\beta \approx 1$), the Cherenkov angle is 42° in water, and 1.4° in air. In water, around 220 photons per centimeter are produced by a singly charged relativistic particle. The corresponding number in air is 30 photons per meter. Figure 4.4 shows the variation of the Cherenkov angle and the photon yield, with the particle velocity for water and air. The atmospheric Cherenkov technique permits the identification of photon-induced electromagnetic showers that develop in the atmosphere, and separates them from the more abundant hadronic cascades. This is possible because the recorded Cherenkov pattern is different for electromagnetic and hadronic cascades; also photons

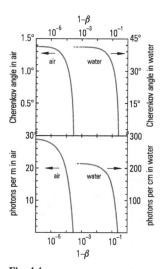

Fig. 4.4
Variation of the Cherenkov angle and photon yield of singly charged particles in water and air

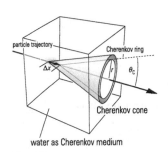

Fig. 4.5
Production of a Cherenkov ring in
a water Cherenkov counter

point back to their sources, while hadrons only produce an isotropic background. The axis of the Cherenkov cone follows the direction of incidence of the primary photon. The Cherenkov cone for γ-induced cascades in air spans only $\pm 1.4°$, therefore the hadronic background in such a small angular range is relatively small.

Apart from the atmospheric Cherenkov technique, the Cherenkov effect is also utilized in large water Cherenkov detectors for neutrino astronomy. The operation principle of a water Cherenkov counter is sketched in Fig. 4.5. Cherenkov radiation is emitted along a distance Δx. The Cherenkov cone projects an image on the detector surface, which is at a distance d from the source. The image is a ring with an average radius

$$r = d \tan \theta_C \; . \tag{4.12}$$

4.4 Special Aspects of Photon Detection

*"Are not the rays of light very
small bodies emitted from shining
substances?"*

Sir Isaac Newton

The detection of photons is more indirect compared to charged particles. Photons first have to create charged particles in an interaction process. These charged particles will then be detected via the processes described, such as ionization, excitation, bremsstrahlung, and the production of Cherenkov radiation.

At comparatively low energies, as in X-ray astronomy, photons can be imaged by reflections at grazing incidence. Photons are detected in the focal plane of an X-ray telescope via the photoelectric effect. Semiconductor counters, X-ray CCDs[2], or multiwire proportional chambers filled with a noble gas of high atomic number (e.g., krypton, xenon) can be used for focal detectors. These types of detectors provide spatial details, as well as energy information.

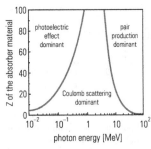

Fig. 4.6
Domains, in which various photon
interactions dominate, shown in
their dependence on the photon
energy and the nuclear charge of
the absorber

The Compton effect dominates for photons at MeV energies (see Fig. 4.6). In *Compton scattering*, a photon of energy E_γ transfers part of its energy ΔE to a target electron, thereby being redshifted. Based on the reaction kinematics,

---

[2] CCD – Charge-Coupled Device (solid-state ionization chamber)

the ratio of the scattered photon energy E'_γ to the incident photon energy E_γ can be derived:

$$\frac{E'_\gamma}{E_\gamma} = \frac{1}{1 + \varepsilon(1 - \cos\theta_\gamma)} \cdot \qquad (4.13)$$

In this equation, $\varepsilon = E_\gamma/m_e c^2$ is the reduced photon energy and θ_γ is the scattering angle of the photon in the γ-electron interaction. With a Compton telescope, not only the energy, but also the direction of incidence of the photons can be determined. In such a telescope, the energy loss of the Compton-scattered photon $\Delta E = E_\gamma - E'_\gamma$ is determined in the upper detector layer by measuring the energy of the Compton electron (see Fig. 4.7). The Compton-scattered photon of reduced energy will subsequently be detected in the lower detector plane, preferentially by the photoelectric effect. Based on the kinematics of the scattering process and using (4.13), the scattering angle θ_γ can be determined. As a consequence of the isotropic emission around the azimuth, the reconstructed photon direction does not point back to a unique position in the sky; it only defines a circle in the sky. If, however, many photons are recorded from the source, the intercepts of these circles define the position of the source. The detection of photons via the Compton effect in such Compton telescopes is usually performed using segmented large-area inorganic or organic scintillation counters that are read out via photomultipliers. Alternatively, for high-resolution telescopes, semiconductor pixel detectors can also be used. This 'ordinary' Compton process is taken advantage of for photon detection. In astrophysical sources the inverse Compton scattering plays an important rôle. In such a process a low-energy photon might gain substantial energy in a collision with an energetic electron, and it can be shifted into the X-ray or γ-ray domain.

At high photon energies, the process of electron–positron pair creation dominates. Similarly to Compton telescopes, the electron and positron tracks enable the direction of the incident photon to be determined. The photon energy is obtained from the sum of the electron and positron energy. This is normally determined in electromagnetic calorimeters, in which electrons and positrons deposit their energy to the detector medium in alternating bremsstrahlung and pair-production processes. These electromagnetic calorimeters can be total-absorption crystal detectors such as NaI or CsI, or they can be constructed from the so-called sandwich

Compton telescope

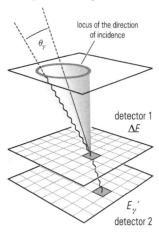

Fig. 4.7
Schematic of a Compton telescope

inverse Compton scattering

electron–positron pair production

electromagnetic calorimeter

sandwich calorimeter

principle. A sandwich calorimeter is a system where absorber and detector layers alternate. Particle multiplication occurs preferentially in the passive absorber sheets, whilst the shower of particles produced is recorded in the active detector layers. Sandwich calorimeters can be compactly constructed and highly segmented, however, they are inferior to crystal calorimeters as far as the energy resolution is concerned.

4.5 Cryogenic Detection Techniques

> *"Ice vendor to his son: 'Stick to it, there is a future in cryogenics.'"*
>
> *Anonymous*

cryogenic detectors

The detection of very small energies can be performed in cryogenic detectors. Cooper bonds in superconductors can already be broken by energy deposits as low as $1\,\mathrm{meV}$ ($= 10^{-3}\,\mathrm{eV}$). The method of classical calorimetry can also be used for particle detection at low temperatures. At low temperatures the specific heat of solids varies with the cube of the temperature ($c_{\mathrm{sp}} \sim T^3$), therefore even the smallest energy deposits ΔE provide a measurable temperature signal. These detectors are mainly used in the search for hypothetical weakly interacting massive particles (WIMPs). To this category also supersymmetric partners of ordinary particles belong. On the other hand, cryogenic detectors are also employed in the realm of high-resolution X-ray spectroscopy where energy resolutions of several eV can be achieved.

4.6 Propagation and Interactions of Astroparticles in Galactic and Extragalactic Space

> *"Space tells matter how to move ... and matter tells space how to curve."*
>
> *John A. Wheeler*

propagation of astroparticles

Now that the principles for the detection of primary and secondary particles have been described, the interactions of astroparticles traveling from their sources to Earth through galactic and extragalactic space shall be briefly discussed.

neutrinos

Neutrinos are only subject to weak interactions with matter, so their range is extremely large. The galactic or intergalactic space does not attenuate the neutrino flux, and magnetic fields do not affect their direction; therefore, neutrinos point directly back to their sources.

The matter density in our galaxy, and particularly in intergalactic space, is very low. This signifies that the ionization energy loss of primary protons traveling from their sources to Earth is extremely small. Protons can, however, interact with cosmic photons. Blackbody photons, in particular, represent a very-high-density target (≈ 400 photons/cm^3). The energy of these photons is very low, typically 250 µeV, and they follow a Planck distribution (Fig. 4.8), see also Chap. 11 on 'The Cosmic Microwave Background'.

The process of pion production by blackbody photons interacting with high-energy protons requires the proton energy to exceed a certain threshold (Greisen–Zatsepin–Kuzmin cutoff). This threshold is reached if photo–pion production via the Δ resonance is kinematically possible in the photon–proton center-of-mass system ($p + \gamma \rightarrow p + \pi^0$). If protons exceed this threshold energy, they quickly lose their energy and fall below the threshold. The GZK cutoff limits the mean free path of the highest-energy cosmic rays (energy $> 6 \times 10^{19}$ eV) to less than a few tens of megaparsecs, quite a small distance in comparison to typical extragalactic scales. Of course, energetic protons lose also energy by inverse Compton scattering on blackbody photons. In contrast to π^0 production via the Δ resonance this process has no threshold. Moreover, the cross section varies like $1/s$, i.e., with the inverse square of the available center-of-mass energy. Compared to the resonant π^0 production the cross section for inverse Compton scattering of protons on blackbody photons is small and therefore has no significant influence on the shape of the primary proton spectrum. A further possible process, $p + \gamma \rightarrow p + e^+ + e^-$, even though it has a lower threshold than $p + \gamma \rightarrow p + \pi^0$, does not proceed through a resonance, and therefore its influence on the propagation of energetic protons in the dense photon field is of little importance. In addition, primary protons (charged particles) naturally interact with the galactic and extragalactic magnetic fields as well as the Earth's magnetic field. Only the most energetic protons (energy $\gg 10^{18}$ eV), which experience a sufficiently small magnetic deflection, can be used for particle astronomy.

Photons are not influenced by magnetic fields. They do, like protons, however, interact with blackbody photons to create electron–positron pairs via the $\gamma\gamma \rightarrow e^+ e^-$ process.

protons and nuclei

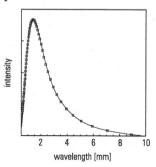

Fig. 4.8
Blackbody spectrum of cosmic microwave background photons

GZK cutoff

magnetic deflection

galactic photon absorption   Owing to the low electron and positron masses, the threshold energy for this process is only about $\approx 10^{15}$ eV. The
attenuation of primary photons (by interactions with blackbody photons), as a function of the primary photon energy, is
shown in Fig. 4.9 for several distances to the γ-ray sources.
A potentially competing process, $\gamma\gamma \rightarrow \gamma\gamma$, is connected
with a very small cross section (it is proportional to the
fourth power of the fine-stucture constant). In addition, the
angular deflection of the photons due to this process is extremely small.

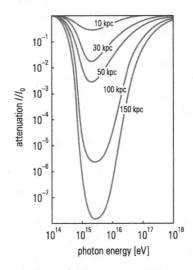

Fig. 4.9
Attenuation of the intensity of
energetic primary cosmic photons
by interactions with blackbody
radiation

The mean free path for energetic photons (energy
$> 10^{15}$ eV) is limited to a few tens of kiloparsecs by this
process. For higher energies, $\gamma\gamma$ processes with different
final states ($\mu^+\mu^-, \ldots$) also occur. For lower energies, photons are attenuated by interactions with infrared or starlight
photons.

4.7 Characteristic Features of Detectors

> *"Detectors can be classified into three*
> *categories – those that don't work, those*
> *that break down, and those that get*
> *lost."*
>
> *Anonymous*

The secondary interaction products of astroparticles are detected in an appropriate device, which can be a detector on

board of a satellite, in a balloon, or at ground level, or even in an underground laboratory. The quality of the measurement depends on the energy and position resolution of the detector. In most cases the ionization energy loss is the relevant detection mechanism.

energy and position resolution

In gaseous detectors an average of typically 30 eV is required to produce an electron–ion pair. The liberated charges are collected in an external electric field and produce an electric signal which can be further processed. In contrast, in solid-state detectors, the average energy for the creation of an electron–hole pair is only ≈ 3 eV, resulting in an improved energy resolution. If, instead, excitation photons produced by the process of scintillation in a crystal detector are recorded, e.g., by photomultipliers, energy deposits of about 25 eV are necessary to yield a scintillation photon in inorganic materials (like NaI(Tl)), while in organic crystals ≈ 100 eV are required to create a scintillation photon. In cryogenic detectors much less energy is needed to produce charge carriers. This substantial advantage which gives rise to excellent energy resolutions is only obtained at the expense of operating the detectors at cryogenic temperatures, mostly in the milli-Kelvin range.

solid-state detectors

scintillators

cryogenic devices

4.8 Problems

1. Show that (4.1) is dimensionally correct.
2. The average energy required for the production of
   a) a photon in a plastic scintillator is 100 eV,
   b) an electron–ion pair in air is 30 eV,
   c) an electron–hole pair in silicon is 3.65 eV,
   d) a quasiparticle (break-up of a Cooper pair in a superconductor) is 1 meV.

   What is the relative energy resolution in these counters for a stopping 10 keV particle assuming Poisson statistics (neglecting the Fano effect[3])?

---

[3] For any specific value of a particle energy the fluctuations of secondary particle production (like electron–ion pairs) are smaller than might be expected according to a Poissonian distribution. This is a simple consequence of the fact that the total energy loss is constrained by the fixed energy of the incident particle. This leads to a standard error of $\sigma = \sqrt{F\,N}$, where N is the number of produced secondaries and F, the Fano factor, is smaller than 1.

3. The simplified energy loss of a muon is parameterized by (4.8). Work out the range of a muon of energy E ($= 100\,\text{GeV}$) in rock ($\varrho_{rock} = 2.5\,\text{g/cm}^3$) under the assumption that a ($= 2\,\text{MeV}/(\text{g/cm}^2)$) and b ($= 4.4 \times 10^{-6}\,\text{cm}^2/\text{g}$ for rock) are energy independent.

4. Show that the mass of a charged particle can be inferred from the Cherenkov angle θ_C and momentum p by

$$m_0 = \frac{p}{c}\sqrt{n^2\cos^2\theta_C - 1}\,,$$

where n is the index of refraction.

5. In a cryogenic argon calorimeter ($T = 1.1$ Kelvin, mass 1 g) a WIMP (weakly interacting massive particle) deposits 10 keV. By how much does the temperature rise? (The specific heat of argon at 1.1 K is $c_{sp} = 8 \times 10^{-5}\,\text{J}/(\text{g\,K})$.)

6. Derive (4.13) using four-momenta.

7. Work out the maximum energy which can be transferred to an electron in a Compton process! As an example use the photon transition energy of 662 keV emitted by an excited ^{137}Ba nucleus after a beta decay from ^{137}Cs,

$$^{137}\text{Cs} \rightarrow {}^{137}\text{Ba}^* + e^- + \bar{\nu}_e$$
$$\hookrightarrow {}^{137}\text{Ba} + \gamma\ (662\,\text{keV})\ \cdot$$

What kind of energy does the electron get for infinitely large photon energies? Is there, on the other hand, a minimum energy for the backscattered photon in this limit?

8. Figure 4.2 shows the energy loss of charged particles as given by the Bethe–Bloch formula. The abscissa is given as momentum and also as product of the normalized velocity β and the Lorentz factor γ. Show that $\beta\gamma = p/m_0 c$ holds.

9. Equation (4.11) shows that the number of emitted Cherenkov photons N is proportional to $1/\lambda^2$. The wavelength for X-ray photons is shorter than that for the visible light region. Why then is Cherenkov light not emitted in the X-ray region?

5 Acceleration Mechanisms

> *"Physics also solves puzzles. However,*
> *these puzzles are not posed by mankind, but*
> *rather by nature."*
>
> *Maria Goeppert-Mayer*

The origin of cosmic rays is one of the major unsolved astrophysical problems. The highest-energy cosmic rays possess macroscopic energies and their origin is likely to be associated with the most energetic processes in the universe. When discussing cosmic-ray origin, one must in principle distinguish between the power source and the acceleration mechanism. Cosmic rays can be produced by particle interactions at the sites of acceleration like in pulsars. The acceleration mechanism can, of course, also be based on conventional physics using electromagnetic or gravitational potentials such as in supernova remnants or active galactic nuclei. One generally assumes that in most cases cosmic-ray particles are not only produced in the sources but also accelerated to high energies in or near the source. Candidate sites for cosmic-ray production and acceleration are supernova explosions, highly magnetized spinning neutron stars, i.e., pulsars, accreting black holes, and the centers of active galactic nuclei. However, it is also possible that cosmic-ray particles powered by some source experience acceleration during the propagation in the interstellar or intergalactic medium by interactions with extensive gas clouds. These gas clouds are created by magnetic-field irregularities and charged particles can gain energy while they scatter off the constituents of these 'magnetic clouds'.

    In *top–down scenarios* energetic cosmic rays can also be produced by the decay of topological defects, domain walls, or cosmic strings, which could be relics of the Big Bang.

    There is a large number of models for cosmic-ray acceleration. This appears to indicate that the actual acceleration mechanisms are not completely understood and identified. On the other hand, it is also possible that different mechanisms are at work for different energies. In the following the most plausible ideas about cosmic-ray-particle acceleration will be presented.

origin of cosmic rays

acceleration mechanisms

supernovae
pulsars
black holes
active galactic nuclei

top–down scenario

5.1 Cyclotron Mechanism

"Happy is he who gets to know the reason for things."

Virgil

Even normal stars can accelerate charged particles up to the GeV range. This acceleration can occur in time-dependent magnetic fields. These magnetic sites appear as star spots **sunspots** or sunspots, respectively. The temperature of sunspots is slightly lower compared to the surrounding regions. They appear darker, because part of the thermal energy has been transformed into magnetic field energy. Sunspots in typical stars can be associated with magnetic field strengths of up to 1000 Gauss (1 Tesla = 10^4 Gauss). The lifetime of such sunspots can exceed several rotation periods. The spatial extension of sunspots on the Sun can be as large as 10^9 cm. The observed Zeeman splitting of spectral lines has shown beyond any doubt that magnetic fields are responsible for the sunspots. Since the Zeeman splitting of spectral lines depends on the magnetic field strength, this fact can also be used to measure the strength of the magnetic fields on stars.

creation of magnetic fields   The magnetic fields in the Sun are generated by turbulent plasma motions where the plasma consists essentially of protons and electrons. The motions of this plasma constitute currents which produce magnetic fields. When these magnetic fields are generated and when they decay, electric fields are created in which protons and electrons can be accelerated.

Figure 5.1 shows schematically a sunspot of extension $A = \pi R^2$ with a variable magnetic field $\boldsymbol{B}$.

The time-dependent change of the magnetic flux ϕ produces a potential U,

$$-\frac{\mathrm{d}\phi}{\mathrm{d}t} = \oint \boldsymbol{E} \cdot \mathrm{d}\boldsymbol{s} = U \tag{5.1}$$

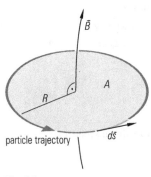

particle trajectory

Fig. 5.1
Principle of particle acceleration by variable sunspots

($\boldsymbol{E}$ – electrical field strength, d$\boldsymbol{s}$ – infinitesimal distance along the particle trajectory). The magnetic flux is given by

$$\phi = \int \boldsymbol{B} \cdot \mathrm{d}\boldsymbol{A} = B\pi R^2 , \tag{5.2}$$

circular acceleration   where d$\boldsymbol{A}$ is the infinitesimal area element. In this equation it is assumed that $\boldsymbol{B}$ is perpendicular to the area, i.e., $\boldsymbol{B} \parallel \boldsymbol{A}$, (the vector $\boldsymbol{A}$ is always perpendicular to the area). One turn of a charged particle around the time-dependent magnetic field leads to an energy gain of

$$E = eU = e\pi R^2 \frac{dB}{dt} \ . \tag{5.3}$$

A sunspot of an extension $R = 10^9$ cm and magnetic field $B = 2000$ Gauss at a lifetime of one day ($\frac{dB}{dt} = 2000$ Gauss/day) leads to

$$E = 1.6 \times 10^{-19} \, \text{A s} \, \pi \, 10^{14} \, \text{m}^2 \, \frac{0.2 \, \text{V s}}{86\,400 \, \text{s m}^2}$$

$$= 1.16 \times 10^{-10} \, \text{J} = 0.73 \, \text{GeV} \ . \tag{5.4}$$

Actually, particles from the Sun with energies up to 100 GeV have been observed. This, however, might also represent the limit for the acceleration power of stars based on the cyclotron mechanism.

The cyclotron model can explain the correct energies, however, it does not explain why charged particles propagate in circular orbits around time-dependent magnetic fields. Circular orbits are only stable in the presence of guiding forces such as they are used in earthbound accelerators.

5.2 Acceleration by Sunspot Pairs

> *"Living on Earth may be expensive, but it includes an annual free trip around the Sun."*
>
> *Ashleigh Brilliant*

Sunspots often come in pairs of opposite magnetic polarity (see Fig. 5.2).

The sunspots normally approach each other and merge at a later time. Let us assume that the left sunspot is at rest and the right one approaches the first sunspot with a velocity v. The moving magnetic dipole produces an electric field perpendicular to the direction of the dipole and perpendicular to its direction of motion v, i.e., parallel to $v \times B$. Typical solar magnetic sunspots can create electrical fields of 10 V/m. In spite of such a low field strength, protons can be accelerated since the collision energy loss is smaller than the energy gain in the low-density chromosphere. Under realistic assumptions (distance of sunspots 10^7 m, magnetic field strengths 2000 Gauss, relative velocity $v = 10^7$ m/day) particle energies in the GeV range are obtained. This shows that the model of particle acceleration in approaching magnetic dipoles can only explain energies which can also be

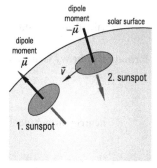

Fig. 5.2
Sketch of a sunspot pair

acceleration to GeV energies

provided by the cyclotron mechanism. The mechanism of approaching sunspots, however, sounds more plausible because in this case no guiding forces (like in the cyclotron model) are required.

5.3 Shock Acceleration

"Basic research is what I am doing when I don't know what I am doing."
Wernher von Braun

gravitational collapse

successive fusion processes

If a massive star has exhausted its hydrogen, the radiation pressure can no longer withstand the gravitational pressure and the star will collapse under its own gravity. The liberated gravitational energy increases the central temperature of a massive star to such an extent that helium burning can start. If the helium reservoir is used up, the process of gravitational infall of matter repeats itself until the temperature is further increased so that the products of helium themselves can initiate fusion processes. These successive fusion processes can lead at most to elements of the iron group (Fe, Co, Ni). For higher nuclear charges the fusion reaction is endotherm, which means that without providing additional energy heavier elements cannot be synthesized. When the fusion process stops at iron, the massive star will implode. In this process part of its mass will be ejected into interstellar space. This material can be recycled for the production of a new star generation which will contain – like the Sun – some heavy elements. As a result of the implosion a compact neutron star will be formed that has a density which is comparable to the density of atomic nuclei. In the course of a supernova explosion some elements heavier than iron are produced if the copiously available neutrons are attached to the elements of the iron group, which – with successive β^- decays – allows elements with higher nuclear charge to be formed:

birth of neutron stars

formation of heavy elements

neutron attachment

$$^{56}_{26}\text{Fe} + n \rightarrow {}^{57}_{26}\text{Fe} \ ,$$
$$^{57}_{26}\text{Fe} + n \rightarrow {}^{58}_{26}\text{Fe} \ ,$$
$$^{58}_{26}\text{Fe} + n \rightarrow {}^{59}_{26}\text{Fe}^* \tag{5.5}$$
$$\hookrightarrow {}^{59}_{27}\text{Co} + e^- + \bar{\nu}_e \ ,$$
$$^{59}_{27}\text{Co} + n \rightarrow {}^{60}_{27}\text{Co}^*$$
$$\hookrightarrow {}^{60}_{28}\text{Ni} + e^- + \bar{\nu}_e \ .$$

The ejected envelope of a supernova represents a *shock front* with respect to the interstellar medium. Let us assume that the shock front moves at a velocity u_1. Behind the shock front the gas recedes with a velocity u_2. This means that the gas has a velocity $u_1 - u_2$ in the laboratory system (see Fig. 5.3).

A particle of velocity v colliding with the shock front and being reflected gains the energy

$$\Delta E = \frac{1}{2}m(v + (u_1 - u_2))^2 - \frac{1}{2}mv^2$$

$$= \frac{1}{2}m(2v(u_1 - u_2) + (u_1 - u_2)^2) . \qquad (5.6)$$

Since the linear term dominates ($v \gg u_1, u_2, u_1 > u_2$), this simple model provides a relative energy gain of

$$\frac{\Delta E}{E} \approx \frac{2(u_1 - u_2)}{v} . \qquad (5.7)$$

A more general, relativistic treatment of shock acceleration including also variable scattering angles leads to

$$\frac{\Delta E}{E} = \frac{4}{3}\frac{u_1 - u_2}{c} , \qquad (5.8)$$

where it has been assumed that the particle velocity v can be approximated by the speed of light c. Similar results are obtained if one assumes that particles are trapped between two shock fronts and are reflected back and forth from the fronts.

Usually the inner front will have a much higher velocity (v_2) compared to the outer front (v_1), which is decelerated already in interactions with the interstellar material (Fig. 5.4). The inner shock front can provide velocities up to 20 000 km/s, as obtained from measurements of the Doppler shift of the ejected gas. The outer front spreads into the interstellar medium with velocities between some 100 km/s up to 1000 km/s. For shock accelerations in active galactic nuclei even superfast shocks with $v_2 = 0.9\,c$ are discussed.

A particle of velocity v being reflected at the inner shock fronts gains the energy

$$\Delta E_1 = \frac{1}{2}m(v+v_2)^2 - \frac{1}{2}mv^2 = \frac{1}{2}m(v_2^2 + 2vv_2) . \quad (5.9)$$

Reflection at the outer shock front leads to an energy loss

$$\Delta E_2 = \frac{1}{2}m(v-v_1)^2 - \frac{1}{2}mv^2 = \frac{1}{2}m(v_1^2 - 2vv_1) . \quad (5.10)$$

shock acceleration

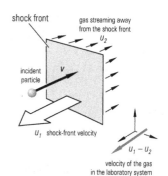

Fig. 5.3
Schematics of shock-wave acceleration

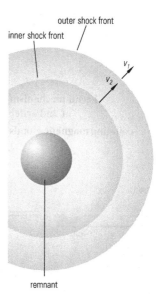

Fig. 5.4
Particle acceleration by multiple reflection between two shock fronts

On average, however, the particle gains an energy

$$\Delta E = \frac{1}{2}m(v_1^2 + v_2^2 + 2v(v_2 - v_1)) \ . \tag{5.11}$$

Since the quadratic terms can be neglected and because of $v_2 > v_1$, one gets

$$\Delta E \approx mv\Delta v \ , \quad \frac{\Delta E}{E} \approx 2\frac{\Delta v}{v} \ . \tag{5.12}$$

This calculation followed similar arguments as in (5.6) and (5.7).

Both presented shock acceleration mechanisms are linear in the relative velocity. Sometimes this type of shock acceleration is called Fermi mechanism of first order. Under plausible conditions using the relativistic treatment, maximum energies of about 100 TeV can be explained in this way.

Fermi mechanism of 1st order acceleration to 100 TeV *(margin note)*

5.4 Fermi Mechanism

> *"Results! Why man, I have gotten a lot of results, I know several thousand things that don't work."*
>
> *Thomas Edison*

Fermi mechanism of 2nd order colliding magnetic clouds *(margin note)*

Fermi mechanism of second order (or more general Fermi mechanism) describes the interaction of cosmic-ray particles with magnetic clouds. At first sight it appears improbable that particles can gain energy in this way. Let us assume that a particle (with velocity v) is reflected from a gas cloud which moves with a velocity u (Fig. 5.5).

If v and u are antiparallel, the particle gains the energy

$$\Delta E_1 = \frac{1}{2}m(v+u)^2 - \frac{1}{2}mv^2 = \frac{1}{2}m(2uv + u^2) \ . \tag{5.13}$$

In case that v and u are parallel, the particle loses an energy

$$\Delta E_2 = \frac{1}{2}m(v-u)^2 - \frac{1}{2}mv^2 = \frac{1}{2}m(-2uv + u^2) \ . \tag{5.14}$$

On average a net energy gain of

$$\Delta E = \Delta E_1 + \Delta E_2 = mu^2 \tag{5.15}$$

results, leading to the relative energy gain of

$$\frac{\Delta E}{E} = 2\frac{u^2}{v^2} \ . \tag{5.16}$$

case 1:

case 2:

Fig. 5.5
Energy gain of a particle by a reflection from a magnetic cloud

Since this acceleration mechanism is quadratic in the cloud velocity, this variant is often called Fermi mechanism of 2nd order. The result of (5.16) remains correct even under relativistic treatment. Since the cloud velocity is rather low compared to the particle velocities ($u \ll v \approx c$), the energy gain per collision ($\sim u^2$) is very small. Therefore, the acceleration of particles by the Fermi mechanism requires a very long time. In this acceleration type one assumes that magnetic clouds act as collision partners – and not normal gas clouds – because the gas density and thereby the interaction probability is larger in magnetic clouds.

slow energy increase

Another important aspect is that cosmic-ray particles will lose some of their gained energy by interactions with the interstellar or intergalactic gas between two collisions. This is why this mechanism requires a minimum injection energy above which particles can only be effectively accelerated. These injection energies could be provided by the Fermi mechanism of 1st order, that is, by shock acceleration.

injection energy

5.5 Pulsars

"Rhythmically pulsating radio source,
Can you not tell us what terrible force
Renders your density all so immense
To account for your signal so sharp and
intense?"

Dietrick E. Thomsen
Jonathan Eberhart

Spinning magnetized neutron stars (pulsars) are remnants of supernova explosions. While stars typically have radii of 10^6 km, they shrink under a gravitational collapse to a size of just about 20 km. This process leads to densities of 6×10^{13} g/cm^3 comparable to nuclear densities. In this process electrons and protons are so closely packed that in processes of weak interactions neutrons are formed:

$$p + e^- \rightarrow n + \nu_e . \tag{5.17}$$

Since the Fermi energy of electrons in such a neutron star amounts to several hundred MeV, the formed neutrons cannot decay because of the Pauli principle, since the maximum energy of electrons in neutron beta decay is only 0.78 MeV and all energy levels in the Fermi gas of electrons up to this energy and even beyond are occupied.

neutron star

gravitational collapse

pulsar periods

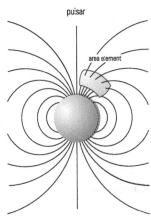

star

area element

contraction

pulsar

area element

Fig. 5.6
Increase of the magnetic field
during the gravitational collapse of
a star

The gravitational collapse of stars conserves the angular momentum. Therefore, because of their small size, rotating neutron stars possess extraordinary short rotational periods.

Assuming orbital periods of a normal star of about one month like for the Sun, one obtains – if the mass loss during contraction can be neglected – pulsar frequencies ω_{pulsar} of (Θ – moment of inertia)

$$\Theta_{\text{star}}\,\omega_{\text{star}} = \Theta_{\text{pulsar}}\,\omega_{\text{pulsar}}\,,$$

$$\omega_{\text{pulsar}} = \frac{R_{\text{star}}^2}{R_{\text{pulsar}}^2}\,\omega_{\text{star}} \tag{5.18}$$

corresponding to pulsar periods of

$$T_{\text{pulsar}} = T_{\text{star}}\,\frac{R_{\text{pulsar}}^2}{R_{\text{star}}^2}\,. \tag{5.19}$$

For a stellar size $R_{\text{star}} = 10^6\,\text{km}$, a pulsar radius $R_{\text{pulsar}} = 20\,\text{km}$, and a rotation period of $T_{\text{star}} = 1$ month one obtains

$$T_{\text{pulsar}} \approx 1\,\text{ms}\,. \tag{5.20}$$

The gravitational collapse amplifies the original magnetic field extraordinarily. If one assumes that the magnetic flux, e.g., through the upper hemisphere of a star, is conserved during the contraction, the magnetic field lines will be tightly squeezed. One obtains (see Fig. 5.6)

$$\int_{\text{star}} \boldsymbol{B}_{\text{star}} \cdot d\boldsymbol{A}_{\text{star}} = \int_{\text{pulsar}} \boldsymbol{B}_{\text{pulsar}} \cdot d\boldsymbol{A}_{\text{pulsar}}\,,$$

$$B_{\text{pulsar}} = B_{\text{star}}\,\frac{R_{\text{star}}^2}{R_{\text{pulsar}}^2}\,. \tag{5.21}$$

For $B_{\text{star}} = 1000$ Gauss magnetic pulsar fields of $2.5 \times 10^{12}\,\text{Gauss} = 2.5 \times 10^8\,\text{T}$ are obtained! These theoretically expected extraordinary high magnetic field strengths have been experimentally confirmed by measuring quantized energy levels of free electrons in strong magnetic fields ('Landau levels'). The rotational axis of pulsars usually does not coincide with the direction of the magnetic field. It is obvious that the vector of these high magnetic fields spinning around the non-aligned axis of rotation will produce strong electric fields in which particles can be accelerated.

For a 30 ms pulsar with rotational velocities of

$$v = \frac{2\pi\,R_{\text{pulsar}}}{T_{\text{pulsar}}} = \frac{2\pi \times 20 \times 10^3\,\text{m}}{3 \times 10^{-2}\,\text{s}} \approx 4 \times 10^6\,\text{m/s}$$

one obtains, using $E = v \times B$ with $v \perp B$, electrical field strengths of

$$|E| \approx v\,B = 10^{15}\,\text{V/m} \,. \qquad (5.22)$$

This implies that singly charged particles can gain 1 PeV = 1000 TeV per meter. However, it is not at all obvious how pulsars manage in detail to transform the rotational energy into the acceleration of particles. Pulsars possess a rotational energy of

$$E_{\text{rot}} = \frac{1}{2}\,\Theta_{\text{pulsar}}\,\omega^2_{\text{pulsar}} = \frac{1}{2}\frac{2}{5}\,m\,R^2_{\text{pulsar}}\,\omega^2_{\text{pulsar}} \qquad (5.23)$$
$$\approx 7 \times 10^{42}\,\text{J} \approx 4.4 \times 10^{61}\,\text{eV}$$

($T_{\text{pulsar}} = 30\,\text{ms}$, $M_{\text{pulsar}} = 2 \times 10^{30}\,\text{kg}$, $R_{\text{pulsar}} = 20\,\text{km}$, $\omega = 2\pi/T$). If the pulsars succeed to convert a fraction of only 1% of this enormous energy into the acceleration of cosmic-ray particles, one obtains an injection rate of

energy production rate

$$\frac{dE}{dt} \approx 1.4 \times 10^{42}\,\text{eV/s} \,, \qquad (5.24)$$

if a pulsar lifetime of 10^{10} years is assumed.

If one considers that our galaxy contains 10^{11} stars and if the supernova explosion rate (pulsar creation rate) is assumed to be 1 per century, a total number of 10^8 pulsars have provided energy for the acceleration of cosmic-ray particles since the creation of our galaxy (age of the galaxy $\approx 10^{10}$ years). This leads to a total energy of 2.2×10^{67} eV for an average pulsar injection time of 5×10^9 years. For a total volume of our galaxy (radius 15 kpc, average effective thickness of the galactic disk 1 kpc) of $2 \times 10^{67}\,\text{cm}^3$ this corresponds to an energy density of cosmic rays of $1.1\,\text{eV/cm}^3$.

energy density of cosmic rays

One has, of course, to consider that cosmic-ray particles stay only for a limited time in our galaxy and are furthermore subject to energy-loss processes. Still, the above presented crude estimate describes the actual energy density of cosmic rays of $\approx 1\,\text{eV/cm}^3$ rather well.

5.6 Binaries

> *"The advantage of living near a binary is to get a tan in half the time."*
>
> *Anonymous*

Binaries consisting of a pulsar or neutron star and a normal star can also be considered as a site of cosmic-ray-particle acceleration. In such a binary system matter is permanently

normal star

matter
transfer

pulsar

Fig. 5.7
Formation of accretion disks in
binaries

dragged from the normal star and whirled into an accretion disk around the compact companion. Due to these enormous plasma motions very strong electromagnetic fields are
produced in the vicinity of the neutron star. In these fields
charged particles can be accelerated to high energies (see
Fig. 5.7).

**acceleration
in gravitational potentials**

The energy gain of infalling protons (mass m_p) in the
gravitational potential of a pulsar (mass M_{pulsar}) is

$$\Delta E = -\int_{\infty}^{R_{\text{pulsar}}} G \frac{m_p \, M_{\text{pulsar}}}{r^2} \, \mathrm{d}r = G \frac{m_p \, M_{\text{pulsar}}}{R_{\text{pulsar}}}$$

$$\approx 1.1 \times 10^{-11} \, \text{J} \approx 70 \, \text{MeV} \qquad (5.25)$$

($m_p \approx 1.67 \times 10^{-27}$ kg, $M_{\text{pulsar}} = 2 \times 10^{30}$ kg, $R_{\text{pulsar}} =$
20 km, $G \approx 6.67 \times 10^{-11}$ m^3 kg^{-1} s^{-2} gravitational constant).

The matter falling into the accretion disk achieves velocities v which are obtained under classical treatment from

$$\frac{1}{2} m v^2 = \Delta E = G \frac{m \, M_{\text{pulsar}}}{R_{\text{pulsar}}} \qquad (5.26)$$

to provide values of

$$v = \sqrt{\frac{2 G M_{\text{pulsar}}}{R_{\text{pulsar}}}} \approx 1.2 \times 10^8 \, \text{m/s} \, . \qquad (5.27)$$

The variable magnetic field of the neutron star which is perpendicular to the accretion disk will produce via the Lorentz force a strong electric field. Using

$$F = e(v \times B) = eE , \qquad (5.28)$$

the particle energy E is obtained, using $v \perp B$, to

$$E = \int F \cdot ds = evB\Delta s . \qquad (5.29)$$

Under plausible assumptions ($v \approx c$, $B = 10^6$ T, $\Delta s = 10^5$ m) particle energies of 3×10^{19} eV are possible. Even more powerful are accretion disks which form around black holes or the compact nuclei of active galaxies. One assumes that in these active galactic nuclei and in jets ejected from such nuclei, particles can be accelerated to the highest energies observed in primary cosmic rays.

The details of these acceleration processes are not yet fully understood. Sites in the vicinity of black holes – a billion times more massive than the Sun – could possibly provide the environment for the acceleration of the highest-energy cosmic rays. Confined highly relativistic jets are a common feature of such compact sources. It is assumed that the jets of particles accelerated near a black hole or the nucleus of a compact galaxy are injected into the radiation field of the source. Electrons and protons accelerated in the jets via shocks initiate electromagnetic and hadronic cascades. High-energy γ rays are produced by inverse Compton scattering off accelerated electrons. High-energy neutrinos are created in the decays of charged pions in the development of the hadronic cascade. It is assumed that one detects emission from these sources only if the jets are beamed into our line of sight. A possible scenario for the acceleration of particles in *beamed jets* from massive compact sources is sketched in Fig. 5.8.

accretion disks

active galactic nuclei

beamed jets from black holes

5.7 Energy Spectra of Primary Particles

> *"Get first your facts right and then you can distort them as much as you please."*
> Mark Twain

At the present time it is not at all clear which of the presented mechanisms contribute predominantly to the acceleration of cosmic-ray particles. There are good arguments to

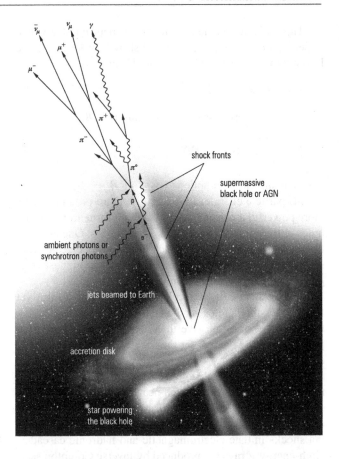

Fig. 5.8
Acceleration model for relativistic
jets powered by a black hole or an
active galactic nucleus (the
reactions are only sketched)

assume that the majority of galactic cosmic rays is produced
by shock acceleration where the particles emitted from the
source are possibly further accelerated by the Fermi mechanism of 2nd order. In contrast, it is likely that the extremely
energetic cosmic rays are predominantly accelerated in pulsars, binaries, or in jets emitted from black holes or active
galactic nuclei. For shock acceleration in supernova explosions the shape of the energy spectrum of cosmic-ray particles can be derived from the acceleration mechanism.

shock-acceleration model

Let E_0 be the initial energy of a particle and εE_0 the
energy gain per acceleration cycle. After the first cycle one
gets

$$E_1 = E_0 + \varepsilon E_0 = E_0(1 + \varepsilon) \tag{5.30}$$

while after the nth cycle (e.g., due to multiple reflection at
shock fronts) one has

$$E_n = E_0(1 + \varepsilon)^n .$$ (5.31)

To obtain the final energy $E_n = E$, a number of

$$n = \frac{\ln(E/E_0)}{\ln(1 + \varepsilon)}$$ (5.32)

cycles is required. Let us assume that the escape probability per cycle is P. The probability that particles still take part in the acceleration mechanism after n cycles is $(1 - P)^n$. This leads to the following number of particles with energies in excess of E:

cyclic energy gain

$$N(> E) \sim \sum_{m=n}^{\infty} (1 - P)^m .$$ (5.33)

Because of $\displaystyle\sum_{m=0}^{\infty} x^m = \frac{1}{1 - x}$ (for $x < 1$), (5.33) can be rewritten as

$$N(> E) \sim (1 - P)^n \sum_{m=n}^{\infty} (1 - P)^{m-n}$$

$$= (1 - P)^n \sum_{m=0}^{\infty} (1 - P)^m = \frac{(1 - P)^n}{P} ,$$ (5.34)

where $m - n$ has been renamed m. Equations (5.32) and (5.34) can be combined to form the integral energy spectrum

integral primary spectra

$$N(> E) \sim \frac{1}{P} \left(\frac{E}{E_0}\right)^{-\gamma} \sim E^{-\gamma} ,$$ (5.35)

where the spectral index γ is obtained from (5.34) and (5.35) with the help of (5.32) to

$$(1 - P)^n = \left(\frac{E}{E_0}\right)^{-\gamma} ,$$

$$n \ln(1 - P) = -\gamma \ln(E/E_0) ,$$

$$\gamma = -\frac{n \ln(1 - P)}{\ln(E/E_0)} = \frac{\ln(1/(1 - P))}{\ln(1 + \varepsilon)} .$$ (5.36)

This simple consideration yields a power law of primary cosmic rays in agreement with observation.

The energy gain per cycle surely is rather small ($\varepsilon \ll 1$). If also the escape probability P is low (e.g., at reflections between two shock fronts), (5.36) is simplified to

$$\gamma \approx \frac{\ln(1 + P)}{\ln(1 + \varepsilon)} \approx \frac{P}{\varepsilon} \ . \tag{5.37}$$

Experimentally one finds that the spectral index up to energies of 10^{15} eV is $\gamma = 1.7$. For higher energies the primary cosmic-ray-particle spectrum steepens with $\gamma = 2$.

5.8 Problems

1. Work out the kinetic energy of electrons accelerated in a betatron for the classical ($v \ll c$) and the relativistic case ($B = 1$ Tesla, $R = 0.2$ m). See also Problem 1.2.
2. A star of 10 solar masses undergoes a supernova explosion. Assume that 50% of its mass is ejected and the other half ends up in a pulsar of 10 km radius. What is the Fermi energy of the electrons in the pulsar? What is the consequence of it?
3. It is assumed that active galactic nuclei are powered by black holes. What is the energy gain of a proton falling into a one-million-solar-mass black hole down to the event horizon?
4. If the Sun were to collapse to a neutron star ($R_{NS} = 50$ km), what would be the rotational energy of such a solar remnant ($M_{\odot} = 2 \times 10^{30}$ kg, $R_{\odot} = 7 \times 10^{8}$ m, $\omega_{\odot} = 3 \times 10^{-6}$ s^{-1})? Compare this rotational energy to the energy which a main-sequence star like the Sun can liberate through nuclear fusion!
5. In a betatron the change of the magnetic flux $\phi = \int B \, dA = \pi R^2 B$ induces an electric field,

$$\int E \, ds = -\dot{\phi} \ ,$$

in which particles can be accelerated,

$$E = -\frac{\dot{\phi}}{2\pi R} = -\frac{1}{2} R \dot{B} \ .$$

The momentum increase is given by

$$\dot{p} = -eE = \frac{1}{2} eR\dot{B} \ . \tag{5.38}$$

What kind of guiding field would be required to keep the charged particles on a stable orbit?

6 Primary Cosmic Rays

"It will be found that everything depends on the composition of the forces with which the particles of matter act upon one another, and from these forces, as a matter of fact, all phenomena of nature take their origin."

R. J. Boscovich

Cosmic rays provide important information about high-energy processes occurring in our galaxy and beyond. Cosmic radiation produced in the sources is usually called *primordial cosmic rays*. This radiation is modified during its propagation in galactic and extragalactic space. Particles of galactic origin pass on average through a column density of $6 \, \text{g/cm}^2$ before reaching the top of the Earth's atmosphere. Of course, the atmosphere does not really have a 'top' but it rather exhibits an exponential density distribution. It has become common practice to understand under the top of the atmosphere an altitude of approximately $40 \, \text{km}$. This height corresponds to a residual column density of $5 \, \text{g/cm}^2$ corresponding to a pressure of $5 \, \text{mbar}$ due to the residual atmosphere above altitudes of $40 \, \text{km}$. Cosmic rays arriving unperturbed at the Earth's atmosphere are usually called *primary cosmic rays*.

primordial radiation

top of the atmosphere

Sources of cosmic rays accelerate predominantly charged particles such as protons and electrons. Since all elements of the periodic table are produced during element formation, nuclei as helium, lithium, and so on can be also accelerated. Cosmic rays represent an extraterrestrial or even extragalactic matter sample whose chemical composition exhibits certain features similar to the elemental abundance in our solar system.

Charged cosmic rays accelerated in sources can produce a number of secondary particles by interactions in the sources themselves.

production of secondary particles

These mostly unstable secondary particles, i.e., pions and kaons, produce stable particles in their decay, i.e., photons from $\pi^0 \rightarrow \gamma\gamma$ and neutrinos from $\pi^+ \rightarrow \mu^+ + \nu_\mu$ decays. Secondary particles also emerge from the sources and can reach Earth. Let us first discuss the originally accelerated charged component of primary cosmic rays.

6.1 Charged Component of Primary Cosmic Rays

"Coming out of space and incident on the high atmosphere, there is a thin rain of charged particles known as primary cosmic rays."

C. F. Powell

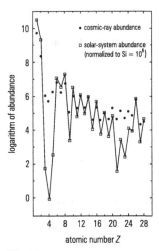

Fig. 6.1
Elemental abundance of primary cosmic rays for $1 \leq Z \leq 28$

The elemental abundance of primary cosmic rays is shown in Figs. 6.1 and 6.2 in comparison to the chemical composition of the solar system. Protons are the dominant particle species ($\approx 85\%$) followed by α particles ($\approx 12\%$). Elements with a nuclear charge $Z \geq 3$ represent only a 3% fraction of charged primary cosmic rays. The chemical composition of the solar system, shown in Figs. 6.1 and 6.2, has many features in common with that of cosmic rays. However, remarkable differences are observed for lithium, beryllium, and boron ($Z = 3$–5), and for the elements below the iron group ($Z < 26$). The larger abundance of Li, Be, and B in cosmic rays can easily be understood by fragmentation of the heavier nuclei carbon ($Z = 6$) and in particular oxygen ($Z = 8$) in galactic matter on their way from the source to Earth.

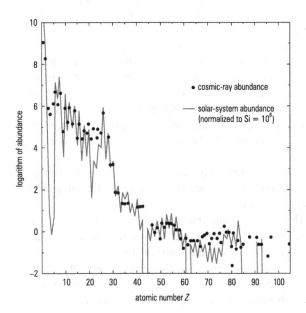

Fig. 6.2
Elemental abundance of primary cosmic rays for $1 \leq Z \leq 100$

In the same way the fragmentation or spallation of the relatively abundant element iron populates elements below the iron group. The general trend of the dependence of the chemical composition of primary cosmic rays on the atomic number can be understood by nuclear physics arguments. In the framework of the shell model it is easily explained that nuclear configurations with even proton and neutron numbers (even–even nuclei) are more abundant compared to nuclei with odd proton and neutron numbers (odd–odd nuclei). As far as stability is concerned, even–odd and odd–even nuclei are associated with abundances between ee and oo configurations. Extremely stable nuclei occur for filled shells ('magic nuclei'), where the magic numbers (2, 8, 20, 50, 82, 126) refer separately to protons and neutrons. As a consequence, doubly magic nuclei (like helium and oxygen) are particularly stable and correspondingly abundant. But nuclei with a large binding energy such as iron which can be produced in fusion processes, are also relatively abundant in charged primary cosmic rays. The energy spectra of primary nuclei of hydrogen, helium, carbon, and iron are shown in Fig. 6.3.

The low-energy part of the primary spectrum is modified by the Sun's and the Earth's magnetic field. The 11-year period of the sunspot cycle modulates the intensity of low-energy primary cosmic rays ($< 1\,$GeV/nucleon). The active Sun reduces the cosmic-ray intensity because a stronger magnetic field created by the Sun prevents galactic charged particles from reaching Earth.

In general, the intensity decreases with increasing energy so that a direct observation of the high-energy component of cosmic rays at the top of the atmosphere with balloons or satellites eventually runs out of statistics. Measurements of the charged component of primary cosmic rays at energies in excess of several hundred GeV must therefore resort to indirect methods. The atmospheric air Cherenkov technique (see Sect. 6.3: Gamma Astronomy) or the measurement of extensive air showers via air fluorescence or particle sampling (see Sect. 7.4: Extensive Air Showers) can in principle cover this part of the energy spectrum, however, a determination of the chemical composition of primary cosmic rays by this indirect technique is particularly difficult. Furthermore, the particle intensities at these high energies are extremely low. For particles with energies in excess of about 10^{19} eV the rate is only 1 particle per km^2 and year.

shell model

magic nuclei

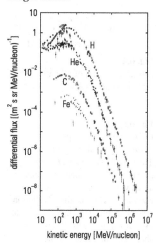

Fig. 6.3
Energy spectra of the main components of charged primary cosmic rays

modification of the low-energy part

decreasing intensity with high energies

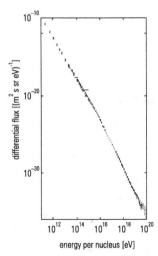

Fig. 6.4
Energy spectrum of all particles of
primary cosmic rays

Fig. 6.5
Energy spectrum of primary
cosmic rays scaled by a factor E^3.
The data from the Japanese
air-shower experiment AGASA
agree well – except at very high
energies – with the air-scintillation
results of the Utah High Resolution
experiment as far as the spectral
shape is concerned, but they
disagree in absolute intensity
(AGASA – Akeno Giant Air
Shower Array, HiRes – High
Resolution Fly's Eye)

The all-particle spectrum of charged primary cosmic
rays (Fig. 6.4) is relatively steep so that practically no details are observable. Only after multiplication of the intensity with a power of the primary energy, structures in the
primary spectrum become visible (Fig. 6.5). The bulk of
cosmic rays up to at least an energy of 10^{15} eV is believed to
originate from within our galaxy. Above that energy which is
associated with the so-called '*knee*' the spectrum steepens.
Above the so-called '*ankle*' at energies around 5×10^{18} eV
the spectrum flattens again. This latter feature is often interpreted as a crossover from a steeper galactic component to a
harder component of extragalactic origin.

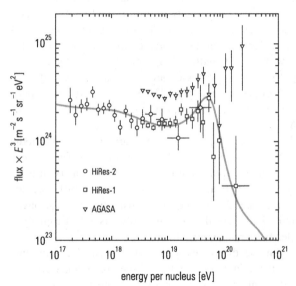

Fig. 6.6
Artist's impression of the different
structures in the primary
cosmic-ray spectrum

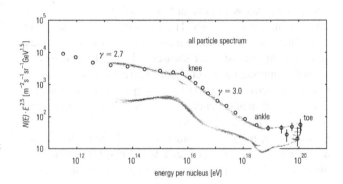

Cosmic rays originate predominantly from within our galaxy. Galactic objects do not in general have such a combination of size and magnetic field strength to contain particles at very high energies.

Because of the equilibrium between the centrifugal and Lorentz force ($v \perp B$ assumed) one has

equilibrium between centrifugal and Lorentz force

$$mv^2/\varrho = Z\,e\,v\,B \qquad (6.1)$$

which yields for the momentum of singly charged particles

$$p = e\,\varrho\,B$$

(p is the particle momentum, B the magnetic field, v the particle velocity, m the particle mass, ϱ the bending radius or gyroradius). For a large-area galactic magnetic field of $B = 10^{-10}$ Tesla in the galaxy (about 10^5 times weaker compared to the magnetic field on the surface of the Earth) and a gyroradius of 5 pc, from which particles start to leak from the galaxy, particles with momenta up to

galactic containment

$$p[\text{GeV}/c] = 0.3\,B[\text{T}]\,\varrho[\text{m}]\;,$$
$$p_{\max} = 4.6 \times 10^6\,\text{GeV}/c = 4.6 \times 10^{15}\,\text{eV}/c \qquad (6.2)$$

can be contained. 1 parsec (pc) is the popular unit of distance in astronomy (1 pc = 3.26 light-years = 3.0857 × 10^{16} m). Particles with energies exceeding 10^{15} eV start to leak from the galaxy. This causes the spectrum to get steeper to higher energies. Since the containment radius depends on the atomic number, see (6.1), the position of the knee should depend on the charge of primary cosmic rays in this scenario, i.e., the knee for iron would be expected at higher energies compared to the proton knee.

1 parsec (pc) = 3.26 LY

proton knee

Another possible reason for the knee in cosmic radiation could be related to the fact that 10^{15} eV is about the maximum energy which can be supplied by supernova explosions. For higher energies a different acceleration mechanism is required which might possibly lead to a steeper energy spectrum. The knee could in principle also have its origin in a possible change of interaction characteristics of high-energy particles. The energy of the knee coincides with the maximum energy presently available at accelerators ($\sqrt{s} = 1.8$ TeV, corresponding to $E_{\text{lab}} \approx 2 \times 10^{15}$ eV) beyond which no direct measurements are available. It is conceivable that the interaction cross section changes with energy giving rise to features at the knee of the primary cosmic-ray spectrum. The flattening of the spectrum above

iron knee

ankle of cosmic rays

10^{19} eV ('ankle') is generally assumed to be due to an extragalactic component.

In 1966 it was realized by Greisen, Zatsepin, and Kuzmin (GZK) that cosmic rays above the energy of approximately 6×10^{19} eV would interact with the cosmic blackbody radiation. Protons of higher energies would rapidly lose energy by this interaction process causing the spectrum to be cut off at energies around 6×10^{19} eV. Primary protons with these energies produce pions on blackbody photons via the Δ resonance according to

$$\gamma + p \; \rightarrow \; p + \pi^0 \,, \;\; \gamma + p \; \rightarrow \; n + \pi^+ \,, \tag{6.3}$$

Greisen–Zatsepin–Kuzmin
cutoff

thereby losing a large fraction of their energy.

The threshold energy for the photoproduction of pions can be determined from four-momentum conservation

$$(q_\gamma + q_p)^2 = (m_p + m_\pi)^2 \tag{6.4}$$

(q_γ, q_p are four-momenta of the photon or proton, respectively; m_p, m_π are proton and pion masses) yielding

$$E_p = (m_\pi^2 + 2m_p m_\pi)/4E_\gamma \tag{6.5}$$

for head-on collisions.

A typical value of the Planck distribution corresponding to the blackbody radiation of temperature 2.7 Kelvin is around 1.1 meV. With this photon energy the threshold energy for the photoproduction of pions is

$$E_p \approx 6 \times 10^{19} \,\text{eV} \,. \tag{6.6}$$

toe of cosmic rays

The observation of several events in excess of 10^{20} eV (the '*toe*' of primary cosmic rays), therefore, represents a certain mystery. The Greisen–Zatsepin–Kuzmin cutoff limits the mean free path of high-energy protons to something like 10 Mpc. Therefore, the experimental verification of the GZK cutoff in the spectrum of primary cosmic rays would be the clearest proof that the high-energy particles beyond several 10^{19} eV are generated at extragalactic sources. Pho-

$\gamma\gamma$ interactions

tons as candidates for primary particles have even shorter mean free paths (≈ 10 kpc) because they produce electron pairs in gamma–gamma interactions with blackbody photons, infrared and starlight photons ($\gamma\gamma \rightarrow e^+e^-$). The hypothesis that primary neutrinos are responsible for the highest-energy events is rather unlikely. The interaction probability for neutrinos in the atmosphere is extremely small ($< 10^{-4}$). Furthermore, the observed zenith-angle distribution of energetic events and the position of primary vertices of the cascade development in the atmosphere are inconsistent with the assumption that primary neutrinos are responsible for these events. Because of their low interaction probability one would expect that the primary vertices for neutrinos would be distributed uniformly in the atmosphere. In contrast, one observes that the first interaction takes place predominantly in the 100 mbar layer which is characteristic of hadron or photon interactions. One way out would be to assume that after all protons are responsible for the events with energies exceeding 6×10^{19} eV. This would support the idea that the sources of the highest-energy cosmic-ray events are relatively close. A candidate source is M87, an elliptic giant galaxy in the Virgo cluster at a distance of 15 Mpc. From the center of M87 a jet of 1500 pc length is ejected that could be the source of energetic particles. M87 coincides with Virgo A (3C274), one of the strongest radio sources in the constellation Virgin.

neutrino origin

M87, a source of cosmic rays?

A close look at the experimental situation (see Fig. 6.5) shows that the two major experiments measuring in the $\leq$ EeV range do not agree in absolute intensity and also not in the shape of the spectrum for energies in excess of $\approx 5 \times 10^{19}$ eV. Even though the HiRes experiment finds events beyond 10^{20} eV, the HiRes spectrum is not in disagreement with the expectation based on the GZK cutoff, quite in contrast to the AGASA findings. One might argue that possibly the energy assignment for the showers from HiRes is superior over that from AGASA since HiRes records the complete longitudinal development of the shower, while AGASA only samples the shower information in one atmospheric layer, i.e., at ground level.

HiRes vs. AGASA

It is to be expected that the Auger experiment will clarify the question concerning the GZK cutoff. In this context it is interesting to note that the Auger-south array has recorded an event with $\approx 10^{20}$ eV already during the construction phase [4]. The full Auger-south detector will be completed in spring 2006.

Auger experiment

Considering the enormous rigidity of these high-energy particles and the weakness of the intergalactic magnetic field one would not expect substantial deflections of these particles over distances of 50 Mpc. This would imply that one can consider to do astronomy with these extremely high-energy cosmic rays. Since there is no correlation of the arrival directions of these high-energy cosmic-ray events with known

exotic particles?

astronomical sources in the immediate neighbourhood of our galaxy, one might resort to the assumption that new and so far unknown elementary particles might be responsible for the events exceeding 6×10^{19} eV, if the AGASA results are confirmed, e.g., by the Auger experiment.

**antiparticles
in primary cosmic rays**

$\bar{p}$ **generation**

Antiparticles are extremely rare in primary cosmic rays. The measured primary antiprotons are presumably generated in interactions of primary charged cosmic rays with the interstellar gas. Antiprotons can be readily produced according to

$$p + p \rightarrow p + p + p + \bar{p} \tag{6.7}$$

e^+ **generation**

(compare Example 1, Chap. 3), while positrons are most easily formed in pair production by energetic photons (compare Example 4, Chap. 3). The flux of primary antiprotons for energies > 10 GeV has been measured to be

$$\left. \frac{N(\bar{p})}{N(p)} \right|_{>10\,\text{GeV}} \approx 10^{-4} . \tag{6.8}$$

primary electrons

The fraction of primary electrons in relation to primary protons is only 1%. Primary positrons constitute only 10% of the electrons at energies around 10 GeV. They are presumably also consistent with secondary origin.

One might wonder whether the continuous bombardment of the Earth with predominantly positively charged particles (only 1% are negatively charged) would lead to a positive charge-up of our planet. This, however, is not true. When the rates of primary protons and electrons are compared, one normally considers only energetic particles. The spectra of protons and electrons are very different with electrons populating mainly low-energy regions. If *all* energies are considered, there are equal numbers of protons and electrons so that there is no charge-up of our planet.

stars of antimatter?

To find out whether there are stars of antimatter in the universe, the existence of primary antinuclei (antihelium, anticarbon) must be established because secondary production of antinuclei with $Z \geq 2$ by cosmic rays is practically excluded. The non-observation of primary antimatter with $Z \geq 2$ is a strong hint that our universe is matter dominated.

**chemical composition
of high-energy cosmic rays**

The chemical composition of high-energy primary cosmic rays ($> 10^{15}$ eV) is to large extent an unknown territory. If the current models of nucleon–nucleon interactions are extrapolated into the range beyond 10^{15} eV (corresponding to a center-of-mass energy of $\gtrsim 1.4$ TeV in proton–proton collisions) and if the muon content and lateral distribution of muons in extensive air showers are taken as a criterion for the identity of the primary particle, then one would arrive at the conclusion that the chemical composition of primary cosmic rays cannot be very different from the composition below the knee ($< 10^{15}$ eV). Some experiments, however, seem to indicate that the iron fraction of primary cosmic rays increases with energy beyond the knee. This could just be a consequence of galactic containment: the iron knee is expected to occur at higher energies compared to the proton knee.

Even though cosmic rays have been discovered about 90 years ago, their origin is still an open question. It is generally assumed that active galactic nuclei, quasars, or supernova explosions are excellent source candidates for high-energy cosmic rays, but there is no direct evidence for this assumption. In the energy range up to 100 TeV individual sources have been identified by primary gamma rays. It is conceivable that gamma rays of these energies are decay products of elementary particles (π^0 decay, Centaurus A?), which have been produced by those particles that have been originally accelerated in the sources. Therefore, it would be interesting to see the sources of cosmic rays in the light of these originally accelerated particles.

origin of cosmic rays

hadronic origin of energetic γs?

This, however, presents a serious problem: photons and neutrinos travel on straight lines in galactic and intergalactic space, therefore pointing directly back to the sources. Charged particles, on the other hand, are subject to the influence of homogeneous or irregular magnetic fields. This causes the accelerated particles to travel along chaotic trajectories thereby losing all directional information before finally reaching Earth. Therefore, it is of very little surprise that the sky for charged particles with energies below 10^{14} eV appears completely isotropic. The level of observed anisotropies lies below 0.5%. There is some hope that for energies exceeding 10^{18} eV a certain directionality could be found. It is true that also in this energy domain the galactic magnetic fields must be taken into account, however, the deflection radii are already rather large. The situation is even more complicated because of a rather uncertain topology of galactic magnetic fields. In addition, one must in principle know the time evolution of magnetic fields over the last ≈ 50 million years because the sources can easily reside at distances of > 10 Mpc ($\hat{=} 32.6$ million light-years). For simultaneous observation of cosmic-ray sources in the

isotropy of charged particles

anisotropy

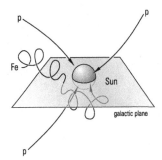

Fig. 6.7
Sketch of proton and iron-nucleus trajectories in our Milky Way at 10^{18} eV

light of charged particles and photons, one must take into account that charged particles are delayed with respect to photons because they travel on trajectories bent by the magnetic field.

Since the magnetic deflection is proportional to the charge of a particle, proton astronomy is more promising than astronomy with heavy nuclei. This idea is outlined in Fig. 6.7 where the trajectories of protons and an iron nucleus ($Z = 26$) at an energy of 10^{18} eV are sketched for our galaxy. This figure clearly shows that one should only use – if experimentally possible – protons for *particle astronomy*. Actually, there are some hints that the origins of the few events with energies $> 10^{19}$ eV could lie in the supergalactic plane, a cluster of relatively close-by galaxies including our Milky Way ('local super galaxy'). It has also been discussed that the galactic center of our Milky Way and in particular the Cygnus region could be responsible for a certain anisotropy at 10^{18} eV. It must, however, be mentioned that claims for such a possible correlation are based on very low statistics and are therefore not unanimously supported. They certainly need further experimental confirmation.

6.2 Neutrino Astronomy

> *"Neutrino physics is largely an art of learning a great deal by observing nothing."*
>
> Haim Harari

limits of classical astronomy   The disadvantage of classical astronomies like observations in the radio, infrared, optical, ultraviolet, X-ray, or γ-ray band is related to the fact that electromagnetic radiation is quickly absorbed in matter. Therefore, with these astronomies one can only observe the surfaces of astronomical objects. In addition, energetic γ rays from distant sources are attenuated via $\gamma\gamma$ interactions with photons of the black-body radiation by the process

$$\gamma + \gamma \rightarrow e^+ + e^- .$$

Energetic photons ($> 10^{15}$ eV), for example, from the Large Magellanic Cloud (LMC, 52 kpc distance) are significantly attenuated by this process (see Fig. 4.9).

charged primaries   Charged primaries can in principle also be used in astroparticle physics. However, the directional information is only conserved for very energetic protons ($> 10^{19}$ eV) because otherwise the irregular and partly not well-known galactic magnetic fields will randomize their original direction. For these high energies the Greisen–Zatsepin–Kuzmin cutoff also comes into play whereby protons lose their energy via the photoproduction of pions off blackbody photons. For protons with energies exceeding 6×10^{19} eV the universe is no longer transparent (attenuation length $\lambda \approx 10$ Mpc). As a consequence of these facts, the requirement for an optimal astronomy can be defined in the following way:

requirement for an optimal astronomy

1. The optimal astroparticles or radiation should not be influenced by magnetic fields.
2. The particles should not decay from source to Earth. This practically excludes neutrons as carriers unless neutrons have extremely high energy ($\tau^0_{\text{neutron}} = 885.7$ s; at $E = 10^{19}$ eV one has $\gamma c\tau^0_{\text{neutron}} \approx 300\,000$ light-years).
3. Particles and antiparticles should be different. This would in principle allow to find out whether particles originate from a matter or antimatter source. This requirement excludes photons because a photon is its own antiparticle, $\gamma = \bar{\gamma}$.
4. The particles must be penetrating so that one can look into the central part of the sources.
5. Particles should not be absorbed by interstellar or intergalactic dust or by infrared or blackbody photons.

These five requirements are fulfilled by neutrinos in an ideal way! One could ask oneself why *neutrino astronomy* has not been a major branch of astronomy all along. The fact that neutrinos can escape from the center of the sources is related to their low interaction cross section. This, unfortunately, goes along with an enormous difficulty to detect these neutrinos on Earth.

neutrino astronomy

For solar neutrinos in the range of several 100 keV the cross section for neutrino–nucleon scattering is

$$\sigma(\nu_e N) \approx 10^{-45} \text{ cm}^2/\text{nucleon} . \qquad (6.9)$$

The interaction probability of these neutrinos with our planet Earth at central incidence is

$$\phi = \sigma N_A d \varrho \approx 4 \times 10^{-12} \qquad (6.10)$$

(N_A is the Avogadro number, d the diameter of the Earth, ϱ the average density of the Earth). Out of the 7×10^{10} neutrinos per cm^2 and s radiated by the Sun and arriving at Earth only one at most is 'seen' by our planet.

neutrino telescopes

As a consequence of this, neutrino telescopes must have an enormous target mass, and one has to envisage long exposure times. However, for high energies the interaction cross section rises with neutrino energy. Neutrinos in the energy range of several 100 keV can be detected by radiochemical methods. For energies exceeding 5 MeV large-volume water Cherenkov counters are an attractive possibility.

neutrino detection

Neutrino astronomy is a very young branch of astroparticle physics. Up to now four different sources of neutrinos have been investigated. The physics results and implications of these measurements will be discussed in the following four sections.

6.2.1 Atmospheric Neutrinos

For real neutrino astronomy neutrinos from atmospheric sources are an annoying background. For the particle physics aspect of astroparticle physics atmospheric neutrinos have turned out to be a very interesting subject. Primary cosmic rays interact in the atmosphere with the atomic nuclei of nitrogen and oxygen. In these proton–air interactions nuclear fragments and predominantly charged and neutral pions are produced. The decay of charged pions (lifetime 26 ns) produces muon neutrinos:

neutrino production

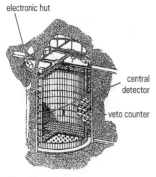

electronic hut

central detector

veto counter

Fig. 6.8
The Super-Kamiokande detector in the Kamioka mine in Japan {8}

$$\pi^+ \rightarrow \mu^+ + \nu_\mu , \quad \pi^- \rightarrow \mu^- + \bar{\nu}_\mu . \tag{6.11}$$

Muons themselves are also unstable and decay with an average lifetime of 2.2 μs according to

$$\mu^+ \rightarrow e^+ + \nu_e + \bar{\nu}_\mu , \quad \mu^- \rightarrow e^- + \bar{\nu}_e + \nu_\mu . \tag{6.12}$$

Therefore, the atmospheric neutrino beam contains electron and muon neutrinos and one would expect a ratio

$$\frac{N(\nu_\mu, \bar{\nu}_\mu)}{N(\nu_e, \bar{\nu}_e)} \equiv \frac{N_\mu}{N_e} \approx 2 , \tag{6.13}$$

expected dominance of ν_μ

as can be easily seen by counting the decay neutrinos in reactions (6.11) and (6.12).

The presently largest experiments measuring atmospheric neutrinos are Super-Kamiokande (see Fig. 6.8) and AMANDA II (see Sect. 6.2.4). Neutrino interactions in the Super-Kamiokande detector are recorded in a tank of approximately 50 000 tons of ultrapure water. Electron neutrinos transfer part of their energy to electrons,

$$\nu_e + e^- \rightarrow \nu_e + e^- , \tag{6.14}$$

or produce electrons in neutrino–nucleon interactions

$$\nu_e + N \rightarrow e^- + N' . \tag{6.15}$$

Muon neutrinos are detected in neutrino–nucleon interactions according to

$$\nu_\mu + N \rightarrow \mu^- + N' . \tag{6.16}$$

Electron antineutrinos and muon antineutrinos produce correspondingly positrons and positive muons. The charged leptons (e^+, e^-, μ^+, μ^-) can be detected via the Cherenkov effect in water. The produced Cherenkov light is measured with 11 200 photomultipliers of 50 cm cathode diameter. In the GeV-range electrons initiate characteristic electromagnetic cascades of short range while muons produce long straight tracks. This presents a basis for distinguishing electron from muon neutrinos. On top of that, muons can be identified by their decay in the detector thereby giving additional evidence concerning the identity of the initiating neutrino species. Figures 6.9 and 6.10 show an electron and muon event in Super-Kamiokande. Muons have a well-defined range and produce a clear Cherenkov pattern with sharp edges while electrons initiate electromagnetic cascades thereby creating a fuzzy ring pattern.

The result of the Super-Kamiokande experiment is that the number of electron-neutrino events corresponds to the theoretical expectation while there is a clear deficit of events initiated by muon neutrinos. The measured electron and muon spectra compared to expectation are shown in Figs. 6.11 and 6.12.

Because of the different acceptance for electrons and muons in the water Cherenkov detector, the ratio of muons to electrons is compared to a Monte Carlo simulation. For the double ratio

$$R = \frac{(N_\mu/N_e)_{\text{data}}}{(N_\mu/N_e)_{\text{Monte Carlo}}} \tag{6.17}$$

one would expect the value $R = 1$ in agreement with the standard interaction and propagation models. However, the Super-Kamiokande experiment obtains

$$R = 0.69 \pm 0.06 , \tag{6.18}$$

which represents a clear deviation from expectation.

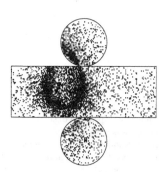

Fig. 6.9
Cherenkov pattern of an energetic electron in the Super-Kamiokande Detector {9}

**distinguishing
electron and muon neutrinos**

deficit of muon neutrinos

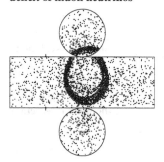

Fig. 6.10
Cherenkov pattern for an energetic muon in the Super-Kamiokande detector {9}

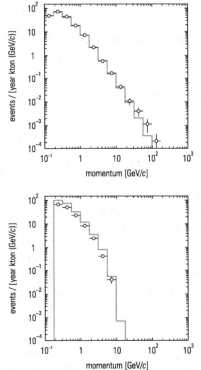

Fig. 6.11
Momentum spectrum of single-ring electron-like events in Super-Kamiokande. The *solid line* represents the Monte Carlo expectation {9}

Fig. 6.12
Momentum spectrum of single-ring muon-like events in Super-Kamiokande. The *solid line* represents the Monte Carlo expectation. The cutoff arround 10 GeV originates from the condition that the muon tracks must be contained in the detector {9}

After careful checks of the experimental results and investigations of possible systematic effects the general opinion prevails that the deficit of muon neutrinos can only be explained by *neutrino oscillations*.

neutrino oscillations

Mixed particle states are known from the quark sector (see Chap. 2). Similarly, it is conceivable that in the lepton sector the eigenstates of weak interactions ν_e, ν_μ, and

eigenstates of the weak interaction

ν_τ are superpositions of mass eigenstates ν_1, ν_2, and ν_3. A muon neutrino ν_μ born in a pion decay could be transformed during the propagation from the source to the observation in the detector into a different neutrino flavour. If the muon neutrino in reality was a mixture of two different mass

mass eigenstates

eigenstates ν_1 and ν_2, these two states would propagate at different velocities if their masses were not identical and so the mass components get out of phase with each other. This could possibly result in a different neutrino flavour at the detector. If, however, all neutrinos were massless, they would all propagate precisely at the velocity of light, and the mass eigenstates can never get out of phase with each other.

For an assumed two-neutrino mixing of v_e and v_μ the **mixing angle**
weak eigenstates could be related to the mass eigenstates by
the following two equations:

$$v_e = v_1 \cos\theta + v_2 \sin\theta \,,$$
$$v_\mu = -v_1 \sin\theta + v_2 \cos\theta \,. \tag{6.19}$$

The *mixing angle* θ determines the degree of mixing.
This assumption requires that the neutrinos have non-zero
mass and, in addition, $m_1 \neq m_2$ must hold.

In the framework of this oscillation model the probability that an electron neutrino stays an electron neutrino, can
be calculated to be (see also Problem 4 in this section):

$$P_{v_e \to v_e}(x) = 1 - \sin^2 2\theta \sin^2\left(\pi \frac{x}{L_v}\right) \,, \tag{6.20}$$

where x is the distance from the source to the detector and
L_v the oscillation length

$$L_v = \frac{2.48 \, E_v[\text{MeV}]}{(m_1^2 - m_2^2) \, [\text{eV}^2/c^4]} \text{ m} \,. \tag{6.21}$$

The expression $m_1^2 - m_2^2$ is usually abbreviated as δm^2. Equations (6.20) and (6.21) can be combined to give

$$P_{v_e \to v_e}(x) = 1 - \sin^2 2\theta \sin^2\left(1.27 \, \delta m^2 \frac{x}{E_v}\right) \tag{6.22}$$

where δm^2 is measured in eV^2, x in km, and E_v in GeV.
The idea of a two-neutrino mixing is graphically presented
in Fig. 6.13.

For the general case of mixing of all three neutrino
flavours one obtains as generalization of (6.19) **mixing matrix**

$$\begin{pmatrix} v_e \\ v_\mu \\ v_\tau \end{pmatrix} = U_N \begin{pmatrix} v_1 \\ v_2 \\ v_3 \end{pmatrix} \,, \tag{6.23}$$

where U_N is the (3×3) neutrino mixing matrix.

The deficit of muon neutrinos can now be explained by **deficit of muon neutrinos**
the assumption that some of the muon neutrinos transform
themselves during propagation from the point of production
to the detector into a different neutrino flavour, e.g., into tau
neutrinos. The sketch shown in Fig. 6.13 demonstrated that
for an assumed mixing angle of 45° all neutrinos of a certain type have transformed themselves into a different neutrino flavour after propagating half the oscillation length.

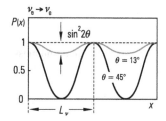

Fig. 6.13
Oscillation model for v_e–v_μ
mixing for different mixing angles

tau production?

If, however, muon neutrinos have oscillated into tau neutrinos, a deficit of muon neutrinos will be observed in the detector because tau neutrinos would only produce taus in the water Cherenkov counter, but not muons. Since, however, the mass of the tau is rather high ($1.77\,\text{GeV}/c^2$), tau neutrinos normally would not meet the requirement to provide the necessary center-of-mass energy for tau production. Consequently, they would escape from the detector without interaction. If the deficit of muon neutrinos would be interpreted by ($\nu_\mu \rightarrow \nu_\tau$) oscillations, the mixing angle and the difference of mass squares δm^2 can be determined from the experimental data. The measured value of the double ratio $R = 0.69$ leads to

$$\delta m^2 \approx 2 \times 10^{-3}\,\text{eV}^2 \tag{6.24}$$

neutrino mass

at maximal mixing ($\sin^2 2\theta = 1$, corresponding to $\theta = 45°$).[1] If one assumes that in the neutrino sector a similar mass hierarchy as in the sector of charged leptons exists ($m_e \ll m_\mu \ll m_\tau$), then the mass of the heaviest neutrino can be estimated from (6.24),

$$m_{\nu_\tau} \approx \sqrt{\delta m^2} \approx 0.045\,\text{eV} \tag{6.25}$$

zenith-angle dependence

(see also parts b,c of Problem 4 in this section). The validity of this conclusion relies on the correctly measured absolute fluxes of electron and muon neutrinos. Because of the different Cherenkov pattern of electrons and muons in the water Cherenkov detector the efficiencies for electron neutrino and muon neutrino detections might be different. To support the oscillation hypothesis one would therefore prefer to have an additional independent experimental result. This is provided in an impressive manner by the ratio of upward- to downward-going muons. Upward-coming atmospheric neutrinos have traversed the whole Earth ($\approx 12\,800\,\text{km}$). They would have a much larger probability to oscillate into tau neutrinos compared to downward-going neutrinos which have traveled typically only $20\,\text{km}$. Actually, according to the experimental result of the Super-Kamiokande collaboration the

---

[1] The 90% coincidence limit for δm^2 given by the Super-Kamiokande experiment is $1.3 \times 10^{-3}\,\text{eV}^2 \leq \delta m^2 \leq 3 \times 10^{-3}\,\text{eV}^2$. The accelerator experiment K2K sending muon neutrinos to the Kamioka mine gets $\delta m^2 = 2.8 \times 10^{-3}\,\text{eV}^2$ [5]. K2K – from KEK to Kamioka, Long-baseline Neutrino Oscillation Experiment

upward-going muon neutrinos which have traveled through the whole Earth are suppressed by a factor of two compared to the downward-going muons. This is taken as a strong indication for the existence of oscillations (see Fig. 6.14). For the ratio of upward- to downward-going muon neutrinos one obtains

$$S = \frac{N(\nu_\mu, \text{up})}{N(\nu_\mu, \text{down})} = 0.54 \pm 0.06 , \qquad (6.26)$$

which presents a clear effect in favour of oscillation.

Details of the observed zenith-angle dependence of atmospheric ν_e and ν_μ fluxes also represent a particularly strong support for the oscillation model.

The comparison of electron and muon events in both the sub-GeV and multi-GeV range, in their dependence on the zenith angle, is shown in Fig. 6.15.

The electron events are in perfect agreement with expectation, while the no-oscillation hypothesis for muons is clearly ruled out by the data.

Since the production altitude L and energy E_ν of atmospheric neutrinos are known ($\approx 20\,\text{km}$ for vertically downward-going neutrinos), the observed zenith-angle dependence of electron and muon neutrinos can also be converted

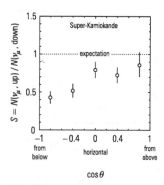

Fig. 6.14

Ratio of ν_μ fluxes as a function of zenith angle as measured in the Super-Kamiokande experiment

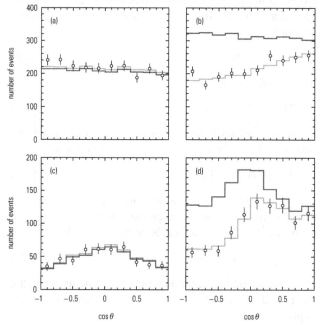

Fig. 6.15

Zenith-angle distribution of electron-like and muon-like events for the sub-GeV range (**a:** electrons, **b:** muons) and the multi-GeV range (**c:** electrons, **d:** muons) in Super-Kamiokande. The *dark grey line* is the expectation for the null hypothesis (no oscillations) while the *light grey histogram* represents the expectation for oscillations with maximal mixing ($\sin^2 2\theta = 1$) and $\delta m^2 = 3 \times 10^{-3}\,\text{eV}^2$ {9}

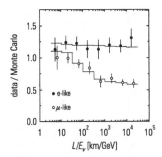

Fig. 6.16
Ratio of fully contained events measured in the Super-Kamiokande detector as a function of the reconstructed value of distance over energy (L/E_ν). The *lower histogram* for μ-like events corresponds to the expectation for $\nu_\mu \leftrightarrow \nu_\tau$ oscillations with $\delta m^2 = 2.2 \times 10^{-3}$ eV and $\sin 2\theta = 1$ {10}

fusion reactions in the Sun

'chemistry' of nuclear fusion

into a dependence of the rate versus the reconstructed ratio of L/E_ν. Figure 6.16 shows the ratio data/Monte Carlo for fully contained events as measured in the Super-Kamiokande experiment. The data exhibit a zenith-angle- (i.e., distance-) dependent deficit of muon neutrinos, while the electron neutrinos follow the expectation for no oscillations. The observed behaviour is consistent with ($\nu_\mu \leftrightarrow \nu_\tau$) oscillations, where a best fit is obtained for $\delta m^2 = 2.2 \times 10^{-3}$ eV for maximal mixing ($\sin 2\theta = 1$).

In the Standard Model of elementary particles neutrinos have zero mass. Therefore, neutrino oscillations represent an important extension of the physics of elementary particles. In this example of neutrino oscillations the synthesis between astrophysics and particle physics becomes particularly evident.

6.2.2 Solar Neutrinos

The Sun is a nuclear fusion reactor. In its interior hydrogen is burned to helium. The longevity of the Sun is related to the fact that the initial reaction

$$p + p \rightarrow d + e^+ + \nu_e \tag{6.27}$$

proceeds via the weak interaction. 86% of solar neutrinos are produced in this proton–proton reaction. Deuterium made according to (6.27) fuses with a further proton to produce helium 3,

$$d + p \rightarrow {}^3\text{He} + \gamma . \tag{6.28}$$

In ${}^3\text{He}$–${}^3\text{He}$ interactions

$$^3\text{He} + {}^3\text{He} \rightarrow {}^4\text{He} + 2p \tag{6.29}$$

the isotope helium 4 can be formed. On the other hand, the isotopes ${}^3\text{He}$ and ${}^4\text{He}$ could also produce beryllium,

$$^3\text{He} + {}^4\text{He} \rightarrow {}^7\text{Be} + \gamma . \tag{6.30}$$

${}^7\text{Be}$ is made of four protons and three neutrons. Light elements prefer symmetry between the number of protons and neutrons. ${}^7\text{Be}$ can capture an electron yielding ${}^7\text{Li}$,

$$^7\text{Be} + e^- \rightarrow {}^7\text{Li} + \nu_e , \tag{6.31}$$

where a proton has been transformed into a neutron. On the other hand, ${}^7\text{Be}$ can react with one of the abundant protons to produce ${}^8\text{B}$,

$$^7\text{Be} + p \rightarrow {}^8\text{B} + \gamma \ . \tag{6.32}$$

^7Li produced according to (6.31) will usually interact with protons forming helium,

$$^7\text{Li} + p \rightarrow {}^4\text{He} + {}^4\text{He} \ , \tag{6.33}$$

while the boron isotope ^8B will reduce its proton excess by β^+ decay,

$$^8\text{B} \rightarrow {}^8\text{Be} + e^+ + \nu_e \ , \tag{6.34}$$

and the resulting ^8Be will disintegrate into two helium nuclei. Apart from the dominant pp neutrinos, reaction (6.27), further 14% are generated in the electron-capture reaction (6.31), while the ^8B decay contributes only at the level of 0.02% albeit yielding high-energy neutrinos. In total, the solar neutrino flux at Earth amounts to about 7×10^{10} particles per cm^2 and second.

The energy spectra of different reactions which proceed in the solar interior at a temperature of 15 million Kelvin are shown in Fig. 6.17. The Sun is a pure electron–neutrino source. It does not produce electron antineutrinos and, in particular, no other neutrino flavours (ν_μ, ν_τ).

Three radiochemical experiments and two water Cherenkov experiments are trying to measure the flux of solar neutrinos.

The historically first experiment for the search of solar neutrinos is based on the reaction

$$\nu_e + {}^{37}\text{Cl} \rightarrow {}^{37}\text{Ar} + e^- \ , \tag{6.35}$$

where the produced ^{37}Ar has to be extracted from a huge tank filled with 380 000 liters of perchlorethylene (C_2Cl_4). Because of the low capture rate of less than one neutrino per day the experiment must be shielded against atmospheric cosmic rays. Therefore, it is operated in a gold mine at 1500 meter depth under the Earth's surface (see Fig. 6.18). After a run of typically one month the tank is flushed with a noble gas and the few produced ^{37}Ar atoms are extracted from the detector and subsequently counted. Counting is done by means of the electron-capture reaction of ^{37}Ar where again ^{37}Cl is produced. Since the electron capture occurs predominantly from the K shell, the produced ^{37}Cl atom is now missing one electron in the innermost shell (in the K shell). The atomic electrons of the ^{37}Cl atom are rearranged under emission of either characteristic X rays or by the emission of Auger electrons. These Auger electrons and, in particu-

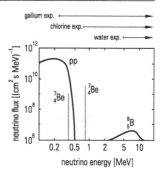

Fig. 6.17
Neutrino spectra from solar fusion processes. The reaction thresholds of the gallium, chlorine, and water Cherenkov experiments are indicated. The line fluxes of beryllium isotopes are given in units of cm^{-2} s^{-1}

Fig. 6.18
The detector of the chlorine experiment of R. Davis for the measurement of solar neutrinos. The detector is installed at a depth of 1400 m in the Homestake Mine in South Dakota. It is filled with 380 000 liters of perchlorethylene {11}

lar, the characteristic X rays are the basis for counting ^{37}Ar atoms produced by solar neutrinos.

Davis experiment

In the course of 30 years of operation a deficit of solar neutrinos has become more and more evident. The experiment led by Davis only finds 27% of the expected solar neutrino flux. To solve this neutrino puzzle, two further neutrino experiments were started. The gallium experiment GALLEX in a tunnel through the Gran Sasso mountains in Italy and the Soviet–American gallium experiment (SAGE) in Caucasus measure the flux of solar neutrinos also in radiochemical experiments. Solar neutrinos react with gallium according to

GALLEX, SAGE

$$\nu_e + {}^{71}\text{Ga} \rightarrow {}^{71}\text{Ge} + e^- \; . \tag{6.36}$$

In this reaction ^{71}Ge is produced and extracted like in the Davis experiment and counted. The gallium experiments have the big advantage that the reaction threshold for the reaction (6.36) is as low as 233 keV so that these experiments are sensitive to neutrinos from the proton–proton fusion while the Davis experiment with a threshold of 810 keV essentially only measures neutrinos from the ^{8}B decay. GALLEX and SAGE have also measured a deficit of solar neutrinos. They only find 52% of the expected rate which presents a clear discrepancy to the prediction on the basis of the standard solar model. However, the discrepancy is not so pronounced as in the Davis experiment. A strong point for the gallium experiments is that the neutrino capture rate and the extraction technique have been checked with neutrinos of an artificial ^{51}Cr source. It could be convincingly shown that the produced ^{71}Ge atoms could be successfully extracted in the expected quantities.

***pp* fusion neutrinos**

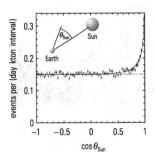

Fig. 6.19
Arrival directions of neutrinos measured in the Super-Kamiokande experiment

The Kamiokande and Super-Kamiokande experiment, respectively, measure solar neutrinos via the reaction

$$\nu_e + e^- \rightarrow \nu_e + e^- \tag{6.37}$$

at a threshold of 5 MeV in a water Cherenkov counter. Since the emission of the knock-on electron follows essentially the direction of the incident neutrinos, the detector can really 'see' the Sun. This directionality gives the water Cherenkov counter a superiority over the radiochemical experiments. Figure 6.19 shows the neutrino counting rate of the Super-Kamiokande experiment as a function of the angle with respect to the Sun. The Super-Kamiokande experiment also measures a low flux of solar neutrinos representing only 40% of the expectation.

Kamiokande, Super-Kamiokande

A reconstructed image of the Sun in the light of neutrinos is shown in Fig. 6.20.

Many proposals have been made to solve the solar neutrino problem. The first thing for elementary particle physicists is to doubt the correctness of the standard solar model. The flux of ^{8}B neutrinos varies with the central temperature of the Sun like $\sim T^{18}$. A reduction by only 5% of the central solar temperature would bring the Kamiokande experiment already in agreement with the now reduced expectation. However, solar astrophysicists consider even a somewhat lower central temperature of the Sun rather improbable.

The theoretical calculation of the solar neutrino flux uses the cross sections for the reactions (6.27) up to (6.34). An overestimate of the reaction cross sections would also lead to a too high expectation for the neutrino flux. A variation of these cross sections in a range which is considered realistic by nuclear physicists is insufficient to explain the discrepancy between the experimental data and expectation.

If neutrinos had a mass, they could also possess a magnetic moment. If their spin is rotated while propagating from the solar interior to the detector at Earth, one would not be able to measure these neutrinos because the detectors are insensitive to neutrinos of wrong helicity.

Finally, solar neutrinos could decay on their way from Sun to Earth into particles which might be invisible to the neutrino detectors.

A drastic assumption would be that the solar fire has gone out. In the light of neutrinos this would become practically immediately evident (more precisely: in 8 minutes). The energy transport from the solar interior to the surface, however, requires a time of several 100 000 years so that the Sun would continue to shine for this period even though the nuclear fusion at its center has come to an end.

Since all mentioned explanations are considered rather unlikely, it is attractive to interpret a deficit of solar neutrinos also by oscillations like in the case of atmospheric neutrinos.

In addition to the vacuum oscillations described by (6.23), solar neutrinos can also be transformed by so-called matter oscillations. The flux of electron neutrinos and its oscillation property can be modified by neutrino–electron scattering when the solar neutrino flux from the interior of the Sun encounters collisions with the abundant number of solar electrons. Flavour oscillations can even be magnified in a resonance-like fashion by matter effects so that certain

Fig. 6.20
Reconstructed image of the Sun in the light of solar neutrinos. Due to the limited spatial and angular resolution of Super-Kamiokande, the image of the Sun appears larger than it really is {12}

cross section

magnetic moment of neutrinos?

neutrino decay?

extinct solar fire?

neutrino oscillations?

matter oscillations

MSW effect

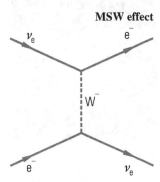

Fig. 6.21
Feynman diagram responsible for
matter oscillation (MSW effect).
Given the energy of solar
neutrinos and the fact that there
are only target electrons in the
Sun, this process can only occur
for v_e, but not for v_μ or v_τ

energy ranges of the solar v_e spectrum are depleted. The
possibility of matter oscillations has first been proposed by
Mikheyev, Smirnov, and Wolfenstein. The oscillation property of the MSW effect is different from that of vacuum
oscillations. It relates to the fact that $v_e e^-$ scattering contributes a term to the mixing matrix that is not present in
vacuum. Due to this charged-current interaction (Fig. 6.21),
which is kinematically not possible for v_μ and v_τ in the Sun,
the interaction Hamiltonian for v_e is modified compared to
the other neutrino flavours. This leads to alterations for the
energy difference of the two neutrino eigenvalues in matter compared to vacuum. Therefore, electron neutrinos are
singled out by this additional interaction process in matter.

Depending on the electron density in the Sun the originally dominant mass eigenstate v_e can propagate into a
different mass eigenstate for which the neutrino detectors
are not sensitive. One might wonder, how such matter oscillations work in the Sun. The probability for a neutrino
to interact in matter is extremely small. The way the solar
electron density affects the propagation of solar neutrinos,
however, depends on amplitudes which are square roots of
probabilities. Therefore, even though the probabilities of
interactions are small, the neutrino flavours can be significantly altered because of the amplitude dependence of the
oscillation mechanism.

If the three neutrino flavours v_e, v_μ, v_τ would completely
mix, only $1/3$ of the original electron neutrinos would arrive at Earth. Since the neutrino detectors, however, are blind
for MeV neutrinos of v_μ and v_τ type, the experimental results could be understood in a framework of oscillations.
Obviously, the solar neutrino problem cannot be solved that
easily. The results of the four so far described experiments
which measure solar neutrinos do not permit a unique solu-
tion in the parameter space $\sin^2 2\theta$ and δm^2, compare (6.20)

mixing angle   and (6.21). If $(v_e \to v_\mu)$ or $(v_e \to v_\tau)$ oscillations are assumed and if it is considered that the MSW effect is respon-
sible for the oscillations, a δm^2 on the order of 4×10^{-4}–
$2 \times 10^{-5}\,\text{eV}^2$ and a large-mixing-angle solution, although
disfavouring maximal mixing, is presently favoured. Assuming a mass hierarchy also in the neutrino sector, this

possible values   would lead to a v_μ or v_τ mass of 0.02–0.004 eV. This is not
for neutrino masses   necessarily in contradiction to the results from atmospheric
neutrinos since solar neutrinos could oscillate into muon
neutrinos and atmospheric muon neutrinos into tau neutri-

nos (or into so far undiscovered sterile neutrinos which are not even subject to weak interactions). If this scenario were correct, one could have $m_{\nu_\mu} \approx 10\,\text{meV}$ and $m_{\nu_\tau} \approx 50\,\text{meV}$.

Recently the Sudbury Neutrino Observatory (SNO) has convincingly confirmed the oscillation picture. The SNO Cherenkov detector installed at a depth of 2000 m underground in a nickel mine in Ontario, Canada, consists of a 1000 ton heavy water target (D_2O) contained in a 12 m diameter acrylic vessel. The interaction target is viewed by 9600 phototubes. This central detector is immersed in a 30 m barrel-shaped cavity containing 7000 tons of normal light water to suppress background reactions from cosmic rays or terrestrial radiation from radioisotopes in the surrounding rock or the mine dust. The SNO experiment can distinguish the charged-current interaction (CC)

SNO experiment

charged currents

(a) $\quad \nu_e + d \rightarrow p + p + e^-$,

which can only be initiated by electron neutrinos, from the neutral-current reaction (NC)

neutral currents

(b) $\quad \nu_x + d \rightarrow p + n + \nu'_x \quad (x = e, \mu, \tau)$,

where an incoming neutrino of any flavour interacts with a deuteron. The neutrons produced in this reaction are captured by deuterons giving rise to the emission of 6.25 MeV photons, which signal the NC interaction. While the ν_e flux as obtained by the CC reaction is only $1/3$ of the predicted solar neutrino flux, the total neutrino flux measured by the NC reaction is in agreement with the expectation of solar models, thereby providing evidence for a non-ν_e component.

This result solves the long-standing neutrino problem. It does not, however, resolve the underlying mechanism of the oscillation process. It is not at all clear, whether the ν_e oscillate into ν_μ or ν_τ. It is considered very likely that matter oscillations via the MSW effect in the Sun is the most likely mechanism for the transmutation of the solar electron neutrinos into other neutrino flavors, which unfortunately cannot directly be measured in a light-water Cherenkov counter.

The oscillation mechanism suggested by the different solar experiments ($\nu_e \rightarrow \nu_\mu$) was confirmed at the end of 2002 by the KamLAND[2] reactor neutrino detector, which removed all doubts about possible uncertainties of the standard solar model predictions.

KamLAND reactor experiment

---

[2] KamLAND – Kamioka Liquid-scintillator Anti-Neutrino Detector

6.2.3 Supernova Neutrinos

discovery of SN 1987A

Fig. 6.22
Supernova 1987A in the Tarantula
Nebula {6}

fusion cycles of a star

gravitational collapse

production of a neutron star

**deleptonization
→ neutrino burst**

The brightest supernova since the observation of Kepler in
the year 1604 was discovered by Ian Shelton at the Las Cam-
panas observatory in Chile on February 23, 1987 (see Fig.
6.22). The region of the sky in the Tarantula Nebula in the
Large Magellanic Cloud (distance 170 000 light-years), in
which the supernova exploded, was routinely photographed
by Robert McNaught in Australia already 20 hours earlier. However, McNaught developed and analyzed the photographic plate only the following day. Ian Shelton was struck
by the brightness of the supernova which was visible to the
naked eye. For the first time a progenitor star of the supernova explosion could be located. Using earlier exposures of
the Tarantula Nebula, a bright blue supergiant, Sanduleak,
was found to have exploded. Sanduleak was an inconspicuous star of 10-fold solar mass with a surface temperature
of 15 000 K. During hydrogen burning Sanduleak increased
its brightness reaching a luminosity 70 000 higher than the
solar luminosity. After the hydrogen supply was exhausted,
the star expanded to become a red supergiant. In this process its central temperature and pressure rose to such values that He burning became possible. In a relatively short
time (600 000 years) the helium supply was also exhausted.
Helium burning was followed by a gravitational contraction
in which the nucleus of the star reached a temperature of
740 million Kelvin and a central density of $240 \, \text{kg/cm}^3$.
These conditions enabled carbon to ignite. In a similar fashion contraction and fusion phases occurred leading via oxygen, neon, silicon, and sulphur finally to iron, the element
with the highest binding energy per nucleon.

The pace of these successive contraction and fusion
phases got faster and faster until finally iron was reached.
Once the star has reached such a state, there is no way to
gain further energy by fusion processes. Therefore, the stability of Sanduleak could no longer be maintained. The star
collapsed under its own gravity. During this process the electrons of the star were forced into the protons and a neutron
star of approximately 20 km diameter was produced. In the
course of this deleptonization a *neutrino burst* of immense
intensity was created,

$$e^- + p \rightarrow n + \nu_e . \tag{6.38}$$

In the hot phase of the collapse corresponding to a temperature of 10 MeV ($\approx 10^{11}$ K), the thermal photons produced electron–positron pairs which, however, were immediately absorbed because of the high density of the surrounding matter. Only the weak-interaction process with a virtual Z,

$$e^+ + e^- \rightarrow Z \rightarrow \nu_\alpha + \bar{\nu}_\alpha , \qquad (6.39)$$

allowed energy to escape from the hot stellar nucleus in the form of neutrinos. In this reaction all three neutrino flavours ν_e, ν_μ, and ν_τ were produced 'democratically' in equal numbers. The total neutrino burst comprised 10^{58} neutrinos and even at Earth the neutrino flux from the supernova was comparable to that of solar neutrinos for a short period.

Actually, the neutrino burst of the supernova was the first signal to be registered on Earth. The large water Cherenkov counters of Kamiokande and IMB (Irvine–Michigan–Brookhaven) recorded a total of 20 out of the emitted 10^{58} neutrinos. The energy threshold of the Kamiokande experiment was as low as 5 MeV. In contrast, the IMB collaboration could only measure neutrinos with energies exceeding 19 MeV. The Baksan liquid scintillator was lucky to record – even though their fiducial mass was only 200 t – five coincident events with energies between 10 MeV and 25 MeV.

measurement of neutrinos

Since the neutrino energies in the range of 10 MeV are insufficient to produce muons or taus, only electron-type neutrinos were recorded via the reactions

$$\begin{aligned} \bar{\nu}_e + p &\rightarrow e^+ + n , \\ \bar{\nu}_e + e^- &\rightarrow \bar{\nu}_e + e^- , \\ \nu_e + e^- &\rightarrow \nu_e + e^- . \end{aligned} \qquad (6.40)$$

In spite of the low number of measured neutrinos on Earth some interesting astrophysical conclusions can be drawn from this supernova explosion. If E_ν^i is the energy of individual neutrinos measured in the detector, ε_1 the probability for the interaction of a neutrino in the detector, and ε_2 the probability to also see this reaction, then the total energy emitted in form of neutrinos can be estimated to be

energy output

$$E_{\text{total}} = \sum_{i=1}^{20} \frac{E_\nu^i}{\varepsilon_1(E_\nu^i)\,\varepsilon_2(E_\nu^i)} \, 4\pi r^2 \, f(\nu_\alpha, \bar{\nu}_\alpha) , \qquad (6.41)$$

where the correction factor f takes into account that the water Cherenkov counters are sensitive not to all neutrino flavours. Based on the 20 recorded neutrino events a total energy of

$$E_{\text{total}} = (6 \pm 2) \times 10^{46} \text{ Joule} \qquad (6.42)$$

is obtained. It is hard to comprehend this enormous energy. (The world energy consumption is 10^{21} Joule per year.) During the 10 seconds lasting neutrino burst Sanduleak radiated more energy than the rest of the universe and hundred times more than the Sun in its total lifetime of about 10 billion years.

limits of neutrino masses   Measurements over the last 40 years have ever tightened the limits for neutrino masses. At the time of the supernova explosion the mass limit for the electron neutrino from measurements of the tritium beta decay ($^3\text{H} \rightarrow {}^3\text{He} + e^- + \bar{\nu}_e$) was about 10 eV. Under the assumption that all supernova neutrinos are emitted practically at the same time, one would expect that their arrival times at Earth would be subject to a certain spread if the neutrinos had mass. Neutrinos of nonzero mass have different velocities depending on their energy. The expected difference of arrival times Δt of two **difference of propagation time** neutrinos with velocities v_1 and v_2 emitted at the same time from the supernova is

$$\Delta t = \frac{r}{v_1} - \frac{r}{v_2} = \frac{r}{c}\left(\frac{1}{\beta_1} - \frac{1}{\beta_2}\right) = \frac{r}{c}\frac{\beta_2 - \beta_1}{\beta_1 \beta_2}. \qquad (6.43)$$

If the recorded electron neutrinos had a rest mass m_0, their energy would be

$$E = mc^2 = \gamma m_0 c^2 = \frac{m_0 c^2}{\sqrt{1 - \beta^2}}, \qquad (6.44)$$

and their velocity

$$\beta = \left(1 - \frac{m_0^2 c^4}{E^2}\right)^{1/2} \approx 1 - \frac{1}{2}\frac{m_0^2 c^4}{E^2}, \qquad (6.45)$$

since one can safely assume that $m_0 c^2 \ll E$. This means that the neutrino velocities are very close to the velocity of light. Obviously, the arrival-time difference Δt depends on the velocity difference of the neutrinos. Using (6.43) and (6.45) one gets

$$\Delta t \approx \frac{r}{c}\frac{\frac{1}{2}\frac{m_0^2 c^4}{E_1^2} - \frac{1}{2}\frac{m_0^2 c^4}{E_2^2}}{\beta_1 \beta_2} \approx \frac{1}{2}m_0^2 c^4 \frac{r}{c}\frac{E_2^2 - E_1^2}{E_1^2 E_2^2}. \qquad (6.46)$$

The experimentally measured arrival-time differences and individual neutrino energies allow in principle to work out the electron neutrino rest mass

$$m_0 = \left\{ \frac{2\Delta t}{r\,c^3} \frac{E_1^2\,E_2^2}{E_2^2 - E_1^2} \right\}^{1/2} . \tag{6.47}$$

Since, however, not all neutrinos are really emitted simultaneously, (6.47) only allows to derive an upper limit for the neutrino mass using pairs of particles of known energy and known arrival-time difference. Using the results of the Kamiokande and IMB experiments a mass limit of the electron neutrino of

neutrino mass limit

$$m_{\nu_e} \leq 10\,\text{eV} \tag{6.48}$$

could be established. This result was obtained in a measurement time of approximately 10 seconds. It demonstrates the potential superiority of astrophysical investigations over laboratory experiments.

Similarly, a possible explanation for the deficit of solar neutrinos by assuming neutrino decay was falsified by the mere observation of electron neutrinos from a distance of 170 000 light-years. For an assumed neutrino mass of $m_0 = 10\,\text{meV}$ the Lorentz factor of 10 MeV neutrinos would be

$$\gamma = \frac{E}{m_0 c^2} \approx 10^9 . \tag{6.49}$$

This would allow to derive a lower limit for the neutrino lifetime from $\tau_\nu^0 = \tau_\nu / \gamma$ to

neutrino lifetime

$$\tau_\nu^0 = 170\,000\,\text{a}\,\frac{1}{\gamma} \approx 5\,000\,\text{s} . \tag{6.50}$$

The supernova 1987A has turned out to be a rich astrophysical laboratory. It has shown that the available supernova models can describe the spectacular death of massive stars on the whole correctly. Given the agreement of the measured neutrinos fluxes with expectation, the supernova neutrinos do not seem to require oscillations. On the other hand, the precision of simulations and the statistical errors of measurements are insufficient to draw a firm conclusion about such a subtle effect for supernova neutrinos. The probability that such a spectacle of a similarly bright supernova in our immediate vicinity will happen again in the near future to clarify whether supernova neutrinos oscillate or not

supernova models confirmed

is extremely small. It is not a surprise that the decision on the oscillation scenario has come from observations of solar and atmospheric neutrinos and accelerator experiments with well-defined, flavour-selected neutrino beams. With the experimental evidence of cosmic-ray-neutrino experiments (Davis, GALLEX, SAGE, Super-Kamiokande, SNO) and recent accelerator and reactor experiments (K2K, Kam-LAND) there is now unanimous agreement that oscillations in the neutrino sector are an established fact.

6.2.4 High-Energy Galactic and Extragalactic Neutrinos

The measurement of high-energy neutrinos ($\geq$ TeV range) represents a big experimental challenge. The arrival direction of such neutrinos, however, would directly point back to the sources of cosmic rays. Therefore, a substantial amount of work is devoted to prototype studies for neutrino detectors in the TeV range and the development of experimental setups for the measurement of galactic and extragalactic high-energy neutrinos. The reason to restrict oneself to high-energy neutrinos is obvious from the inspection of Fig. 6.23. The neutrino echo of the Big Bang has produced energies below the meV range. About a second after the Big Bang weak interactions have transformed protons into neutrons and neutrons into protons thereby producing neutrinos ($p + e^- \rightarrow n + \nu_e$, $n \rightarrow p + e^- + \bar{\nu}_e$). The temperature of these primordial neutrinos should be at 1.9 K at present time.

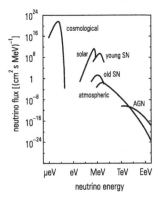

Fig. 6.23
Comparison of cosmic neutrino fluxes in different energy domains

'blackbody' neutrinos

Blackbody photons have a slightly higher temperature (2.7 K) because, in addition, electrons and positrons have transformed their energy by annihilation into photon energy. The Big Bang neutrinos even originate from an earlier cosmological epoch than the blackbody photons since the universe was much earlier transparent for neutrinos. In so far these cosmological neutrinos are very interesting as far as details of the creation and the development of the early universe are concerned. Unfortunately, at present it is almost inconceivable that neutrinos of such low energies in the meV range can be measured at all.

The observation of solar ($\approx$ MeV range) and supernova neutrinos ($\approx$ 10 MeV) is experimentally established. Atmospheric neutrinos represent a background for neutrinos from astrophysical sources. Atmospheric neutrinos originate essentially from pion and muon decays. Their production spectra can be inferred from the measured atmospheric muon spectra. However, they are also directly measured (see Figs. 6.11 and 6.12). Their intensity is only known with an accuracy of about 30%. The spectral shape and intensity of neutrinos from extragalactic sources (AGN – Active Galactic Nuclei) shown in Fig. 6.23 represents only a very rough estimate.

It is generally assumed that binaries are good candidates for the production of energetic neutrinos. A binary consisting of a pulsar and a normal star could represent a strong neutrino source (Fig. 6.24).

The pulsar and the star rotate around their common center of mass. If the stellar mass is large compared to the pulsar mass, one can assume for illustration purposes of the neutrino production mechanism that the pulsar orbits the companion star on a circle. There are models which suggest that the pulsar can manage to accelerate protons to very high energies. These accelerated protons collide with the gas of the atmosphere of the companion star and produce predominantly secondary pions in the interactions. The neutral pions decay relatively fast ($\tau_{\pi^0} = 8.4 \times 10^{-17}$ s) into two energetic γ rays, which would allow to locate the astronomical object in the light of γ rays. The charged pions produce energetic neutrinos by their ($\pi \to \mu\nu$) decay. Whether such a source radiates high-energy γ quanta or neutrinos depends crucially on subtle parameters of the stellar atmosphere. If pions are produced in a proton interaction such as

$$p + \text{nucleus} \to \pi^+ + \pi^- + \pi^0 + \text{anything}, \quad (6.51)$$

equal amounts of neutrinos and photons would be produced by the decays of charged and neutral pions ($\pi^+ \to \mu^+ + \nu_\mu$, $\pi^- \to \mu^- + \bar{\nu}_\mu$, $\pi^0 \to \gamma + \gamma$). With increasing column density of the stellar atmosphere, however, photons would be reabsorbed, and for densities of stellar atmospheres of $\varrho \leq 10^{-8}$ g/cm^3 and column densities of more than 250 g/cm^2 this source would be only visible in the light of neutrinos (Fig. 6.25).

The source would shine predominantly in muon neutrinos (ν_μ or $\bar{\nu}_\mu$). These neutrinos can be recorded in a detector via the weak charged current in which they produce muons (Fig. 6.26).

Muons created in these interactions follow essentially the direction of the incident neutrinos. The energy of the muon is measured by its energy loss in the detector. For energies exceeding the TeV range, muon bremsstrahlung and

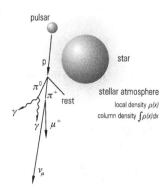

Fig. 6.24
Production mechanism of high-energy neutrinos in a binary system

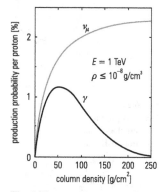

Fig. 6.25
Competition between production and absorption of photons and neutrinos in a binary system

competition:
neutrino, gamma emission

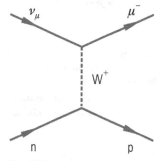

Fig. 6.26
Reaction for muon neutrino detection

direct electron pair production by muons dominate. The energy loss by these two processes is proportional to the muon energy and therefore allows a calorimetric determination of the muon energy (compare Sect. 7.3, Fig. 7.17).

neutrino detectors

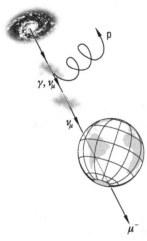

Fig. 6.27
Neutrino production, propagation in intergalactic space, and detection at Earth

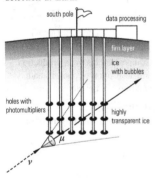

Fig. 6.28
Sketch of a neutrino detector for high-energy extragalactic neutrinos

instrumentation of a large volume

Because of the low interaction probability of neutrinos and the small neutrino fluxes, neutrino detectors must be very large and massive. Since the whole detector volume has to be instrumented to be able to record the interactions of neutrinos and the energy loss of muons, it is necessary to construct a simple, cost-effective detector. The only practicable candidates which meet this condition are huge water or ice Cherenkov counters. Because of the extremely high transparency of ice at large depths in Antarctica and the relatively simple instrumentation of the ice, ice Cherenkov counters are presently the most favourable choice for a realistic neutrino telescope. To protect the detector against the relatively high flux of atmospheric particles, it has become common practice to use the Earth as an absorber and concentrate on neutrinos which enter the detector 'from below'. The principle of such a setup is sketched in Fig. 6.27. Protons from cosmic-ray sources produce pions on a target (e.g., stellar atmosphere, galactic medium) which provide neutrinos and γ quanta in their decay. Photons are frequently absorbed in the galactic medium or disappear in $\gamma\gamma$ interactions with blackbody photons, infrared radiation, or starlight photons. The remaining neutrinos traverse the Earth and are detected in an underground detector. The neutrino detector itself consists of a large array of photomultipliers which record the Cherenkov light of muons produced in ice (or in water), Fig. 6.28. In such neutrino detectors the photomultipliers will be mounted in a suitable distance on strings and many of such strings will be deployed in ice (or water). The mutual distance of the photomultipliers on the strings and the string spacing depends on the absorption and scattering length of Cherenkov light in the detector medium. The direction of incidence of neutrinos can be inferred from arrival times of the Cherenkov light at the photomultipliers. In a water Cherenkov counter in the ocean bioluminescence and potassium-40 activity presents an annoying background which is not present in ice. In practical applications it became obvious that the installation of photomultiplier strings in the antarctic ice is much less problematic compared to the deployment in the ocean.

The setup of a prototype detector in the antarctic ice is shown in Fig. 6.29. In a first installation of photomultiplier strings at depths of 810 up to 1000 meters it turned out that the ice still contained too many bubbles. Only for depths below 1500 m the pressure is sufficiently large (≥ 150 bar) that the bubbles disappear yielding an excellent transparency with absorption lengths of about 300 m. The working principle of the detector could be demonstrated by measuring atmospheric muons and muon neutrinos. For a 'real' neutrino telescope, however, the present AMANDA detector is still too small. To be able to record the fluxes of extragalactic neutrinos, a volume of 1 km^3 is needed. The original AMANDA detector is continuously upgraded so that finally such a big volume will be instrumented (IceCube). Figure 6.30 shows a muon event in AMANDA. The following example will confirm that volumes of this size are really required.

It is considered realistic that a point source in our galaxy produces a neutrino spectrum according to

interaction rate in a ν detector

$$\frac{dN}{dE_\nu} = 2 \times 10^{-11} \frac{100}{E_\nu^2 \, [\text{TeV}^2]} \, \text{cm}^{-2} \, \text{s}^{-1} \, \text{TeV}^{-1} \, . \quad (6.52)$$

This leads to an integral flux of neutrinos of

$$\Phi_\nu(E_\nu > 100 \, \text{TeV}) = 2 \times 10^{-11} \, \text{cm}^{-2} \, \text{s}^{-1} \quad (6.53)$$

(see also Fig. 6.23 for extragalactic sources).

The interaction cross section of high-energy neutrinos was measured at accelerators to be

$$\sigma(\nu_\mu N) = 6.7 \times 10^{-39} \, E_\nu \, [\text{GeV}] \, \text{cm}^2/\text{nucleon} \, . \quad (6.54)$$

For 100 TeV neutrinos one would arrive at a cross section of 6.7×10^{-34} cm^2/nucleon. For a target thickness of one kilometer an interaction probability W per neutrino of

$$W = N_A \, \sigma \, d \, \varrho = 4 \times 10^{-5} \quad (6.55)$$

is obtained ($d = 1 \, \text{km} = 10^5 \, \text{cm}$, $\varrho(\text{ice}) \approx 1 \, \text{g/cm}^3$).

The total interaction rate R is obtained from the integral neutrino flux Φ_ν, the interaction probability W, the effective collection area $A_{\text{eff}} = 1 \, \text{km}^2$, and a measurement time t. This leads to an event rate of

$$R = \Phi_\nu \, W \, A_{\text{eff}} \quad (6.56)$$

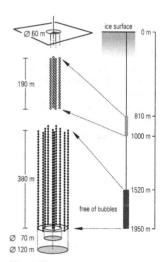

Fig. 6.29
Setup of the AMANDA detector at the South Pole (AMANDA – Antarctic Muon And Neutrino Detector Array)

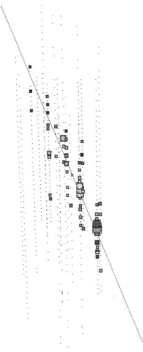

Fig. 6.30
A neutrino-induced upward-going
muon recorded in AMANDA. The
size of the symbols is proportional
to the measured Cherenkov light
{13}

neutrino telescopes

corresponding to 250 events per year. For large absorption
lengths of the produced Cherenkov light the effective collection area of the detector is even larger than the cross section
of the instrumented volume. Assuming that there are about
half a dozen sources in our galaxy, the preceding estimate
would lead to a counting rate of about four events per day.
In addition to this rate from point sources one would also expect to observe events from the diffuse neutrino background
which, however, carries little astrophysical information.

Excellent candidates within our galaxy are the supernova
remnants of the Crab Nebula and Vela, the galactic center,
and Cygnus X3. Extragalactic candidates could be represented by the Markarian galaxies Mrk 421 and Mrk 501,
by M87, or by quasars (e.g., 3C273).

Neutrino astronomy was pioneered with the Baikal telescope installed in the lake Baikal in Siberia. The most advanced larger telescopes are AMANDA and ANTARES[3].
AMANDA is taking data in Antarctica since several years,
while the ANTARES detector which is installed in the
Mediterranean offshore Toulon started data taking early in
2003. The NESTOR[4] detector has seen first results at the
end of 2003. It is also operated in the Mediterranean off
the coast of Greece. Real neutrino astronomy, however, will
require larger detectors like IceCube which is presently installed in Antarctica. It is expected that first results from
IceCube might become available from the year 2005 on.

6.3 Gamma Astronomy

"Let there be light."
The Bible; Genesis 1:3

6.3.1 Introduction

The observation of stars in the optical spectral range belongs
to the field of classical astronomy. Already the Chinese,
Egyptians, and Greeks performed numerous systematic observations and learned a lot about the motion of heavenly
bodies. The optical range, however, covers only a minute
range of the total electromagnetic spectrum (Fig. 6.31).

---

[3] ANTARES – Abyssal Neutrino Telescope And Research Environment of deep Sea

[4] NESTOR – Neutrino Extended Submarine Telescope with
Oceanographic Research

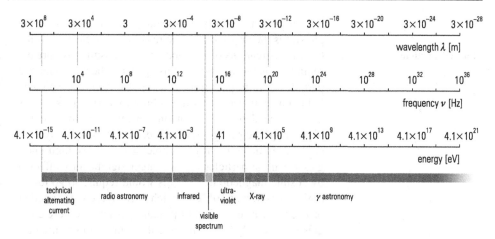

Fig. 6.31
Spectral range of electromagnetic radiation

All parts of this spectrum have been used for astronomical observations. From large wavelengths (radio astronomy), the sub-optical range (infrared astronomy), the classical optical astronomy, the ultraviolet astronomy, and X-ray astronomy one arrives finally at the *gamma-ray astronomy*.

Gamma-ray astronomers are used to characterize gamma quanta not by their wavelength λ or frequency ν, but rather by their energy,

$$E = h\nu . \qquad (6.57)$$

Planck's constant in practical units is

$$h = 4.136 \times 10^{-21}\,\mathrm{MeV\,s} . \qquad (6.58)$$

The frequency ν is measured in $\mathrm{Hz} = 1/\mathrm{s}$. The wavelength λ is obtained to be

$$\lambda = c/\nu , \qquad (6.59)$$

where c is the speed of light in vacuum ($c = 299\,792\,458$ m/s).

In atomic and nuclear physics one distinguishes gamma rays from X rays by the production mechanism. X rays are emitted in transitions of electrons in the atomic shell while gamma rays are produced in transformations of the atomic nucleus. This distinction also results naturally in a classification of X rays and gamma rays according to their energy. X rays typically have energies below 100 keV. Electromagnetic radiation with energies in excess of 100 keV is called γ rays. There is no upper limit for the energy of γ rays. Even cosmic γ rays with energies of 10^{15} eV $= 1$ PeV have been observed.

distinction between γ and X rays

charged-particle astronomy?

An important, so far unsolved problem of astroparticle physics is the origin of cosmic rays (see also Sects. 6.1 and 6.2). Investigations of charged primary cosmic rays are essentially unable to answer this question because charged particles have to pass through extended irregular magnetic fields on their way from the source to Earth. This causes them to be deflected in an uncontrolled fashion thereby 'forgetting' their origin. Therefore, particle astronomy with charged particles is only possible at extremely high energies when the particles are no longer significantly affected by cosmic magnetic fields. This would require to go to energies in excess of 10^{19} eV which, however, creates another problem because the flux of primary particles at these energies is extremely low. Whatever the sources of cosmic rays are, they will also be able to emit energetic penetrating γ rays which are not deflected by intergalactic or stellar magnetic fields and therefore point back to the sources. It must, however, be kept in mind that also X and γ rays from distant sources might be subject to time dispersions. Astronomical objects in the line of sight of these sources can distort their trajectory by gravitational lensing thus making them look blurred and causing time-of-flight dispersions in the arrival time also for electromagnetic radiation.

arrival-time dispersions

6.3.2 Production Mechanisms for γ Rays

sources of γ rays

Possible sources for cosmic rays and thereby also for γ rays are supernovae and their remnants, rapidly rotating objects like pulsars and neutron stars, active galactic nuclei, and matter-accreting black holes. In these sources γ rays can be produced by different mechanisms.

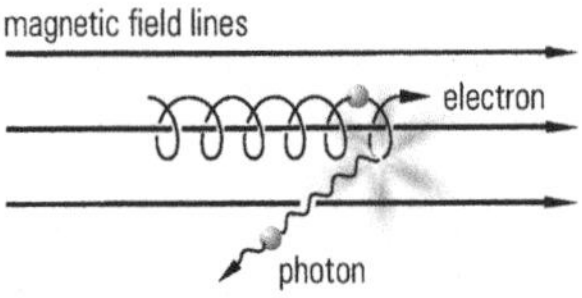

Fig. 6.32
Production of synchrotron radiation by deflection of charged particles in a magnetic field

a) Synchrotron radiation:

The deflection of charged particles in a magnetic field gives rise to an accelerated motion. An accelerated electrical charge radiates electromagnetic waves (Fig. 6.32). This 'bremsstrahlung' of charged particles in magnetic fields is called *synchrotron radiation*. In circular earth-bound accelerators the production of synchrotron radiation is generally considered as an undesired energy-loss mechanism. On the other hand, synchrotron radiation from accelerators is widely used for structure investigations in atomic and solid state physics as well as in biology and medicine.

applications of synchrotron radiation

Synchrotron radiation produced in cosmic magnetic fields is predominantly emitted by the lightweight electrons.

The energy spectrum of synchrotron photons is continuous. The power P radiated by an electron of energy E in a magnetic field of strength B is

$$P \sim E^2 B^2 . \tag{6.60}$$

b) Bremsstrahlung:

A charged particle which is deflected in the Coulomb field of a charge (atomic nucleus or electron) emits bremsstrahlung photons (Fig. 6.33). This mechanism is to a certain extent similar to synchrotron radiation, only that in this case the deflection of the particle occurs in the Coulomb field of a charge rather than in a magnetic field.

The probability for *bremsstrahlung* ϕ varies with the square of the projectile charge z and also with the square of the target charge Z, see also (4.7). ϕ is proportional to the particle energy E and it is inversely proportional to the mass squared of the deflected particle:

$$\phi \sim \frac{z^2 Z^2 E}{m^2} . \tag{6.61}$$

Because of the smallness of the electron mass bremsstrahlung is predominantly created by electrons. The energy spectrum of bremsstrahlung photons is continuous and decreases like $1/E_\gamma$ to high energies.

c) Inverse Compton Scattering:

In the twenties of the last century Compton discovered that energetic photons can transfer part of their energy to free electrons in a collision, thereby losing a certain amount of energy. In astrophysics the *inverse Compton effect* plays an important rôle. Electrons accelerated to high energies in the source collide with the numerous photons of the blackbody radiation ($E_\gamma \approx 250\,\mu\text{eV}$, photon density $N_\gamma \approx 400/\text{cm}^3$) or starlight photons ($E_\gamma \approx 1\,\text{eV}$, $N_\gamma \approx 1/\text{cm}^3$) and transfer part of their energy to the photons which are 'blueshifted' (Fig. 6.34).

d) π^0 Decay:

Protons accelerated in the sources can produce charged and neutral pions in proton–proton or proton–nucleus interactions (Fig. 6.35). A possible process is

$$p + \text{nucleus} \rightarrow p' + \text{nucleus}' + \pi^+ + \pi^- + \pi^0 . \tag{6.62}$$

Charged pions decay with a lifetime of 26 ns into muons and neutrinos, while neutral pions decay rapidly ($\tau = 8.4 \times 10^{-17}$ s) into two γ quanta,

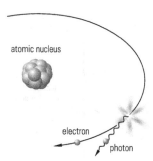

Fig. 6.33
Production of bremsstrahlung by deflection of charged particles in the Coulomb field of a nucleus

bremsstrahlung probability

spectrum of bremsstrahlung photons

inverse Compton scattering

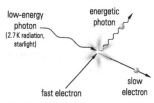

Fig. 6.34
Collision of an energetic electron with a low-energy photon. The electron transfers part of its energy to the photon and is consequently slowed down

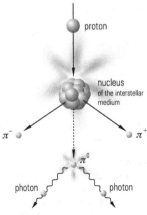

Fig. 6.35
π^0 production in proton interactions and π^0 decay into two photons

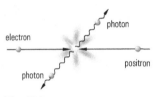

Fig. 6.36
e^+e^- pair annihilation into two photons

γ-ray lines

neutralino annihilation

$$\pi^0 \to \gamma + \gamma \ . \tag{6.63}$$

If the neutral pion decays at rest, both photons are emitted back to back. In this decay they get each half of the π^0 rest mass ($m_{\pi^0} = 135\,\text{MeV}$). In the π^0 decay in flight the photons get different energies depending on their direction of emission with respect to the direction of flight of the π^0 (see Example 8, Chap. 3). Since most pions are produced at low energies, photons from this particular source have energies of typically 70 MeV.

e) Photons from Matter–Antimatter Annihilation:

In the same way as photons can produce particle pairs (pair production), charged particles can annihilate with their antiparticles into energy. The dominant sources for this production mechanism are electron–positron and proton–antiproton annihilations,

$$e^+ + e^- \to \gamma + \gamma \ . \tag{6.64}$$

Momentum conservation requires that at least two photons are produced. In e^+e^- annihilation at rest the photons get 511 keV each corresponding to the rest mass of the electron or positron, respectively (Fig. 6.36). An example for a proton–antiproton annihilation reaction is

$$p + \bar{p} \to \pi^+ + \pi^- + \pi^0 \ , \tag{6.65}$$

where the neutral pion decays into two photons.

f) Photons from Nuclear Transformations:

Heavy elements are 'cooked' in supernova explosions. In these processes not only stable but also radioactive isotopes are produced. These radioisotopes will emit, mostly as a consequence of a beta decay, photons in the MeV range like, e.g.,

$$^{60}\text{Co} \to \ ^{60}\text{Ni}^{**} + e^- + \bar{\nu}_e$$
$$\hookrightarrow \ ^{60}\text{Ni}^* + \gamma\,(1.17\,\text{MeV}) \tag{6.66}$$
$$\hookrightarrow \ ^{60}\text{Ni} + \gamma\,(1.33\,\text{MeV}) \ .$$

g) Annihilation of Neutralinos:

In somewhat more exotic scenarios, energetic γ rays could also originate from annihilation of neutralinos, the neutral supersymmetric partners of ordinary particles, according to

$$\chi + \bar{\chi} \to \gamma + \gamma \ . \tag{6.67}$$

6.3.3 Measurement of γ Rays

In principle the inverse production mechanisms of γ rays can be used for their detection (see also Chap. 4). For γ rays with energies below several hundred keV the photoelectric effect dominates,

photoelectric effect,
$E_\gamma \lesssim 100\,\text{keV}$

$$\gamma + \text{atom} \rightarrow \text{atom}^+ + e^- . \qquad (6.68)$$

The photoelectron can be recorded, e.g., in a scintillation counter. For energies in the MeV range as it is typical for nuclear decays, Compton scattering has the largest cross section,

Compton effect, $E_\gamma \lesssim 1\,\text{MeV}$

$$\gamma + e^-_{\text{at rest}} \rightarrow \gamma' + e^-_{\text{fast}} . \qquad (6.69)$$

In this case the material of a scintillation counter can also act as an electron target which records at the same time the scattered electron. For higher energies ($\gg 1\,\text{MeV}$) electron–positron pair creation dominates,

electron–positron pair production, $E_\gamma \gg 1\,\text{MeV}$

$$\gamma + \text{nucleus} \rightarrow e^+ + e^- + \text{nucleus}' . \qquad (6.70)$$

Figure 6.37 shows the dependence of the mass attenuation coefficient μ for the three mentioned processes in a NaI scintillation counter.

This coefficient is defined through the photon intensity attenuation in matter according to

$$I(x) = I_0\, e^{-\mu x} \qquad (6.71)$$

(I_0 – initial intensity, $I(x)$ – photon intensity after attenuation by an absorber of thickness x).

Since pair production dominates at high energies, this process is used for photon detection in the GeV range. Figure 6.38 shows a typical setup of a satellite experiment for the measurement of γ rays in the GeV range.

Energetic photons are converted into e^+e^- pairs in a modular tracking-chamber system (e.g., in a multiplate spark chamber or a stack of semiconductor silicon counters). The energies E_{e^+} and E_{e^-} are measured in an electromagnetic calorimeter (mostly a crystal-scintillator calorimeter, NaI(Tl) or CsI(Tl)) so that the energy of the original photon is

crystal calorimeter

$$E_\gamma = E_{e^+} + E_{e^-} . \qquad (6.72)$$

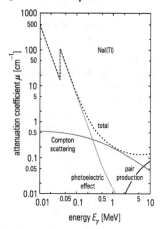

Fig. 6.37
Mass attenuation coefficient for photons in a sodium-iodide scintillation counter

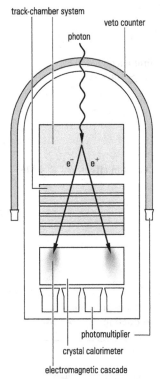

track-chamber system

veto counter

photon

e^- e^+

photomultiplier

crystal calorimeter

electromagnetic cascade

Fig. 6.38
Sketch of a satellite experiment for
the measurement of γ rays in the
GeV range

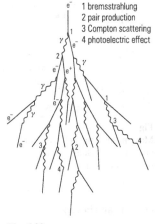

e^- 1 bremsstrahlung
2 pair production
3 Compton scattering
4 photoelectric effect

Fig. 6.39
Schematic representation of an
electron cascade

Cherenkov radiation

The direction of incidence of the photon is derived from the electron and positron momenta where the photon momentum is determined to be $\boldsymbol{p}_\gamma = \boldsymbol{p}_{e^+} + \boldsymbol{p}_{e^-}$. For high energies ($E \gg m_e c^2$) the approximations $|\boldsymbol{p}_{e^+}| = E_{e^+}/c$ and $|\boldsymbol{p}_{e^-}| = E_{e^-}/c$ are well satisfied.

The detection of electrons and positrons in the crystal calorimeter proceeds via electromagnetic cascades. In these showers the produced electrons initially radiate bremsstrahlung photons which convert into e^+e^- pairs. In alternating processes of bremsstrahlung and pair production the initial electrons and photons decrease their energy until absorptive processes like photoelectric effect and Compton scattering for photons on the one hand and ionization loss for electrons and positrons on the other hand halt further particle multiplication (Fig. 6.39).

The anticoincidence counter in Fig. 6.38 serves the purpose of identifying incident charged particles and rejecting them from the analysis.

For energies in excess of $100\,\text{GeV}$ the photon intensities from cosmic-ray sources are so small that other techniques for their detection must be applied, since sufficiently large setups cannot be installed on board of satellites. In this context the detection of photons via the *atmospheric Cherenkov technique* plays a special rôle.

When γ rays enter the atmosphere they produce – like already described for the crystal calorimeter – a cascade of electrons, positrons, and photons which are generally of low energy. This shower does not only propagate longitudinally but it also spreads somewhat laterally in the atmosphere (Fig. 6.40). For initial photon energies below $10^{13}\,\text{eV}$ ($= 10\,\text{TeV}$) the shower particles, however, do not reach sea level. Relativistic electrons and positrons of the cascade which follow essentially the direction of the original incident photon emit blue light in the atmosphere which is known as Cherenkov light. Charged particles whose velocities exceed the speed of light emit this characteristic electromagnetic radiation (see Chap. 4). Since the speed of light in atmospheric air is

$$c_n = c/n \tag{6.73}$$

(n is the index of refraction of air; $n = 1.000\,273$ at $20°C$ and 1 atm), electrons with velocities

$$v \geq c/n \tag{6.74}$$

will emit Cherenkov light. This threshold velocity of

$$v = c_n = 299\,710\,637\,\text{m/s} \tag{6.75}$$

corresponds to a kinetic electron energy of

$$
\begin{aligned}
E_{\text{kin}} &= E_{\text{total}} - m_0 c^2 = \gamma m_0 c^2 - m_0 c^2 \\
&= (\gamma - 1) m_0 c^2 = \left(\frac{1}{\sqrt{1 - v^2/c^2}} - 1 \right) m_0 c^2 \\
&= \left(\frac{1}{\sqrt{1 - 1/n^2}} - 1 \right) m_0 c^2 \\
&= \left(\frac{n}{\sqrt{n^2 - 1}} - 1 \right) m_0 c^2 \approx 21.36 \,\text{MeV} .
\end{aligned} \tag{6.76}
$$

The production of Cherenkov radiation in an optical shock wave (Fig. 6.41) is the optical analogue to sound shock waves which are created when aeroplanes exceed the velocity of sound.

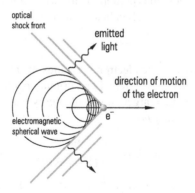

In this way energetic primary γ quanta can be recorded at ground level via the produced Cherenkov light even though the electromagnetic shower does not reach sea level. The Cherenkov light is emitted under a characteristic angle of

Cherenkov angle

$$\theta_C = \arccos \left(\frac{1}{n\beta} \right) . \tag{6.77}$$

For electrons in the multi-GeV range the opening angle of the Cherenkov cone is only $1.4°$.[5]

---

[5] Actually the Cherenkov angle is somewhat smaller ($\approx 1°$), since the shower particles are produced at large altitudes where the density of air and thereby also the index of refraction is smaller.

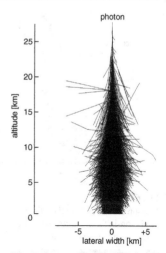

Fig. 6.40
Monte Carlo simulation of an electromagnetic shower in the atmosphere initiated by a photon of energy 10^{14} eV. All secondaries with energies $E \geq 3$ MeV are shown {14}

Fig. 6.41
Emission of Cherenkov radiation in an optical shock wave by particles traversing a medium of refractive index n with a velocity exceeding the velocity of light in that medium $(v > c/n)$

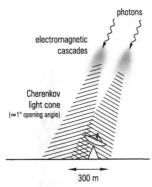

photons

electromagnetic cascades

Cherenkov light cone (≈1° opening angle)

300 m

Fig. 6.42
Measurement of Cherenkov light of photon-induced electromagnetic cascades in the atmosphere

Fig. 6.43
Photograph of the air Cherenkov telescope CANGAROO (CANGAROO – Collaboration of Australia and Nippon (Japan) for a GAmma-Ray Observatory in the Outback) {15}

γγ interactions

A simple Cherenkov detector, therefore, consists of a parabolic mirror which collects the Cherenkov light and a set of photomultipliers which record the light collected at the focal point of the mirror. Figure 6.42 shows the principle of photon measurements via the atmospheric Cherenkov technique. Large Cherenkov telescopes with mirror diameters ≥ 10 m allow to measure comparatively low-energy photons (< 100 GeV) with correspondingly small shower size even in the presence of light from the night sky (see Fig. 6.43). For even higher energies (> 10^{15} eV) the electromagnetic cascades initiated by the photons reach sea level and can be recorded with techniques like those which are used for the investigations and measurements of extensive air showers (particle sampling, air scintillation, cf. Sect. 7.4). At these energies it is anyhow impossible to explore larger regions of the universe in the light of γ rays. The intensity of energetic primary photons is attenuated by photon–photon interactions predominantly with numerous ambient photons of the 2.7 Kelvin blackbody radiation. For the process

$$\gamma + \gamma \rightarrow e^+ + e^- \tag{6.78}$$

twice the electron mass must be provided in the γγ center-of-mass system. For a primary photon of energy E colliding with a target photon of energy ε at an angle θ the threshold energy is

$$E_{\text{threshold}} = \frac{2m_e^2}{\varepsilon(1 - \cos\theta)} . \tag{6.79}$$

For a central collision ($\theta = 180°$) and a typical blackbody photon energy of $\varepsilon \approx 250\,\mu\text{eV}$ the threshold is

$$E_{\text{threshold}} \approx 10^{15}\,\text{eV} . \tag{6.80}$$

The cross section rises rapidly above threshold, reaches a maximum of 200 mb at twice the threshold energy, and decreases thereafter. For even higher energies further absorptive processes with infrared or starlight photons occur ($\gamma\gamma \rightarrow \mu^+\mu^-$) so that distant regions of the universe (> 100 kpc) are inaccessible for energetic photons (> 100 TeV). Photon–photon interactions, therefore, cause a horizon for γ astronomy, which allows us to explore the nearest neighbours of our local group of galaxies in the light of high-energy γ quanta, but they attenuate the γ intensity for larger distances so strongly that a meaningful observation becomes impossible (cf. Chap. 4, Fig. 4.9).

The relation between the threshold energy for absorptive $\gamma\gamma$ interactions and the energy of target photons is shown in Fig. 6.44. Distant γ-ray sources whose high-energy photons are absorbed by blackbody and infrared photons can still be observed in the energy range $< 1\,\mathrm{TeV}$.

6.3.4 Observation of γ-Ray Point Sources

First measurements of galactic γ rays were performed in the seventies with satellite experiments. The results of these investigations (Fig. 6.45) clearly show the galactic center, the Crab Nebula, the Vela X1 pulsar, Cygnus X3, and Geminga as γ-ray point sources. Recent satellite measurements with the Compton gamma-ray observatory (CGRO) show a large number of further γ-ray sources. There were four experiments on board of the CGRO satellite (BATSE, OSSE, EGRET, COMPTEL)[6]. These four telescopes cover an energy range from $30\,\mathrm{keV}$ up to $30\,\mathrm{GeV}$. Apart from γ-ray bursts (see Sect. 6.3.5) numerous galactic pulsars and a large number of extragalactic sources (AGN)[7] have been discovered. It was found out that old pulsars can transform their rotational energy more efficiently into γ rays compared to young pulsars. One assumes that the observed γ radiation is produced by synchrotron radiation of energetic electrons in the strong magnetic fields of the pulsars.

	OSSE	– Oriented Scintillation Spectroscopy Experiment
6	COMPTEL	– COMpton TELescope
	EGRET	– Energy Gamma Ray Telescope Experiment
	BATSE	– Burst And Transient Source Experiment
7	AGN – Active Galactic Nuclei	

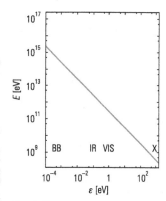

Fig. 6.44
Dependence of the threshold energy E for $\gamma\gamma$ absorption on the energy of the target photons (BB – blackbody radiation, IR – infrared, VIS – visible spectral range, X – X rays

Compton gamma-ray observatory

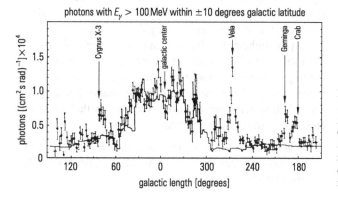

photons with $E_\gamma > 100\,\mathrm{MeV}$ within ± 10 degrees galactic latitude

Fig. 6.45
Measurement of the intensity of galactic γ radiation for photon energies $> 100\,\mathrm{MeV}$. The *solid line* represents the expected γ-ray intensity on the basis of the column density of interstellar gas in that direction

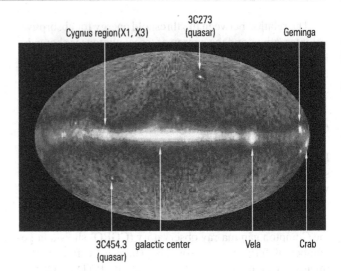

Fig. 6.46
All-sky survey in the light of γ rays {16}

Among the discovered active galaxies highly variable blazars (extremely variable objects on short time scales with strong radio emission) were found which had their maximum of emission in the gamma range. In addition, gamma quasars at high redshifts ($z > 2$) were observed. In this case the gamma radiation could have been produced by inverse Compton scattering of energetic electrons off photons.

Figure 6.46 shows a complete *all-sky survey* in the light of γ rays in galactic coordinates. Apart from the galactic center and several further point sources, the galactic disk is **possible point sources** clearly visible. Candidates for point sources of galactic γ rays are pulsars, binary pulsar systems, and supernovae. Extragalactic sources are believed to be compact active galactic nuclei (AGN), quasi-stellar radio sources (quasars), blazars, **black holes** and accreting black holes. According to common belief black holes could be the 'powerhouses' of quasars. Black holes are found at the center of galaxies where the matter density is highest thereby providing sufficient material for the formation of accretion disks arround black holes. According to the definition of a black hole no radiation can escape from it,[8] however, the infalling matter heats up already before reaching the event horizon so that intensive energetic γ rays can be emitted.

[8] In this context the Hawking radiation, which has no importance for γ-ray astronomy, will not be considered, since black holes have a very low 'temperature'.

Future space-based gamma-ray telescopes like INTE-GRAL[9] (15 keV–10 MeV) and GLAST[10] (30 MeV–300 GeV) will certainly supplement the results from earlier γ-ray missions.

In the TeV range γ-ray point sources have been discovered with the air Cherenkov technique. Recently, a supernova remnant as a source of high-energy photons has been observed by the HESS[11] experiment. The γ-ray spectrum of this object, SNR RX J1713.7-3946 in the galactic plane 7 kpc off the galactic center, near the solar system (distance 1 kpc) can best be explained by the assumption that the photons with energies near 10 TeV are produced by π^0 decays, i.e., this source is a good candidate for a hadron accelerator. The imaging air Cherenkov telescope MAGIC under construction on the Canary Island La Palma with its 17 m diameter mirror will soon compete with the HESS telescope in Namibia. Apart from galactic sources (Crab Nebula) extragalactic objects emitting TeV photons (Markarian 421, Markarian 501, and M87) have also been unambiguously identified. Markarian 421 is an elliptic galaxy with a highly variable galactic nucleus. The luminosity of Markarian 421 in the light of TeV photons would be 10^{10} times higher than that of the Crab Nebula if isotropic emission were assumed. One generally believes that this galaxy is powered by a massive black hole which emits jets of relativistic particles from its poles. It is conceivable that this approximately 400 million light-years distant galaxy beams the high-energy particle jets – and thereby also the photon beam – exactly into the direction of Earth.

hadron accelerator

TeV photons

The highest γ energies from cosmic sources have been recorded by earthbound air-shower experiments, but also by air Cherenkov telescopes. It has become common practice to consider the Crab Nebula, which emits photons with energies up to 100 TeV, as a standard candle. γ-ray sources found in this energy regime are mostly characterized by an extreme variability. In this context the X-ray source Cygnus X3 plays a special rôle. In the eighties γ rays from this source with energies up to 10^{16} eV (10 000 TeV) were claimed to be seen. These high-energy γ rays appeared to show the same variability (period 4.8 hours) as the X rays

highest γ energies

---

[9] INTEGRAL – INTErnational Gamma-Ray Astrophysics Laboratory

[10] GLAST – Gamma Ray Large Area Telescope

[11] HESS – High Energy Stereoscopic System

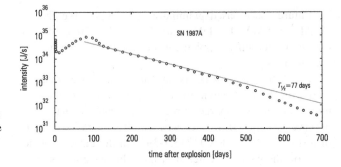

Fig. 6.47
Light curve of SN 1987A. The
solid line corresponds to complete
conversion of ^{56}Co γ rays into the
infrared, optical, and ultraviolet
spectral range

coming from this object. It has to be noted, however, that a
high-energy gamma outburst of this source has never been
seen again.

γ line emission   Apart from the investigation of cosmic sources in the
light of high-energy γ rays the sky is also searched for
γ quanta of certain fixed energy. This γ-ray line emission
hints at radioactive isotopes, which are formed in the process of nucleosynthesis in supernova explosions. It could be
shown beyond any doubt that the positron emitter ^{56}Ni was
produced in the supernova explosion 1987A in the Large
Magellanic Cloud. This radioisotope decays into ^{56}Co with

light curve of SN 1987A   a half-life of 6.1 days. The light curve of this source showed
a luminosity maximum followed by an exponential bright-
ness decay. This could be traced back to the radioactive decay of the daughter ^{56}Co to the stable isotope ^{56}Fe with a
half-life of 77.1 days (see Fig. 6.47).

Interesting results are also expected from an all-sky sur-
vey in the light of the $511\,\mathrm{keV}$ line from e^+e^- annihilation.
This γ-ray line emission could indicate the presence of antimatter in our galaxy. The observation of the distribution of
cosmic antimatter could throw some light on the problem
why our universe seems to be matter dominated.

6.3.5 γ Burster

Cosmic objects which emit sudden single short outbursts of
γ rays have been discovered in the early seventies by Ameri-

discovery of γ-ray bursts   can reconnaissance satellites. The purpose of these satellites
was to check the agreement on the stop of nuclear weapon
tests in the atmosphere. The recorded γ rays, however, did
not come from the surface of the Earth or the atmosphere,
but rather from outside sources and, therefore, were not related to explosions of nuclear weapons which also are a
source of γ rays.

γ-ray bursts occur suddenly and unpredictably with a rate of approximately one burst per day. The durations of the γ-ray bursts are very short ranging from fractions of a second up to 100 seconds. There appear to be two distinct classes of γ-ray bursts, one with short ($\approx$ 0.5 s), the other with longer durations ($\approx$ 50 s), indicating the existence of two different populations of γ-ray bursters. Figure 6.48 shows the γ-ray light curve of a typical short burst. Within only one second the γ-ray intensity increases by a factor of nearly 10. The γ-ray bursters appear to be uniformly distributed over the whole sky. Because of the short burst duration it is very difficult to identify a γ-ray burster with a known object. At the beginning of 1997 researchers succeeded for the first time to associate a γ-ray burst with a rapidly fading object in the optical regime. From the spectral analysis of the optical partner one could conclude that the distance of this γ-ray burster was about several billion light-years.

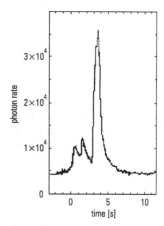

Fig. 6.48
Light curve of a typcial γ-ray burst

The angular distribution of 2000 γ-ray bursters recorded until the end of 1997 is shown in Fig. 6.49 in galactic coordinates. From this graph it is obvious that there is no clustering of γ-ray bursters along the galactic plane. Therefore, the most simple assumption is that these exotic objects are at cosmological distances which means that they are extragalactic. Measurements of the intensity distributions of bursts show that weak bursts are relatively rare. This could imply that the weak (i.e., distant) bursts exhibit a lower spatial density compared to the strong (near) bursts.

angular distribution

Even though violent supernova explosions are considered to be excellent candidates for γ-ray bursts, it is not obvious whether also other astrophysical objects are responsible for this enigmatic phenomenon. The observed spatial

spatial distribution

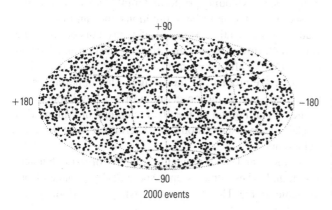

2000 events

Fig. 6.49
Angular distribution of 2000 γ-ray bursts in galactic coordinates recorded with the BATSE detector (Burst And Transient Source Experiment) on board the CGRO satellite (Compton Gamma Ray Observatory) {17}

distribution of γ-ray bursters suggests that they are at extra-galactic distances. In this context the deficit of weak bursts in the intensity distribution could be explained by the red-shift of spectral lines associated with the expansion of the universe. This would also explain why weaker bursts have softer energy spectra.

supernova, hypernova explosions

A large fraction of γ-ray bursts is believed to be caused by violent supernova explosions (e.g., hypernova explosions with a collapse into a rotating black hole). This seems to have been confirmed by the association of the γ-ray burst GRB030329 with the supernova explosion SN2003dh. The observation of the optical afterglow of this burst allowed to measure its distance to be at 800 Mpc. The afterglow of the burst, i.e., the optial luminosity of the associated supernova reached a magnitude of 12 in the first observations after the γ-ray burst. Such a bright supernova may have been visi-ble to the naked eye in the first minutes after the explosion.

Wolf–Rayet star

Such hypernova explosions are considered to be rare events which are probably caused by stars of the 'Wolf–Rayet' type. Wolf–Rayet stars are massive objects ($M > 20\,M_\odot$) which initially consist mainly of hydrogen. During their burning phase they strip themselves off their outer layers thus consisting mainly of helium, oxygen, and heavy ele-ments. When they run out of fuel, the core collapses and forms a black hole surrounded by an accretion disk. It is be-lieved that in this moment a jet of matter is ejected from the black hole which represents the γ-ray burst ('collapsar model').

candidates for γ-ray bursts

As alternative candidates for γ-ray bursts, namely for short-duration bursts, collisions of neutron stars, collisions of neutron stars with black holes, coalescence of two neu-tron stars forming a black hole, asteroid impacts on neu-tron stars, or exploding primordial mini black holes are dis-cussed. From the short burst durations one can firmly con-clude that the spatial extension of γ-ray bursters must be very small. However, only the exact localization and detailed observation of afterglow partners of γ-ray bursts will allow to clarify the problem of their origin. One important input to this question is the observation of a γ-ray burster also in the optical regime by the 10 m telescope on Mauna Kea in Hawaii. Its distance could be determined to be 9 billion light-years.

quasi-periodic γ burster

In the year 1986 a variant of a family of γ-ray bursters was found. These objects emit sporadically γ bursts from the same source. The few so far known quasi-periodic γ-ray bursters all reside in our galaxy or in the nearby Magellanic Clouds. Most of these objects could be identified with young supernova remnants. These 'soft gamma-ray repeaters' appear to be associated with enormous magnetic fields. If such a magnetar rearranges its magnetic field to reach a more favourable energy state, a star quake might occasionally occur in the course of which γ bursts are emitted. The γ-ray burst of the magnetar SGR-1900+14 was recorded by seven research satellites on August 27, 1998. From the observed slowing down of the rotational period of this magnetar one concludes that this object possesses a superstrong magnetic field of 10^{11} Tesla exceeding the magnetic fields of normal neutron stars by a factor of 1000.

soft gamma-ray repeater

magnetars

With these properties γ bursters are also excellent candidates as sources of cosmic rays. It is frequently discussed that the birth or the collapse of neutron stars could be associated with the emission of narrowly collimated particle jets. If this were true, we would only be able to see a small fraction of γ bursters. The total number of γ bursters would then be sufficiently large to explain the observed particle fluxes of cosmic rays. The enormous time-dependent magnetic fields would also produce strong electric fields in which cosmic-ray particles could be accelerated up to the highest energies.

6.4 X-Ray Astronomy

"Light brings the news of the universe."
Sir William Bragg

6.4.1 Introduction

X rays differ from γ rays by their production mechanism and their energy. X rays are produced if electrons are decelerated in the Coulomb field of atomic nuclei or in transitions between atomic electron levels. Their energy ranges between approximately 1–100 keV. In contrast, γ rays are usually emitted in transitions between nuclear levels, nuclear transformations, or in elementary-particle processes.

production of X rays

After the discovery of X rays in 1895 by Wilhelm Conrad Röntgen, X rays were mainly used in medical applications because of their high penetration power. X rays with energies exceeding 50 keV can easily pass through 30 cm of tissue (absorption probability $\approx 50\%$). The column density

**energy dependence
of X-ray absorption**

of the Earth's atmosphere, however, is too large to allow extraterrestrial X rays to reach sea level. In the keV energy region, corresponding to the brightness maximum of most X-ray sources, the range of X rays in air is 10 cm only. To be able to observe X rays from astronomical objects one therefore has to operate detectors at the top of the atmosphere or in space. This would imply balloon experiments, rocket flights, or satellite missions.

balloon experiments   Balloon experiments can reach a flight altitude of 35 to 40 km. Their flight duration amounts to typically between 20 to 40 hours. At these altitudes, however, a substantial fraction of X rays is already absorbed. Balloons, therefore, can only observe X-ray sources at energies exceeding 50 keV

rocket flights   without appreciable absorption losses. In contrast, rockets normally reach large altitudes. Consequently, they can measure X-ray sources unbiased by absorption effects. However, their flight time of typically several minutes, before they fall

satellites   back to Earth, is extremely short. Satellites have the big advantage that their orbit is permanently outside the Earth's atmosphere allowing observation times of several years.

discovery   In 1962 X-ray sources were discovered by chance when
of cosmic X-ray sources   an American Aerobee rocket with a detector consisting of three Geiger counters searched for X rays from the Moon. No lunar X radiation was found, but instead extrasolar X rays from the constellations Scorpio and Sagittarius were observed. This was a big surprise because it was known that our Sun radiates a small fraction of its energy in the X-ray range and – because of solid angle arguments – one did not expect X-ray radiation from other stellar objects. This is because the distance of the nearest stars is more than 100 000

bright X-ray sources   times larger than the distance of our Sun. The brightness of such distant sources must have been enormous compared to the solar X-ray luminosity to be able to detect them with detectors that were in use in the sixties. The mechanism what made the sources Scorpio and Sagittarius shine so bright in the X-ray range, therefore, was an interesting astrophysical question.

6.4.2 Production Mechanisms for X Rays

The sources of X rays are similar to those of gamma rays. Since the energy spectrum of electromagnetic radiation usually decreases steeply with increasing energy, X-ray sources outnumber gamma-ray sources. In addition to the processes already discussed in Sect. 6.3 (gamma-ray astronomy) like synchrotron radiation, bremsstrahlung, and inverse Compton scattering, a further production mechanism for X rays like thermal radiation from hot cosmic sources has to be considered. The Sun with its effective surface temperature of about 6000 Kelvin emits predominantly in the eV range. Sources with a temperature of several million Kelvin would also emit X rays as blackbody radiation.

X rays
as blackbody radiation

The measured spectra of many X-ray sources exhibit a steep intensity drop to very small energies which can be attributed to absorption by cold material in the line of sight. At higher energies a continuum follows which can be described either by a power law ($\sim E^{-\gamma}$) or by an exponential, depending on the type of the dominant production mechanism for X rays. Sources in which relativistic electrons produce X rays by synchrotron radiation or inverse Compton scattering can be characterized by a power-law spectrum like $E^{-\gamma}$. On the other hand, one obtains an exponential decrease to high energies if thermal processes dominate. A bremsstrahlung spectrum is usually relatively flat at low energies. In most cases more than one production process contributes to the generation of X rays. According to the present understanding, the X rays of most X-ray sources appear to be of thermal origin. For thermal X rays one has to distinguish two cases.

shape of spectra

dominance
of thermal production

1. In a hot gas ($\approx 10^7$ K) the atoms are ionized. Electrons of the thermal gas produce in an optically thin medium (practically no self absorption) X rays by bremsstrahlung and by atomic level transitions. The second mechanism requires the existence of atoms which still have at least one bound electron. At temperatures exceeding $\approx 10^7$ K, however, the most abundant atoms like hydrogen and helium are completely ionized so that in this case bremsstrahlung is the dominant source. In this context one understands under bremsstrahlung the emission of X rays which are produced by interactions of electrons in the Coulomb field of positive ions of the plasma in continuum transitions (*thermal bremsstrahlung*). For energies $h\nu > kT$ the spectrum decreases exponentially like $e^{-h\nu/kT}$ (k: Boltzmann constant). On the other hand, if $h\nu \ll kT$, the spectrum is nearly flat. The assumption of low optical density of the source leads to the fact that the emission spectrum and production spectrum are practically identical.

production
in an optical thin medium

2. A hot optical dense body produces a *blackbody spectrum* independently of the underlying production pro-

production
in an optical thick medium

cess because both emission and absorption processes are involved. Therefore, an optically dense bremsstrahlung source which absorbs its own radiation would also produce a blackbody spectrum. The emission P of a blackbody is given by Planck's law

energy spectra

$$P \sim \frac{\nu^3}{e^{h\nu/kT} - 1} .$$ (6.81)

For high energies ($h\nu \gg kT$) P can be described by an exponential

$$P \sim e^{-h\nu/kT} ,$$ (6.82)

while at low energies ($h\nu \ll kT$), because of

$$e^{h\nu/kT} = 1 + \frac{h\nu}{kT} + \cdots ,$$ (6.83)

the spectrum decreases to low frequencies like

$$P \sim \nu^2 .$$ (6.84)

The total radiation S of a hot body is described by the Stefan–Boltzmann law,

$$S = \sigma T^4 ,$$ (6.85)

where σ is the Stefan–Boltzmann constant.

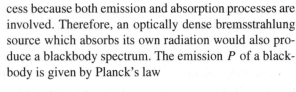

Fig. 6.50
Standard X-ray spectra originating from various production processes

Typical energy spectra for various production mechanisms are sketched in Fig. 6.50.

6.4.3 Detection of X Rays

The observation of X-ray sources is more demanding compared to optical astronomy. X rays cannot be imaged with lenses since the index of refraction in the keV range is very close to unity. If X rays are incident on a mirror they will be absorbed rather than reflected. Therefore, the direction of incidence of X rays has to be measured by different techniques. The most simple method for directional observation is based on the use of slit or wire collimators which are mounted in front of an X-ray detector. In this case the observational direction is given by the alignment of the space probe. Such a geometrical system achieves resolutions on the order of 0.5°. By combining various types of collimators angular resolutions of one arc minute can be obtained.

passive collimators

In 1952 Wolter had already proposed how to build X-ray telescopes based on total reflection. To get reflection at grazing incidence rather than absorption or scattering, the imaging surfaces have to be polished to better than a fraction of 10^{-3} of the optical wavelength. Typically, systems of stacked assemblies of paraboloids or combinations of *parabolic* and *hyperbolic mirrors* are used (see Fig. 6.51).

Wolter telescope

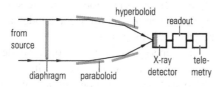

Fig. 6.51
Cross section through an X-ray telescope with parabolic and hyperbolic mirrors

To be able to image X rays in the range between 0.5 nm and 10 nm with this technique, the angles of incidence must be smaller than 1.5° (Fig. 6.52). The wavelength λ[nm] is obtained from the relation

$$\lambda = \frac{c}{\nu} = \frac{hc}{h\nu} = \frac{1240}{E[\text{eV}]}\,\text{nm}\,. \qquad (6.86)$$

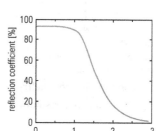

Fig. 6.52
Angular-dependent reflection power of metal mirrors

The mirror system images the incident X rays onto the common focal point. In X-ray satellites commonly several X-ray devices are installed as focal-point detectors. They are usually mounted on a remotely controllable device. Depending on the particular application an appropriate detector can be moved into the focal point. In multimirror systems angular resolutions of one arc second are obtained. The requirement of grazing incidence, however, considerably limits the acceptance of X-ray telescopes.

As detectors for X rays crystal spectrometers (Bragg reflection), proportional counters, photomultipliers, single-channel electron multipliers (channeltrons), semiconductor counters, or X-ray CCDs (charge-coupled devices) are in use.

detector for X rays

In proportional counters the incident photon first creates an electron via the photoelectric effect which then produces an avalanche in the strong electrical field (Fig. 6.53).

In the proportional domain gas amplifications of 10^3 up to 10^5 are achieved. Since the absorption cross section for the photoelectric effect varies proportional to Z^5, a heavy noble gas (Xe, $Z = 54$) with a quencher should be used as counting gas. Thin foils ($\approx 1\,\mu\text{m}$) made from beryllium ($Z = 4$) or carbon ($Z = 6$) are used as entrance windows.

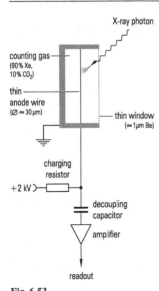

X-ray photon

counting gas
(90% Xe,
10% CO$_2$)

thin
anode wire
(∅ ≈ 30 μm)

thin window
(≈ 1 μm Be)

charging
resistor

+2 kV

decoupling
capacitor

amplifier

readout

Fig. 6.53
Principle of operation of a
proportional counter

photomultiplier, channeltron

angular resolution

charge-coupled device

The incident photon transfers its total energy to the photo-electron. This energy is now amplified during avalanche formation in a proportional fashion. Therefore, this technique not only allows to determine the direction of the incidence of the X-ray photon but also its energy.

With photomultipliers or channeltrons the incident photon is also converted via the photoelectric effect into an electron. This electron is then amplified by ionizing collisions in the discrete or continuous electrode system. The amplified signal can be picked up at the anode and further processed by electronic amplifiers.

The energy measurement of X-ray detectors is based on the number of charge carriers which are produced by the photoelectron. In gas proportional chambers typically 30 eV are required to produce an electron–ion pair. Semiconductor counters possess the attractive feature that only approximately 3 eV are needed to produce an electron–hole pair. Therefore, the energy resolution of semiconductor counters is better by a factor of approximately $\sqrt{10}$ compared to proportional chambers. As solid-state materials, silicon, germanium, or gallium arsenide can be considered. Because of the easy availability and the favourable noise properties mostly silicon semiconductor counters are used.

If a silicon counter is subdivided in a matrix-like fashion into many quadratic elements (pixels), which are shielded against each other by potential wells, the produced energy depositions can be read out line by line. Because of the charge coupling of the pixels this type of silicon image sensor is also called charge-coupled device. Commercial CCDs with areas of 1 cm × 1 cm at a thickness of 300 μm have about 10^5 pixels. Even though the shifting of the charge in the CCD is a serial process, these counters have a relatively high rate capability. Presently, time resolutions of 1 ms up to 100 μs have been obtained. This allows rate measurements in the kHz range which is extremely interesting for the observation of X-ray sources with high variability.

6.4.4 Observation of X-Ray Sources

The Sun was the first star of which X rays were recorded (Friedmann et al. 1951). In the range of X rays the Sun is characterized by a strong variability. In strong flares its intensity can exceed the X-ray brightness of the quiet Sun by a factor of 10 000.

In 1959 the first X-ray telescope was built (R. Giacconni, Nobel Prize 2002) and flown on an Aerobee rocket in 1962. During its six minutes flight time it discovered the first extrasolar X-ray sources in the constellation Scorpio. The observation time could be extended with the first X-ray satellite UHURU (meaning 'freedom' in Swahili), which was launched from a base in Kenya 1970. Every week in orbit it produced more results than all previous experiments combined.

extrasolar X rays

In the course of time a large number of X-ray satellites has provided more and more precision information about the X-ray sky. The satellite with the highest resolution up to 1999 was a common German–British–American project: the ROentgen SATellite ROSAT (see Fig. 6.54). ROSAT measured X rays in the range of 0.1 up to 2.5 keV with a Wolter telescope of 83 cm diameter. As X-ray detectors, multiwire proportional chambers (PSPC)[12] with 25 arc second resolution and a channel-plate multiplier (HRI)[13] with 5 arc second resolution were in use. One of the PSPCs was permanently blinded by looking by mistake into the Sun. The second PSPC stopped operation after a data-taking period of 4 years because its gas supply was exhausted. Since then only the channel-plate multiplier was available as an X-ray detector. Compared to earlier X-ray satellites, ROSAT had a much larger geometrical acceptance, better angular and energy resolution, and a considerably increased signal-to-noise ratio: per angular pixel element the background rate was only one event per day.

Fig. 6.54
Photograph of the X-ray satellite ROSAT {18}

In a sky survey ROSAT has discovered about 130 000 X-ray sources. For comparison: the earlier flown Einstein Observatory HEAO[14] had only found 840 sources. The most frequent type of X-ray sources are nuclei of active galaxies ($\approx 65\,000$) and normal stars ($\approx 50\,000$). About 13 000 galactic clusters and 500 normal galaxies were found to emit X rays. The smallest class of X-ray sources are supernova remnants with approximately 300 identified objects.

130 000 discovered X-ray sources

HEAO

Supernova remnants (SNR) represent the most beautiful X-ray sources in the sky. ROSAT found that the Vela pulsar also emits X rays with a period of 89 ms known from its optical emission. It appears that the X-ray emission of Vela X1 is partially of thermal origin. The supernova rem-

---

[12] PSPC – Position Sensitive Proportional Chamber

[13] HRI – High Resolution Imager

[14] HEAO – High Energy Astronomy Observatory

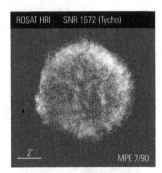

Fig. 6.55
Supernova remnant SNR 1572
recorded with the HRI detector
(High Resolution Instrument) on
board the ROSAT satellite {18}

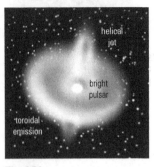

Fig. 6.56
Sketch of X-ray emission from the
Crab pulsar

binaries

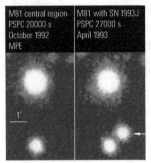

Fig. 6.57
Sprial galaxy M81 with the
supernova SN 1993J. The image
was recorded with the PSPC
detector (Position Sensitive
Proportional Counter) of the
ROSAT satellite {18}

nant SNR 1572 that was observed by the Danish astronomer
Tycho Brahe shows a nearly spherically expanding shell in
the X-ray range (Fig. 6.55). The shell expands into the interstellar medium with a velocity of about 50 km/s and it heats
up in the course of this process to several million degrees.

The topology of X-ray emission from the Crab Pulsar
allows to identify different components: the pulsar itself is
very bright in X rays compared to the otherwise more diffuse emission. The main component consists of a toroidal
configuration which is caused by synchrotron radiation of
energetic electrons and positrons in the magnetic field of the
pulsar. In addition, electrons and positrons escape along the
magnetic field lines at the poles where they produce X rays
in a helical wind (see Fig. 6.56).

Only six days after the explosion of a supernova in the
spiral galaxy M81, ROSAT has measured its X-ray emission. In the right-hand part of Fig. 6.57 the X-ray source SN
1993J south of the center of M81 is visible. In an earlier exposure of the same sky region (left-hand part of the figure)
this source is absent.

A large number of X-ray sources are binaries. In these
binaries mostly a compact object – a white dwarf, a neutron star, or a black hole – accretes matter from a nearby
companion. The matter flowing to the compact object frequently forms an accretion disk (see Fig. 5.7), however, the
matter can also be transported along the magnetic field lines
landing directly on the neutron star. In such cataclysmic
variables a mass transfer from the companion, e.g., to a
white dwarf, can be sufficient to maintain a permanent hydrogen burning. If the ionized hydrogen lands on a neutron
star also thermonuclear X-ray flashes can occur. Initially,
the incident hydrogen fuses in a thin layer at the surface of
the neutron star to helium. If a sufficient amount of matter
is accreted, the helium produced by fusion can achieve such
high densities and temperatures that it can be ignited in a
thermonuclear explosion forming carbon.

The observation of thermal X rays from galactic clusters
allows a mass determination of the hot plasma and the total gravitational mass of the cluster. This method is based
on the fact that the temperature is a measure for the gravitational attraction of the cluster. A high gas temperature –
characterized by the energy of the emitted X rays – represents via the gas pressure the counterforce to gravitation and
prevents the gas from falling into the center of the cluster.

Measurements of X rays from galactic clusters have established that the hot plasma between the galaxies is five times more massive than the galaxies themselves. The discovery of X-ray-emitting massive hot plasmas between the galaxies is a very important input for the understanding of the dynamics of the universe.

In the present understanding of the evolution of the universe all structures are hierarchically formed from objects of the respective earlier stages: stellar clusters combine to galaxies, galaxies form groups of galaxies which grow to galactic clusters which in turn produce superclusters. Distant, i.e., younger galactic clusters, get more massive while close-by galactic clusters hardly grow at all. This allows to conclude that nearby galactic clusters have essentially collected all available matter gravitationally. The mass of these clusters appears to be dominated by gas clouds into which stellar systems are embedded like raisins in a cake. Therefore, the X-ray-emitting gas clouds allow to estimate the matter density in the universe. The present X-ray observations from ROSAT suggest a value of approximately 30% of the critical mass density of the universe. If this were all, this would mean that the universe expands eternally (see Chap. 8).

hierarchical structure of the universe

density of the universe

In 1999 the X-ray satellite AXAF (Advanced X-ray Astrophysics Facility) was launched successfully. To honour the contributions to astronomy and astrophysics the satellite was renamed Chandra after the Indian–American astrophysicist Subrahmanyan Chandrasekhar. At the end of 1999 the X-ray satellite XMM (X-ray Multi-Mirror mission) was brought into orbit (Fig. 6.58). It was also renamed in the year 2000 to honour Newton and its name is now XMM Newton or Newton Observatory, respectively. Both Chandra and the Newton Observatory have better angular and energy resolutions compared to ROSAT, and therefore a better understanding of X-ray sources and the non-luminous matter in the universe is expected. From the experimental point of view one has to take extreme care that the sensitive focal pixel detectors do not suffer radiation damage from low-energy solar particles (p, α, e) emitted during solar eruptions.

AXAF, XMM, ASTRO-E

Fig. 6.58
Photo of the XMM X-ray satellite {19}

It is anticipated that Chandra and XMM Newton will discover $\approx 50\,000$ new X-ray sources per year. Of particular interest are active galactic nuclei which are supposed to be powered by black holes. Many or most of the black holes

expected discoveries of X-ray sources

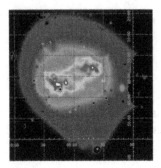

Fig. 6.59
Collision of two merging galactic
clusters, each containing hundreds
of galaxies. The site of this
catastophic event is Abell 754 at a
distance of about 9 million
light-years. The photo shows a
coded pressure map of this region,
where the galaxies themselves are
confined around the white spots
which correspond to regions of
high pressure, followed by
decreasing pressure as one goes
away from the centers {20}

diffuse background X rays

Fig. 6.60
X-ray emission from the Moon
recorded with the PSPC detector
on board of ROSAT. The dark side
of the Moon shields the cosmic
X-ray background {18}

iron line in X-ray spectra

X rays from the Moon

residing at the centers of distant galaxies may be difficult to
find since they are hidden deep inside vast amounts of absorbing dust so that only energetic X rays or γ rays can escape. Already now these new X-ray satellites have observed
galaxies at high redshifts emitting huge amounts of energy
in the form of X rays, far more than can ever expected to be
produced by star formation. Therefore, it is conjectured that
these galaxies must contain actively accreting supermassive
black holes.

Even though presently known classes of sources will
probably dominate the statistics of Chandra and XMM Newton discoveries, the possibility of finding completely new
and exciting populations of X-ray sources also exists.

A very recent result is the observation of the collision
of two galactic clusters in Abell 754 at a distance of about
9 million light-years in the light of X rays. These clusters
with millions of galaxies merge in a catastrophic collision
into one single very large cluster.

On the other hand, the X-ray mission ASTRO-E, started
early in the year 2000, had to be abandoned because the
booster rocket did not carry the satellite into an altitude
required for the intended orbit. The satellite presumably
burned up during reentry into the atmosphere.

The diffuse X-ray background which was discovered relatively early consists to a large extent (75%) of resolved
extragalactic sources. It could easily be that the remaining
diffuse part of X rays consists of so far non-resolved distant
X-ray sources.

The largest fraction of the earlier unresolved X-ray
sources are constituted by active galactic nuclei and quasars.
A long-term exposure of ROSAT (40 hours) revealed more
than 400 X-ray sources per square degree.

In the spectra of many X-ray sources the iron line
(5.9 keV) is observed. This is a clear indication that either
iron is synthesized directly in supernova explosions or the X
rays originate from older sources whose material has been
processed through several stellar generations.

A surprising result of the most recent investigations was
that practically all stars emit X rays. A spectacular observation was also the detection of X rays from the Moon. However, the Moon does not emit these X rays itself. It is rather
reflected corona radiation from our Sun in the same way as
the Moon also does not shine in the optical range but rather
reflects the sunlight (Fig. 6.60).

6.5 Gravitational-Wave Astronomy

> *"What would physics look like without gravitation?"*
>
> *Albert Einstein*

Finally, it is appropriate to mention the new field of gravitational-wave astronomy. Gravitational waves have been predicted by Einstein als early as 1916. Apart from the observation of Taylor and Hulse concerning the energy loss of a binary pulsar (PSR 1913+16) due to the emission of gravitational waves over a period starting from 1974 (Nobel Prize 1993) there is no direct evidence of the existence of gravitational waves. Nobody doubts the correctness of Einstein's prediction, especially since the results of Taylor and Hulse on the energy loss of the binary pulsar system by emission of gravitational radiation agree with the theoretical expectation of general relativity impressibly well (to better than 0.1%).

energy loss by gravitational radiation

Taylor and Hulse have observed the binary system PSR 1913+16 consisting of a pulsar and a neutron star over a period of more than 20 years. The two massive objects rotate around their common center of mass on elliptical orbits. The radio emission from the pulsar can be used as precise clock signal. When the pulsar and neutron star are closest together (periastron), the orbital velocities are largest and the gravitational field is strongest. For high velocities and in a strong gravitational field time is slowed down. This relativistic effect can be checked by looking for changes in the arrival time of the pulsar signal. In this massive and compact pulsar system the periastron time changes in a single day by the same amount for which the planet Mercury needs a century in our solar system. Space-time in the vicinity of the binary is greatly warped.

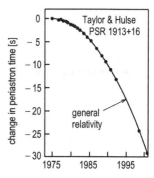

Fig. 6.61
Observed changes in periastron time of the binary system PSR 1913+16 over more than 20 years in comparison to the expectation based on Einstein's theory of general relativity. The agreement between theory and observation is better than 0.1%

The theory of relativity predicts that the binary system will lose energy with time as the orbital rotation energy is converted into gravitational radiation. Fig. 6.61 shows the prediction based on Einstein's theory of general relativity in comparison to the experimental data. The excellent agreement between theory and experiment presents so far the best – albeit indirect – evidence for gravitational waves.

evidence for gravitational radiation

The direct detection of gravitational radiation would open a new window onto violent astrophysical events and it may give a clue to processes where dark matter or dark energy is involved.

However, as far as the direct observation is concerned, the situation is to a certain extent similar to neutrino physics around 1950. At that time nobody really doubted the existence of the neutrino, but there was no strong neutrino source available to test the prediction. Only the oncoming nuclear reactors were sufficiently powerful to provide a large enough neutrino flux to be observed. The problem was related to the low interaction cross section of neutrinos with matter.

low interaction probability of gravitational radiation

Compared to gravitational waves the neutrino interaction with matter can be called 'strong'. The extremely low interaction probability of gravitational waves ensures that they provide a new window to cataclysmic processes in the universe at the expense of a very difficult detection. For neutrinos most astrophysical sources are almost transparent. Because of the feeble interaction of gravitational waves they can propagate out of the most violent cosmological sources even more freely compared to neutrinos.

imaging with gravity waves?

With electromagnetic radiation in various spectral ranges astronomical objects can be imaged. This is because the wavelength of electromagnetic radiation is generally very small compared to the size of astrophysical objects. The wavelength of gravitational waves is much larger so that imaging with this radiation is almost out of question. In electromagnetic radiation time-dependent electric and magnetic fields propagate through space-time. For gravitational waves one is dealing with oscillations of space-time itself.

quadrupole radiation

Electromagnetic radiation is emitted when electric charges are accelerated or decelerated. In a similar way gravitational waves are created whenever there is a non-spherical acceleration of mass–energy distributions. There is, however, one important difference. Electromagnetic radiation is dipole radiation, while gravitational waves present quadrupole radiation. This is equivalent to saying that the quantum of gravitational waves, the graviton, has spin $2\hbar$.

oscillation modes of antennae

Electromagnetic waves are created by time-varying dipole moments where a dipole consists of one positive and one negative charge. In contrast, gravity has no charge, there is only positive mass. Negative mass does not exist. Even antimatter has the same positive mass just as ordinary matter. Therefore, it is not possible to produce an oscillating mass dipole. In a two-body system one mass accelerated to the left creates, because of momentum conservation, an equal and opposite action on the second mass which moves it to the right. For two equal masses the spacing may change but the center of mass remains unaltered. Consequently, there is no monopole or dipole moment. Therefore, the lowest order of oscillation generating gravitational waves is due to a time-varying quadrupole moment. In the same way as a distortion of test masses creates gravitational waves where, e.g., the simplest non-spherical motion is one in which horizontal masses move inside and vertical masses move apart, a gravitational wave will distort an antenna analogously by compression in one direction and elongation in the other (see Fig. 6.62).

The *quadrupole character of gravitational radiation* therefore leads to an action like a tidal force: it squeezes the antenna along one axis while stretching it along the other. Due to the weakness of the gravitational force the relative elongation of an antenna will be at most on the order of $h \approx 10^{-21}$ even for the most violent cosmic catastrophes. There is, however, one advantage of gravitational waves compared to the measurement of electromagnetic radiation: Electromagnetic observables like the energy flux from astrophysical sources are characterized by a $1/r^2$ dependence due to solid-angle reasons. By contrast, the direct observable of gravitational radiation (h) decreases with distance only like $1/r$. h depends linearly on the second derivative of the quadrupole moment of the astrophysical object and it is inversely proportional to the distance r,

$$h \sim \frac{G}{c^4} \frac{\ddot{Q}}{r} \quad (G: \text{Newton's constant}) .$$

Consequently, an improvement of the sensitivity of a gravitational detector by a factor of 2 increases the measurable volume where sources of gravitational waves may reside by a factor of 8. The disadvantage of not being able to image with gravitational radiation goes along with the advantage that gravitational-wave detectors have a nearly 4π steradian sensitivity over the sky.

The most promising candidates as sources for gravitational radiation are mergers of binary systems, accreting black holes, collisions of neutron stars, or special binaries consisting of two black holes orbiting around their common center of mass (like in the radio galaxy 3C66B, which appears to be the result of a merger of two galaxies).

The suppression of noise in these antennae is the most difficult problem. There are stand-alone gravitational-wave detectors, mostly in the form of optical interferometers where the elongation of one lever arm and the compression of the other can be monitored by the technique of Michelson interferometry. These detectors can be operated at ground level or in space. A convincing signal of gravitational waves

Fig. 6.62
Oscillation modes of a spherical antenna upon the impact of a gravitational wave causing it to undergo quadrupole oscillations

**attenuation
of gravitational waves**

noise suppression

would probably require a coincidence of signals from several independent detectors.

6.6 Problems

Problems for Sect. 6.1

1. What is the reason that primary cosmic-ray nuclei like carbon, oxygen, and neon are more abundant than their neighbours in the periodic table of elements (nitrogen, fluorine, sodium)?
2. In Sect. 6.1 it is stated that primary cosmic rays consist of protons, α particles, and heavy nuclei. Only 1% of the primary particles are electrons. Does this mean that the planet Earth will be electrically charged in the course of time because of the continuous bombardment by predominantly positively charged primary cosmic rays? What kind of positive excess charge could have been accumulated during the period of existence of our planet if this were true?
3. The chemical abundance of the sub-iron elements in primary cosmic rays amounts to about 10% of the iron flux.
   a) Estimate the fragmentation cross section for collisions of primary iron nuclei with interstellar/intergalactic nuclei.
   b) What is the chance that a primary iron nucleus will survive to sea level?

Problems for Sect. 6.2

1. The Sun converts protons into helium according to the reaction

   $$4p \rightarrow {}^4\text{He} + 2e^+ + 2\nu_e .$$

   The solar constant describing the power of the Sun at Earth is $P \approx 1400\,\text{W/m}^2$. The energy gain per reaction corresponds to the binding energy of helium ($E_B({}^4\text{He}) = 28.3\,\text{MeV}$). How many solar neutrinos arrive at Earth?
2. If solar electron neutrinos oscillate into muon or tau neutrinos they could in principle be detected via the reactions

   $$\nu_\mu + e^- \rightarrow \mu^- + \nu_e , \quad \nu_\tau + e^- \rightarrow \tau^- + \nu_e .$$

   Work out the threshold energy for these reactions to occur.

3. Radiation exposure due to solar neutrinos.
   a) Use (6.9) to work out the number of interactions of solar neutrinos in the human body (tissue density $\varrho \approx 1\,\mathrm{g\,cm^{-3}}$).
   b) Neutrinos interact in the human body by

   $$\nu_e + N \rightarrow e^- + N' ,$$

   where the radiation damage is caused by the electrons. Estimate the annual dose for a human under the assumption that on average 50% of the neutrino energy is transferred to the electron.
   c) The equivalent dose is defined as

   $$H = (\Delta E/m)\, w_R \qquad (6.87)$$

   (m is the mass of the human body, w_R the radiation weighting factor ($= 1$ for electrons), $[H] = 1\,\mathrm{Sv} = 1 w_R\,\mathrm{J\,kg^{-1}}$), and ΔE the energy deposit in the human body). Work out the annual equivalent dose due to solar neutrinos and compare it with the normal natural dose of $H_0 \approx 2\,\mathrm{mSv/a}$.

4. Neutrino oscillations.[15]
   In the most simple case neutrino oscillations can be described in the following way (as usual $\hbar$ and c will be set to unity): in this scenario the lepton flavour eigenstates are superpositions

   $$|\nu_e\rangle = \cos\theta |\nu_1\rangle + \sin\theta |\nu_2\rangle ,$$
   $$|\nu_\mu\rangle = -\sin\theta |\nu_1\rangle + \cos\theta |\nu_2\rangle$$

   of mass eigenstates $|\nu_1\rangle$ and $|\nu_2\rangle$. All these states are considered as wave packets with well-defined momentum. In an interaction, e.g., a ν_e is assumed to be generated with momentum p, which then propagates as free particle, $|\nu_e; t\rangle = e^{-iHt}|\nu_e\rangle$. For the mass eigenstates one has $e^{-iHt}|\nu_i\rangle = e^{-iE_{\nu_i}t}|\nu_i\rangle$ with $E_{\nu_i} = \sqrt{p^2 + m_i^2}$, $i = 1, 2$. The probability to find a muon neutrino after a time t is

   $$P_{\nu_e \rightarrow \nu_\mu}(t) = |\langle \nu_\mu | \nu_e; t \rangle|^2 .$$

   a) Work out $P_{\nu_e \rightarrow \nu_\mu}(t)$. After a time t the particle is at $x \approx vt = pt/E_{\nu_i}$. Show that — under the assumption of small neutrino masses — the oscillation probability as given in (6.22) can be derived.

---

[15] This problem is difficult and its solution is mathematically demanding.

b) Estimate for ($v_\mu \leftrightarrow v_\tau$) oscillations with maximum mixing ($\sin 2\theta = 1$), $E_v = 1\,\text{GeV}$ and under the assumption of $m_{v_\mu} \ll m_{v_\tau}$ (that is, $m_{v_\tau} \approx \sqrt{\delta m^2}$) the mass of the τ neutrino, which results if the ratio of upward-to-downward-going atmospheric muon neutrinos is assumed to be 0.54.

c) Does the assumption $m_{v_\mu} \ll m_{v_\tau}$ make sense here?

Problems for Sect. 6.3

1. Estimate the detection efficiency for 1 MeV photons in a NaI(Tl) scintillation counter of 3 cm thickness.
   (Hint: Use the information on the mass attenuation coefficient from Fig. 6.37.)
2. Estimate the size of the cosmological object that has given rise to the γ-ray light curve shown in Fig. 6.48.
3. What is the energy threshold for muons to produce Cherenkov light in air ($n = 1.000\,273$) and water ($n = 1.33$)?
4. The Crab Pulsar emits high-energy γ rays with a power of $P = 3 \times 10^{27}\,\text{W}$ at 100 GeV. How many photons of this energy will be recorded by the planned GLAST experiment (Gamma-ray Large Area Space Telescope) per year if isotropic emission is assumed? What is the minimum flux from the Crab that GLAST will be able to detect (in $\text{J}/(\text{cm}^2\,\text{s})$)? (Collecting area of GLAST: $A = 8000\,\text{cm}^2$, distance of the Crab: $R = 3400$ light-years.)
5. The solar constant describing the solar power arriving at Earth is $P_S \approx 1400\,\text{W}/\text{m}^2$.
   a) What is the total power radiated by the Sun?
   b) Which mass fraction of the Sun is emitted in 10^6 years?
   c) What is the daily mass transport from the Sun delivered to Earth?

Problems for Sect. 6.4

1. The radiation power emitted by a blackbody in its dependence on the frequency is given by (6.81). Convert this radiation formula into a function depending on the wavelength.
2. The total energy emitted per second by a star is called its luminosity. The luminosity depends both on the radius of

the star and its temperature. What would be the luminosity of a star ten times larger than our Sun ($R = 10\,R_\odot$) but at the same temperature? What would be the luminosity of a star of the size of our Sun but with ten times higher temperature?

3. The power radiated by a relativistic electron of energy E in a transverse magnetic field B through synchrotron radiation can be worked out from classical electrodynamics to be

$$P = \frac{e^2 c^3}{2\pi} C_\gamma E^2 B^2 \,,$$

where

$$C_\gamma = \frac{4}{3}\pi \frac{r_e}{(m_e c^2)^3} \approx 8.85 \times 10^{-5}\,\mathrm{m\,GeV^{-3}} \,.$$

Work out the energy loss due to synchrotron emission per turn of a 1 TeV electron in a circular orbit around a pulsar at a distance of 1000 km. What kind of magnetic field had the pulsar at this distance?

4. Consider the energy loss by radiation of a particle moving in a transverse homogeneous time-independent magnetic field, see also Problem 3 for this section. The radiated power is given by

$$P = \frac{2}{3}\frac{e^2}{m_0^2 c^3} \gamma^2 |\dot{p}|^2 \,.$$

Calculate the time dependence of the particle energy and of the bending radius of the trajectory for the
   a) ultrarelativistic case,
   b) general case.[16]

5. An X-ray detector on board of a satellite with collection area of $1\,\mathrm{m}^2$ counts 10 keV photons from a source in the Large Magellanic Cloud with a rate of 1/hour. How many 10 keV photons are emitted from the source if isotropic emission is assumed? The detector is a Xe-filled proportional counter of 1 cm thickness (the attenuation coefficient for 10 keV photons is $\mu = 125\,(\mathrm{g/cm}^2)^{-1}$, density $\varrho_{Xe} = 5.8 \times 10^{-3}\,\mathrm{g/cm}^3$).

6. X rays can be produced by inverse Compton scattering of energetic electrons (E_i) off blackbody photons (energy ω_i). Show that the energy of the scattered photon ω_f is related to the scattering angles φ_i and φ_f by

---

[16] The solution to this problem is mathematically demanding.

$$\omega_f \approx \omega_i \, \frac{1 - \cos \varphi_i}{1 - \cos \varphi_f} \,, \tag{6.88}$$

where φ_i and φ_f are the angles between the incoming electron and the incoming and outgoing photon. The above approximation holds if $E_i \gg m_i \gg \omega_i$.

7. What is the temperature of a cosmic object if its maximum blackbody emission occurs at an energy of $E = 50$ keV?

   (Hint: The solution of this problem leads to a transcendental equation which needs to be solved numerically.)

Problems for Sect. 6.5

1. A photon propagating to a celestial object of mass M will gain momentum and will be shifted towards the blue. Work out the relative gain of a photon approaching the Sun's surface from a height of $H = 1$ km. Analogously, a photon escaping from a massive object will be gravitationally redshifted.

   (Radius of the Sun $R_\odot = 6.9635 \times 10^8$ m, Mass of the Sun $M_\odot = 1.993 \times 10^{30}$ kg, acceleration due to Sun's gravity $g_\odot = 2.7398 \times 10^2$ m/s^2.)

2. Accelerated masses radiate gravitational waves. The emitted energy per unit time is worked out to be

$$P = \frac{G}{5c^2} \, \dddot{Q}^2 \,,$$

where Q is the quadrupole moment of a certain mass configuration (e.g., the system Sun–Earth). For a rotating system with periodic time dependence ($\sim \sin \omega t$) each time derivative contributes a factor ω, hence

$$P \approx \frac{G}{5c^2} \, \omega^6 \, Q^2 \,.$$

For a system consisting of a heavy-mass object like the Sun (M) and a low-mass object, like Earth (m), the quadrupole moment is on the order of mr^2. Neglecting numerical factors of order unity, one gets

$$P \approx \frac{G}{c^2} \, \omega^6 \, m^2 \, r^4 \,.$$

Work out the power radiated from the system Sun–Earth and compare it with the gravitational power emitted from typical fast-rotating laboratory equipments.

7 Secondary Cosmic Rays

"There are more things in heaven and earth, Horatio, than are dreamt of in your philosophy."

Shakespeare, Hamlet

For the purpose of astroparticle physics the influence of the Sun and the Earth's magnetic field is a perturbation, which complicates a search for the sources of cosmic rays. The solar activity produces an additional magnetic field which prevents part of galactic cosmic rays from reaching Earth. Figure 7.1, however, shows that the influence of the Sun is limited to primary particles with energies below $10\,\text{GeV}$. The flux of low-energy primary cosmic-ray particles is anti-correlated to the solar activity.

solar modulation

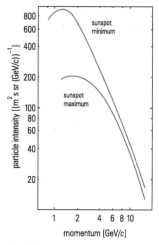

Fig. 7.1
Modulation of the primary spectrum by the 11-year cycle of the Sun

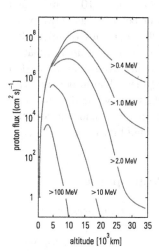

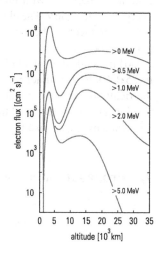

Fig. 7.2
Flux densities of protons and electrons in the radiation belts of the Earth

On the other hand, the solar wind, whose magnetic field modulates primary cosmic rays, is a particle stream in itself, which can be measured at Earth. The particles constituting the solar wind (predominantly protons and electrons) are of low energy (MeV region). These particles are captured to a large extent by the Earth's magnetic field in the Van Allen belts or they are absorbed in the upper layers of the Earth's atmosphere (see Fig. 7.9). Figure 7.2 shows the flux den-

solar wind

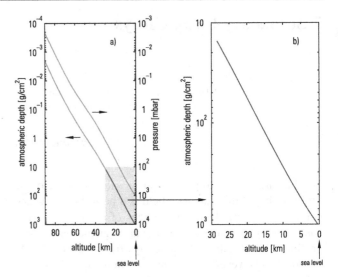

Fig. 7.3
(a) Relation between atmospheric depth (column density) and pressure
(b) column density of the atmosphere as a function of altitude up to 28 km

sities of protons and electrons in the Van Allen belts. The proton belt extends over altitudes from 2 000 to 15 000 km. It contains particles with intensities up to $10^8/(cm^2 s)$ and

radiation belts   energies up to 1 GeV. The electron belt consists of two parts. The inner electron belt with flux densities of up to 10^9 particles per cm^2 and s is at an altitude of approximately 3 000 km, while the outer belt extends from about 15 000 km to 25 000 km. The inner part of the radiation belts is symmetrically distributed around the Earth while the outer part is subject to the influence of the solar wind and consequently deformed by it (see also Fig. 1.9 and Fig. 1.13).

7.1 Propagation in the Atmosphere

"Astroparticles are messengers from different worlds."

Anonymous

interaction in the atmosphere   Primary cosmic rays are strongly modified by interactions with atomic nuclei in the atmospheric air. The column density of the atmosphere amounts to approximately 1000 g/cm^2, corresponding to the atmospheric pressure of about 1000 hPa. Figure 7.3 (a) shows the relation between column

column density   density, altitude in the atmosphere, and pressure. Figure 7.3 (b) shows this relation in somewhat more detail for altitudes below 28 km. The residual atmosphere for flight altitudes of scientific balloons ($\approx$ 35–40 km) corresponds to approximately several g/cm^2. For inclined directions the thickness of the atmosphere increases strongly (approximately like $1/\cos\theta$, with θ – zenith angle). Figure 7.4 shows the variation of atmospheric depth with zenith angle at sea level.

For the interaction behaviour of primary cosmic rays the thickness of the atmosphere in units of the characteristic interaction length for the relevant particles species in question is important. The *radiation length* for photons and electrons in air is $X_0 = 36.66$ g/cm^2. The atmosphere therefore corresponds to a depth of 27 radiation lengths. The relevant *interaction length* for hadrons in air is $\lambda = 90.0$ g/cm^2, corresponding to 11 interaction lengths per atmosphere. This means that practically not a single particle of original primary cosmic rays arrives at sea level. Already at altitudes of 15 to 20 km primary cosmic rays interact with atomic nuclei of the air and initiate – depending on energy and particle species – electromagnetic and/or hadronic cascades.

The momentum spectrum of the singly charged component of primary cosmic rays at the top of the atmosphere is shown in Fig. 7.5. In this diagram the particle velocity $\beta = v/c$ is shown as a function of momentum. Clearly visible are the bands of hydrogen isotopes as well as the low flux of primary antiprotons. Even at these altitudes several muons have been produced via pion decays. Since muon and pion mass are very close, it is impossible to separate them out in this scatter diagram. Also relativistic electrons and positrons would populate the bands labeled μ^+ and μ^-. One generally assumes that the measured antiprotons are not of primordial origin, but are rather produced by interactions in interstellar or interplanetary space or even in the residual atmosphere above the balloon.

The transformation of primary cosmic rays in the atmosphere is presented in Fig. 7.6. Protons with approximately 85% probability constitute the largest fraction of primary cosmic rays. Since the interaction length for hadrons is 90 g/cm^2, primary protons initiate a hadron cascade already in their first interaction approximately at an altitude corresponding to the 100 mbar layer. The secondary particles most copiously produced are pions. Kaons on the other hand are only produced with a probability of 10% compared to pions. Neutral pions initiate via their decay ($\pi^0 \to \gamma + \gamma$) electromagnetic cascades, whose development is characterized by the shorter radiation length ($X_0 \approx \frac{1}{3}\lambda$ in air). This

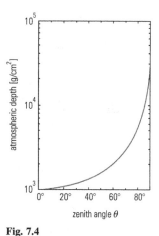

Fig. 7.4
Relation between zenith angle and atmospheric depth at sea level

radiation length

interaction length

electromagnetic and hadronic cascades

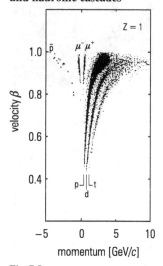

Fig. 7.5
Identification of singly charged particles in cosmic rays at a flight altitude of balloons ($\hat{=}$ 5 g/cm^2 residual atmosphere) {21}

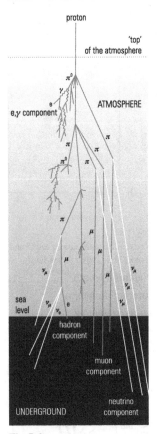

Fig. 7.6
Transformation of primary cosmic rays in the atmosphere

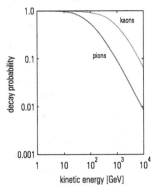

Fig. 7.7
Decay probabilities for charged pions and kaons in the atmosphere as a function of their kinetic energy

shower component is absorbed relatively easily and is therefore also named a soft component. Charged pions and kaons can either initiate further interactions or decay.

The competition between decay and interaction probability is a function of energy. For the same Lorentz factor charged pions (lifetime 26 ns) have a smaller decay probability compared to charged kaons (lifetime 12.4 ns). The decay probability of charged pions and kaons in the atmosphere is shown in Fig. 7.7 as a function of their kinetic energy. The leptonic decays of pions and kaons produce the penetrating muon and neutrino components ($\pi^+ \rightarrow \mu^+ + \nu_\mu$, $\pi^- \rightarrow \mu^- + \bar{\nu}_\mu$; $K^+ \rightarrow \mu^+ + \nu_\mu$, $K^- \rightarrow \mu^- + \bar{\nu}_\mu$). Muons can also decay and contribute via their decay electrons to the soft component and neutrinos to the neutrino component ($\mu^+ \rightarrow e^+ + \nu_e + \bar{\nu}_\mu$, $\mu^- \rightarrow e^- + \bar{\nu}_e + \nu_\mu$).

The energy loss of relativistic muons not decaying in the atmosphere is low (≈ 1.8 GeV). They constitute with 80% of all charged particles the largest fraction of secondary particles at sea level.

Some secondary mesons and baryons can also survive down to sea level. Most of the low-energy charged hadrons observed at sea level are locally produced. The total fraction of hadrons at ground level, however, is very small.

Apart from their longitudinal development electromagnetic and hadronic cascades also spread out laterally in the atmosphere. The lateral size of an electromagnetic cascade is caused by *multiple scattering* of electrons and positrons, while in hadronic cascades the *transverse momenta* at production of secondary particles are responsible for the lateral width of the cascade. Figure 7.8 shows a comparison of the shower development of 100 TeV photons and 100 TeV protons in the atmosphere. It is clearly visible that transverse momenta of secondary particles fan out the hadron cascade.

The intensity of protons, electrons, and muons of all energies as a function of the altitude in the atmosphere is plotted in Fig. 7.9. The absorption of protons can be approximately described by an exponential function.

The electrons and positrons produced through π^0 decay with subsequent pair production reach a maximum intensity at an altitude of approximately 15 km and soon after are relatively quickly absorbed while, in contrast, the flux of muons is attenuated only relatively weakly.

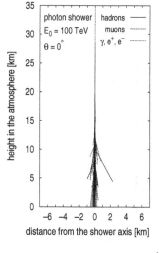

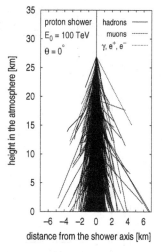

Fig. 7.8
Comparison of the development of electromagnetic (100 TeV photon) and hadronic cascades (100 TeV proton) in the atmosphere. Only secondaries with $E \geq 1$ GeV are shown {22}

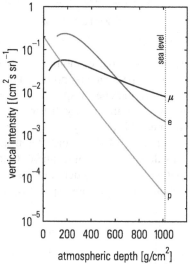

Fig. 7.9
Particle composition in the atmosphere as a function of atmospheric depth

Because of the steepness of the energy spectra the particle intensities are of course dominated by low-energy particles. These low-energy particles, however, are mostly of secondary origin. If only particles with energies in excess of 1 GeV are counted, a different picture emerges (Fig. 7.10).

Primary nucleons (protons and neutrons) with the initial high energies dominate over all other particle species down to altitudes of 9 km, where muons take over. Because of the low interaction probability of neutrinos these particles are practically not at all absorbed in the atmosphere.

sea-level composition

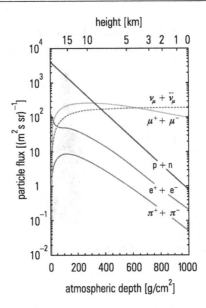

Fig. 7.10

Intensities of cosmic-ray particles with energies > 1 GeV in the atmosphere

Their flux increases monotonically because additional neutrinos are permanently produced by particle decays.

Since the energy spectrum of primary particles is relatively steep, the energy distribution of secondaries also has to reflect this property.

proton and muon spectra

Figure 7.11 shows the proton and muon spectra for various depths in the atmosphere. Clearly visible is the trend that with increasing depth in the atmosphere muons start to dominate over protons especially at high energies.

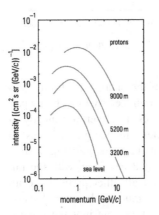
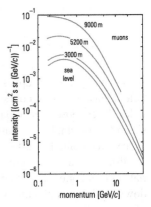

Fig. 7.11

Momentum spectra of protons and muons at various altitudes in the atmosphere

7.2 Cosmic Rays at Sea Level

"The joy of discovery is certainly the liveliest that the mind of man can ever feel."

Claude Bernard

A measurement of charged particles at sea level clearly shows that, apart from some protons, muons are the dominant component (Fig. 7.12).

Approximately 80% of the charged component of secondary cosmic rays at sea level are muons. Their flux through a horizontal area amounts to roughly one particle per cm^2 and minute. These muons originate predominantly from pion decays, since pions as lightest mesons are produced in large numbers in hadron cascades. The muon spectrum at sea level is therefore a direct consequence of the pion source spectrum. There are, however, several modifications. Figure 7.13 shows the parent pion spectrum at the location of production in comparison to the observed sea-level muon spectrum. The shape of the muon spectrum agrees relatively well with the pion spectrum for momenta between 10 and 100 GeV/c. For energies below 10 GeV and above 100 GeV the muon intensity, however, is reduced compared to the pion source spectrum. For low energies the muon decay probability is increased. A muon of 1 GeV with a Lorentz factor of $\gamma = E/m_\mu c^2 = 9.4$ has a mean decay length of

$$s_\mu \approx \gamma \tau_\mu c = 6.2 \,\text{km} \,. \tag{7.1}$$

Since pions are typically produced at altitudes of 15 km and decay relatively fast (for $\gamma = 10$ the decay length is only $s_\pi \approx \gamma \tau_\pi c = 78$ m), the decay muons do not reach sea level but rather decay themselves or get absorbed in the atmosphere. At high energies the situation is changed. For pions of 100 GeV ($s_\pi = 5.6$ km, corresponding to a column density of 160 g/cm^2 measured from the production altitude) the interaction probability dominates ($s_\pi > \lambda$). Pions of these energies will therefore produce further, tertiary pions in subsequent interactions, which will also decay eventually into muons, but providing muons of lower energy. Therefore, the muon spectrum at high energies is always steeper compared to the parent pion spectrum.

If muons from inclined horizontal directions are considered, a further aspect has to be taken into account. For large zenith angles the parent particles of muons travel relatively

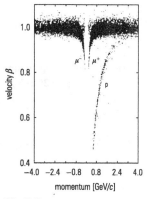

Fig. 7.12
Measurement and identification of charged particles at sea level {21}

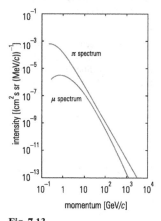

Fig. 7.13
Sea-level muon spectrum in comparison to the pion parent source spectrum at production

muons
from inclined directions

long distances in rare parts of the atmosphere. Because of the low area density at large altitudes for inclined directions the decay probability is increased compared to the interaction probability. Therefore, for inclined directions pions will produce predominantly high-energy muons in their decay.

The result of these considerations is in agreement with observation (Fig. 7.14). For about $170\,\mathrm{GeV}/c$ the muon intensity at $83°$ zenith angle starts to outnumber that of the vertical muon spectrum. The intensity of muons from horizontal directions at low energies is naturally reduced because of muon decays and absorption effects in the thicker atmosphere at large zenith angles.

**sea-level muon spectrum
up to 20 TeV/c**

The sea-level muon spectrum for inclined directions has been measured with solid-iron momentum spectrometers up to momenta of approximately $20\,\mathrm{TeV}/c$ (Fig. 7.15). For higher energies the muon intensity decreases steeply.

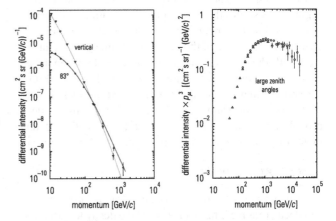

Fig. 7.14
Sea-level muon momentum spectra for vertical and inclined directions

Fig. 7.15
Momentum spectrum of muons at sea level for large zenith angles. In this figure the differential intensity is multiplied by p_μ^3

The total intensity of muons, however, is dominated by low-energy particles. Because of the increased decay probability and the stronger absorption of muons from inclined directions, the total muon intensity at sea level varies like

$$I_\mu(\theta) = I_\mu(\theta = 0)\,\cos^n\theta \tag{7.2}$$

for not too large zenith angles θ. The exponent of the zenith-angle distribution is obtained to be $n = 2$. This exponent varies very little, even at shallow depths underground, if only muons exceeding a fixed energy are counted.

charge ratio of muons

An interesting quantity is the *charge ratio of muons* at sea level. Since primary cosmic rays are positively charged, this positive charge excess is eventually also transferred to muons. If one assumes that primary protons interact with protons and neutrons of atomic nuclei in the atmosphere where the multiplicity of produced pions is usually quite large, the charge ratio of muons, $N(\mu^+)/N(\mu^-)$, can be estimated by considering the possible charge exchange reactions:

charge exchange reactions

$$p + N \rightarrow p' + N' + k\pi^+ + k\pi^- + r\pi^0 ,$$
$$p + N \rightarrow n + N' + (k+1)\pi^+ + k\pi^- + r\pi^0 . \quad (7.3)$$

In this equation k and r are the multiplicities of the produced particle species and N represents a target nucleon. If one assumes that for the reactions in (7.3) the cross sections are the same, the charge ratio of pions is obtained to be

$$R = \frac{N(\pi^+)}{N(\pi^-)} = \frac{2k+1}{2k} = 1 + \frac{1}{2k} . \quad (7.4)$$

For low energies $k = 2$ and thereby $R = 1.25$. Since this ratio is transferred to muons by the pion decay, one would expect a similar value for muons. Experimentally one observes that the charge ratio of muons at sea level is constant over a wide momentum range and takes on a value of

$$N(\mu^+)/N(\mu^-) \approx 1.27 . \quad (7.5)$$

In addition to 'classical' production mechanisms of muons by pion and kaon decays, they can also be produced in semileptonic decays of charmed mesons (for example, $D^0 \rightarrow K^- \mu^+ \nu_\mu$ and $D^+ \rightarrow \bar{K}^0 \mu^+ \nu_\mu$, $D^- \rightarrow K^0 \mu^- \bar{\nu}_\mu$). Since these charmed mesons are very short-lived ($\tau_{D^0} \approx 0.4\,\mathrm{ps}$, $\tau_{D^\pm} \approx 1.1\,\mathrm{ps}$), they decay practically immediately after production without undergoing interactions themselves. Therefore, they are a source of high-energy muons. Since the production cross section of charmed mesons in proton–nucleon interactions is rather small, D decays contribute significantly only at very high energies.

muons from semileptonic decays

Figure 7.12 already showed that apart from muons also some nucleons can be observed at sea level. These nucleons are either remnants of primary cosmic rays, which, however, are reduced in their intensity and energy by multiple interactions, or they are produced in atmospheric hadron cascades. About one third of the nucleons at sea level are neutrons. The proton/muon ratio varies with the momentum of the particles. At low momenta ($\approx 500\,\mathrm{MeV}/c$) a p/μ ratio $N(p)/N(\mu)$ of about 10% is observed decreasing to larger momenta ($N(p)/N(\mu) \approx 2\%$ at $1\,\mathrm{GeV}/c$, $N(p)/N(\mu) \approx 0.5\%$ at $10\,\mathrm{GeV}/c$).

nucleon component

positrons, electrons, and photons from electromagnetic cascades

In addition to muons and protons, one also finds electrons, positrons, and photons at sea level as a consequence of the electromagnetic cascades in the atmosphere. A certain fraction of electrons and positrons originates from muon decays. Electrons can also be liberated by secondary interactions of muons ('knock-on electrons').

pions and kaons at sea level

The few pions and kaons observed at sea level are predominantly produced in local interactions.

production of ν_e and ν_μ

Apart from charged particles, electron and muon neutrinos are produced in pion, kaon, and muon decays. They constitute an annoying background, in particular, for neutrino astronomy. On the other hand, the propagation of atmospheric neutrinos has provided new insights for elementary particle physics, such as neutrino oscillations. A comparison of vertical and horizontal neutrino spectra (Fig. 7.16) shows a similar tendency as for muon spectra.

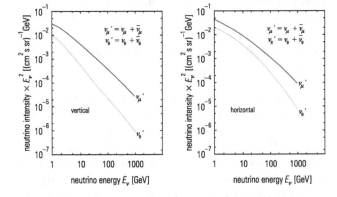

Fig. 7.16
Energy spectra of muon and electron neutrinos for vertical and horizontal directions

neutrino parents

Since the parent particles of neutrinos are dominantly pions and kaons and their decay probability is increased compared to the interaction probability at inclined directions, the horizontal neutrino spectra are also harder in comparison to the spectra from vertical directions. Altogether, muon neutrinos would appear to dominate, since the $(\pi \rightarrow e\nu)$ and $(K \rightarrow e\nu)$ decays are strongly suppressed due to *helicity conservation*. Therefore, pions and kaons almost exclusively produce muon neutrinos only. Only in muon decay equal numbers of electron and muon neutrinos are produced. At high energies also semileptonic decays of charmed mesons constitute a source for neutrinos.

dominance of ν_μ

Based on these 'classical' considerations the integral neutrino spectra yield a neutrino-flavour ratio of

$$\frac{N(\nu_\mu + \bar{\nu}_\mu)}{N(\nu_e + \bar{\nu}_e)} \approx 2 . \tag{7.6}$$

This ratio, however, is modified by propagation effects like neutrino oscillations (see Sect. 6.2: Neutrino Astronomy).

neutrino-flavour ratio

7.3 Cosmic Rays Underground

> *"If your experiment needs statistics, then you ought to have done a better experiment."*
>
> Ernest Rutherford

Particle composition and energy spectra of secondary cosmic rays underground are of particular importance for neutrino astronomy. Experiments in neutrino astronomy are usually set up at large depths underground to provide a sufficient shielding against the other particles from cosmic rays. Because of the rarity of neutrino events even low fluxes of residual cosmic rays constitute an annoying background. In any case it is necessary to know precisely the identity and flux of secondary cosmic rays underground to be able to distinguish a possible signal from cosmic-ray sources from statistical fluctuations or systematical uncertainties of the atmospheric cosmic-ray background.

particle composition underground

Long-range atmospheric muons, secondary particles locally produced by muons, and the interaction products created by atmospheric neutrinos represent the important background sources for neutrino astronomy.

background sources for neutrino astrophysics

Muons suffer energy losses by ionization, direct electron–positron pair production, bremsstrahlung, and nuclear interactions. These processes have been described in rather detail in Chap. 4. While the ionization energy loss at high energies is essentially constant, the cross sections for the other energy-loss processes increase linearly with the energy of the muon,

energy loss of muons

$$-\frac{dE}{dx} = a + b\,E . \tag{7.7}$$

The energy loss of muons as a function of their energy is shown in Fig. 7.17 for iron as absorber material. The energy loss of muons in rock in its dependence on the muon energy was already shown earlier (Fig. 4.3).

range of muons

Equation (7.7) allows to work out the range R of muons by integration,

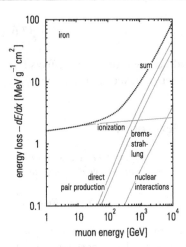

Fig. 7.17
Contributions to the energy loss of
muons in iron

$$R = \int_E^0 \frac{\mathrm{d}E}{-\mathrm{d}E/\mathrm{d}x} = \frac{1}{b}\ln(1 + \frac{b}{a}E) , \qquad (7.8)$$

if it is assumed that the parameters a and b are energy independent.

For not too large energies ($E < 100\,\text{GeV}$) the ionization
energy loss dominates. In this case $bE \ll a$ and therefore

$$R = \frac{E}{a} . \qquad (7.9)$$

The energy loss of a *minimum-ionizing muon* in the atmosphere is

$$\frac{\mathrm{d}E}{\mathrm{d}x} = 1.82\,\text{MeV}/(\text{g/cm})^2 . \qquad (7.10)$$

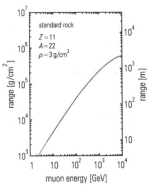

Fig. 7.18
Range of muons in rock

A muon of energy $100\,\text{GeV}$ has a range of about $40\,000$
g/cm^2 in rock corresponding to 160 meter (or 400 meter water equivalent). An energy–range relation for standard rock
is shown in Fig. 7.18. Because of the stochastic character of
muon interaction processes with large energy transfers (e.g.,
bremsstrahlung) muons are subject to a considerable range
straggling.

**determination of the
depth–intensity relation**

The knowledge of the sea-level muon spectrum and the
energy-loss processes of muons allow one to determine the
depth–intensity relation for muons. The integral sea-level
muon spectrum can be approximated by a power law

$$N(> E) = A\,E^{-\gamma} . \qquad (7.11)$$

Using the energy–range relation (7.8), the depth–intensity
relation is obtained,

$$N(> E, R) = A \left[\frac{a}{b}(e^{bR} - 1)\right]^{-\gamma} . \qquad (7.12)$$

For high energies ($E_\mu > 1\,\text{TeV}$, $bE \gg a$) the exponential dominates and one obtains

$$N(> E, R) = A \left(\frac{a}{b}\right)^{-\gamma} e^{-\gamma bR} . \qquad (7.13)$$

For inclined directions the absorbing ground layer increases like $1/\cos\theta = \sec\theta$ (θ – zenith angle) for a flat overburden, so that for muons from inclined directions one obtains a depth–intensity relation of

inclined muon directions

$$N(> E, R, \theta) = A \left(\frac{a}{b}\right)^{-\gamma} e^{-\gamma bR \sec\theta} . \qquad (7.14)$$

For shallower depths (7.12), or also (7.9), however, leads to a power law

$$N(> E, R) = A\,(aR)^{-\gamma} . \qquad (7.15)$$

The measured depth–intensity relation for vertical directions is plotted in Fig. 7.19. From depths of 10 km water equivalent ($\approx 4000\,\text{m}$ rock) onwards muons induced by atmospheric neutrinos dominate the muon rate. Because of the low interaction probability of neutrinos the neutrino-induced muon rate does not depend on the depth. At large depths ($> 10\,\text{km}$ w.e.) a neutrino telescope with a collection area of $100 \times 100\,\text{m}^2$ and a solid angle of π would still measure a background rate of 10 events per day.

The zenith-angle distributions of atmospheric muons for depths of 1500 and 7000 meter water equivalent are shown in Fig. 7.20. For large zenith angles the flux decreases steeply, because the thickness of the overburden increases

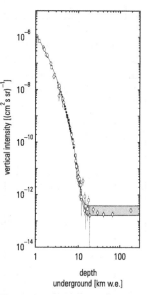

Fig. 7.19
Depth–intensity relation for muons from vertical directions. The *grey-hatched band* at large depths represents the flux of neutrino-induced muons with energies above 2 GeV (*upper line:* horizontal, *lower line:* vertical upward neutrino-induced muons) [2]

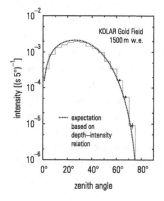

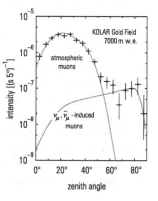

Fig. 7.20
Zenith-angle distribution of atmospheric muons at depths of 1500 and 7000 m w.e.

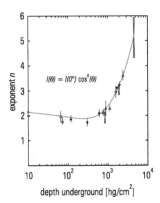

Fig. 7.21
Variation of the exponent n of the zenith-angle distribution of muons with depth

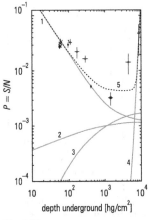

Fig. 7.22
Ratio of stopping to penetrating muons as a function of depth in comparison to some experimental results. (*1*) Stopping atmospheric muons, (*2*) stopping muons from nuclear interactions, (*3*) stopping muons locally produced by photons, (*4*) neutrino-induced stopping muons, and (*5*) sum of all contributions

lateral spread of muons
underground

like $1/\cos\theta$. Therefore, at large depths and from inclined directions neutrino-induced muons dominate.

For not too large zenith angles and depths the zenith-angle dependence of the integral muon spectrum can still be represented by

$$I(\theta) = I(\theta = 0)\cos^n\theta \qquad (7.16)$$

(Fig. 7.21). For large depths the exponent n in this distribution, however, gets very large, so that it is preferable to use (7.14) instead.

The average energy of muons at sea level is in the range of several GeV. Absorption processes in rock reduce predominantly the intensity at low energies. Therefore, the average muon energy of the muon spectrum increases with increasing depth. Muons of high energy can also produce other secondary particles in local interactions. Since low-energy muons can be identified by their ($\mu \rightarrow e\nu\nu$) decay with the characteristic decay time in the microsecond range, the measurement of stopping muons underground provides an information about local production processes. The flux of stopping muons is normally determined for a detector thickness of $100\,\mathrm{g/cm^2}$ and the ratio P of stopping to penetrating muons is presented (Fig. 7.22).

A certain fraction of stopping muons is produced locally by low-energy pions which decay relatively fast into muons. Since the flux of penetrating muons decreases strongly with increasing depth, the ratio P of stopping to penetrating muons is dominated by neutrino interactions for depths larger than 5000 m w.e.

The knowledge of the particle composition at large depths below ground represents an important information for neutrino astrophysics.

Also remnants of extensive air showers, which developed in the atmosphere, are measured underground. Electrons, positrons, photons, and hadrons are completely absorbed already in relatively shallow layers of rock. Therefore, only muons and neutrinos of extensive air showers penetrate to larger depths. The primary interaction vertex of particles which initiate the air showers is typically at an atmospheric altitude of 15 km. Since secondary particles in hadronic cascades have transverse momenta of about $300\,\mathrm{MeV}/c$ only, the high-energy muons essentially follow the shower axis. For primaries of energy around 10^{14} eV lateral displacements of energetic muons ($\approx 1\,\mathrm{TeV}$)

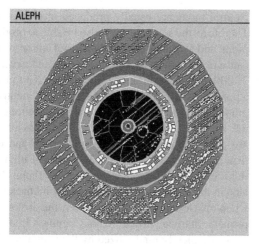

Fig. 7.23
Muon shower in the ALEPH experiment. Muon tracks are seen in the central time-projection chamber and in the surrounding hadron calorimeter. Even though there is a strong 1.5 Tesla magnetic field perpendicular to the projection shown, the muon tracks are almost straight indicating their high momenta. Only a knock-on electron produced in the time-projection chamber by a muon is bent on a circle {23}

at shallow depths underground of typically several meters exclusively caused by transferred transverse momenta are obtained. Typical multiple-scattering angles for energetic muons ($\approx 100\,\mathrm{GeV}$) in thick layers of rock (50–100 m) are on the order of a few mrad.

The multiplicity of produced secondary particles increases with energy of the initiating particle (for a 1 TeV proton the charged multiplicity of particles for proton–proton interactions is about 15). Since the secondaries produced in these interactions decay predominantly into muons, one observes bundles of nearly parallel muons underground in the cores of extensive air showers. Figure 7.23 shows such a shower with more than 50 parallel muons observed by the ALEPH experiment at a depth of 320 m w.e.

High-energy muons are produced by high-energy primaries and, in particular, muon showers correlate with even higher primary energies. Therefore, one is tempted to localize extraterrestrial sources of high-energy cosmic rays via the arrival directions of single or multiple muons. Since Cygnus X3 has been claimed to emit photons with energies up to 10^{16} eV, this astrophysical source also represents an excellent candidate for the acceleration of high-energy charged primary cosmic rays. Cygnus X3 at a distance of approximately 33 000 light-years is an X-ray binary consisting of a superdense pulsar and a stellar companion. The material flowing from the companion into the direction of the pulsar forms an accretion disk around the pulsar. If apparently photons of very high energy can be produced, one would expect them to originate from the π^0 decay ($\pi^0 \rightarrow \gamma\gamma$).

muon bundles

**ALEPH
as cosmic-ray detector**

Cygnus X3

Neutral pions are usually produced in proton interactions. Therefore, the source should also be able to produce charged pions and via their decay muons and muon neutrinos. Because of their short lifetime, muons would never survive the 33 000 light-year distance from Cygnus X3 to Earth, so that a possible muon signal must be caused by neutrino-induced muons. Unfortunately, muons and multi-muons observed in the Frejus experiment from the directions of Cygnus X3 are predominantly of atmospheric origin and do not confirm that Cygnus X3 is a strong source of high-energy particles (Fig. 7.24). The primary particles themselves accelerated in the source could in principle point back to the source when measured on Earth. However, the arrival direction of primary charged particles from Cygnus X3 could also have been completely randomized by the irregular galactic magnetic field. Muon production by neutrinos from Cygnus X3 would have been a rare event which would have required an extremely massive detector to obtain a significant rate.

Cygnus X3, a hadron accelerator?

muon astronomy

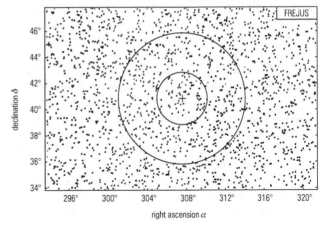

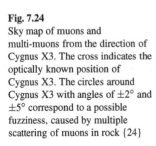

Fig. 7.24
Sky map of muons and multi-muons from the direction of Cygnus X3. The cross indicates the optically known position of Cygnus X3. The circles around Cygnus X3 with angles of $\pm 2°$ and $\pm 5°$ correspond to a possible fuzziness, caused by multiple scattering of muons in rock {24}

7.4 Extensive Air Showers

"Science never solves a problem without creating ten more."

George Bernard Shaw

Extensive air showers are cascades initiated by energetic primary particles which develop in the atmosphere. An extensive air shower (EAS) has an electromagnetic, a muonic, a

components of an extensive air shower

hadronic, and a neutrino component (see Fig. 7.6). The air shower develops a shower nucleus consisting of energetic hadrons, which permanently inject energy into the electromagnetic and the other shower components via interactions and decays. Neutral pions, which are produced in nuclear interactions and whose decay photons produce electrons and positrons via pair production, supply the electron, positron, and photon component. Photons, electrons, and positrons initiate electromagnetic cascades through alternating processes of pair production and bremsstrahlung. The muon and neutrino components are formed by the decay of charged pions and kaons (see also Fig. 7.6).

hadron, electromagnetic, muon, neutrino component

The inelasticity in hadron interactions is on the order of 50%, i.e., 50% of the primary energy is transferred into the production of secondary particles. Since predominantly pions are produced ($N(\pi) : N(K) = 9 : 1$) and all charge states of pions (π^+, π^-, π^0) are produced in equal amounts, one third of the inelasticity is invested into the formation of the electromagnetic component. Since most of the charged hadrons and the hadrons produced in hadron interactions also undergo multiple interactions, the largest fraction of the primary energy is eventually transferred into the electromagnetic cascade. Therefore, in terms of the number of particles, electrons and positrons constitute the main shower component. The particle number increases with shower depth t until absorptive processes like ionization for charged particles and Compton scattering and photoelectric effect for photons start to dominate and cause the shower to die out.

inelasticity

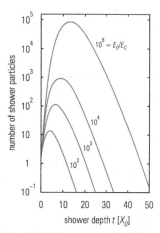

The development of electromagnetic cascades is shown in Fig. 7.25 for various primary energies. The particle intensity increases initially in a parabolical fashion and decays exponentially after the maximum of the shower has been reached. The longitudinal profile of the particle number can be parameterized by

Fig. 7.25
Longitudinal shower development of electromagnetic cascades. (The critical energy in air is $E_c = 84\,\text{MeV}$)

$$N(t) \sim t^\alpha e^{-\beta t} \,, \tag{7.17}$$

where $t = x/X_0$ is the shower depth in units of the radiation length and α and β are free fit parameters. The position of the shower maximum varies only logarithmically with the primary energy, while the total number of shower particles increases linearly with the energy. The latter can therefore be used for the energy determination of the primary particle. One can imagine that the Earth's atmosphere represents a combined hadronic and electromagnetic calorimeter,

longitudinal particle-number profile

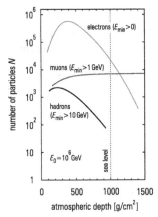

Fig. 7.26
Average longitudinal development
of the various components of an
extensive air shower in the
atmosphere

**longitudinal profile
of a shower**

lateral distribution

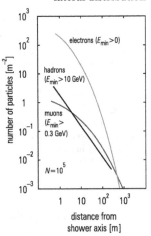

Fig. 7.27
Average lateral distribution of the
shower components for $N = 10^5$
corresponding to $E \approx 10^{15}$ eV

Auger project

in which the extensive air shower develops. The atmosphere constitutes approximately a target of 11 interaction lengths and 27 radiation lengths. The minimum energy for a primary particle to be reasonably well measured at sea level via the particles produced in the air shower is about 10^{14} eV $= 100$ TeV. As a rough estimate for the particle number N at sea level in its dependence on the primary energy E_0, one can use the relation

$$N = 10^{-10} E_0[\text{eV}] . \tag{7.18}$$

Only about 10% of the charged particles in an extensive air shower are muons. The number of muons reaches a plateau already at an atmospheric depth of 200 g/cm^2 (see also Fig. 7.9 and Fig. 7.10). Its number is hardly reduced to sea level, since the probability for catastrophic energy-loss processes, like bremsstrahlung, is low compared to electrons because of the large muon mass. Muons also lose only a small fraction of their energy by ionization. Because of the relativistic time dilation the decay of energetic muons ($E_\mu > 3$ GeV) in the atmosphere is strongly suppressed.

Figure 7.26 shows schematically the longitudinal development of the various components of an extensive air shower in the atmosphere for a primary energy of 10^{15} eV. The lateral spread of an extensive air shower is essentially caused by the transferred transverse momenta in hadronic interactions and by multiple scattering of low-energy shower particles. The muon component is relatively flat compared to the lateral distribution of electrons and hadrons. Figure 7.27 shows the lateral particle profile for the various shower components. Neutrinos essentially follow the shape of the muon component.

Even though an extensive air shower initiated by primary particles with energies below 100 TeV does not reach sea level, it can nevertheless be recorded via the Cherenkov light emitted by the shower particles (see Sect. 6.3 on gamma-ray astronomy). At higher energies one has the choice of various detection techniques.

The classical technique for the measurement of extensive air showers is the sampling of shower particles at sea level with typically 1 m^2 large scintillators or water Cherenkov counters. This technique is sketched in Fig. 7.28. In the Auger project in Argentina 3000 sampling detectors will be used for the measurement of the sea-level component of extensive air showers. However, the energy assignment for the primary particle using this technique is not very

precise. The shower develops in the atmosphere which acts
as a calorimeter of 27 radiation lengths thickness. The information on this shower is sampled in only *one*, the last layer
of this calorimeter and the coverage of this layer is typically
on the order of only 1%. The direction of incidence of the
primary particle can be obtained from the arrival times of
shower particles in the different sampling counters.

energy measurement

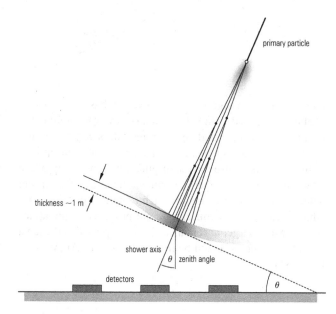

primary particle

thickness ~1 m

shower axis
θ | zenith angle

detectors
θ

Fig. 7.28
Air-shower measurement with
sampling detectors

It would be much more advantageous to measure the
total longitudinal development of the cascade in the atmosphere. This can be achieved using the technique of the Fly's
Eye (Fig. 7.29). Apart from the directional Cherenkov radiation the shower particles also emit an isotropic scintillation
light in the atmosphere.

Fly's Eye

For particles with energies exceeding 10^{17} eV the fluorescence light of nitrogen is sufficiently intense to be
recorded at sea level in the presence of the diffuse background of starlight. The actual detector consists of a system
of mirrors and photomultipliers, which view the whole sky.
An air shower passing through the atmosphere near such
a Fly's Eye detector activates only those photomultipliers
whose field of view is hit. The fired photomultipliers allow
to reconstruct the longitudinal profile of the air shower.
The total recorded light intensity is used to determine the
shower energy. Such a type of detector allows much more

fluorescence technique

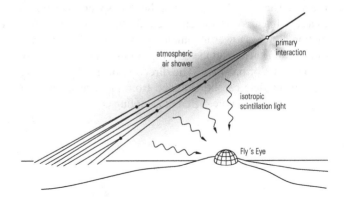

Fig. 7.29
Principle of the measurement of
the scintillation light of extensive
air showers

precise energy assignments, however, it has a big disadvantage compared to the classical air-shower technique that it can only be operated in clear moonless nights. Figure 7.30 shows an arrangement of mirrors and photomultipliers, as they have been used in the original Fly's Eye setup of the Utah group. In the Auger experiment the array of sampling detectors is complemented by such a number of telescopes which measure the scintillation light produced in the atmo-
Air Watch   sphere. Much larger acceptances could be provided if such a Fly's Eye detector would be installed in orbit ('Air Watch', Fig. 7.31).

Fig. 7.30
Arrangement of mirrors and
photomultipliers in the original
Fly's Eye experiment of the Utah
group {25}

radio detection of showers   Apart from these detection techniques it has also been tried to observe air showers via the electromagnetic radiation emitted in the radio band. It is generally believed that this radio signal is caused by shower electrons deflected in

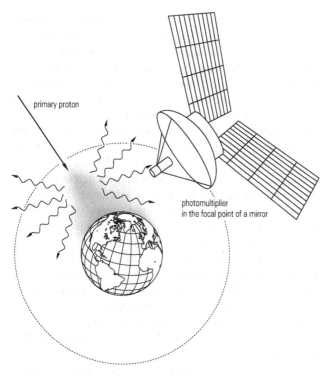

primary proton

photomultiplier
in the focal point of a mirror

Fig. 7.31
Measurement of the isotropic
scintillation light of extensive air
showers by Fly's Eye detectors on
board of satellites ('Air Watch')

the Earth's magnetic field thereby creating synchrotron radiation. Because of the strong background in practically all wavelength ranges these attempts have not been particularly successful so far. The possibility to detect large air showers via their muon content in underground experiments has been followed up in recent experiments.

muon showers underground

Apart from elementary particle physics aspects the purpose of the measurement of extensive air showers is the determination of the chemical composition of primary cosmic rays and the search for the sites of cosmic accelerators.

The arrival directions of the highest-energy particles ($> 10^{19}$ eV) which for intensity reasons can only be recorded via air-shower techniques, practically show no correlation to the galactic plane. This clearly indicates that their origin must be extragalactic. If the highest-energy primary cosmic-ray particles are protons, then their energies must be below 10^{20} eV, if they originate from distances of more than 50 Mpc. Even if their original energy were much higher, they would lose energy by photoproduction of pions on photons of the blackbody radiation until they fall below the threshold of the Greisen–Zatsepin–Kuzmin cutoff

proton horizon

($\approx 6 \times 10^{19}$ eV). Protons of this energy would point back to the sources, because galactic and intergalactic magnetic fields only cause angular distortions on the order of one degree at these high energies. The irregularities of magnetic fields, however, could lead to significant time delays between neutrinos and photons on one hand and protons, on the other hand, from such distant sources. This comes about because the proton trajectories are somewhat longer, even though their magnetic deflection is rather small. Depending on the distance from the source, time delays of months and **extensive air showers** even years can occur. This effect is of particular importance, **and γ-ray bursts** if γ-ray bursters are also able to accelerate the highest-energy particles and if one wants to correlate the arrival times of photons from γ-ray bursts with those of extensive air showers initiated by charged primaries.

energies $> 10^{20}$ TeV The few measured particles with energies in excess of 10^{20} eV show a non-uniform distribution with a certain clustering near the local supergalactic plane. The fact that the attenuation length of protons with energies $> 10^{20}$ eV in the intergalactic space is approximately 10 Mpc would make an origin in the local supercluster (maximum size 30 Mpc) plausible.

particle astronomy? Out of the six measured showers with primary energy exceeding 10^{20} eV the directions of origin for two events are identical within the measurement accuracy. This direction coincides with the position of a radio galaxy (3C134), whose distance unfortunately is unknown, since it lies in the direction of the galactic plane, where optical measurements of extragalactic objects are difficult because of interstellar absorption. The coincidence of the radio galaxy 3C134 with the arrival directions of the two highest-energy particles can, of course, also be an accident.

coincidences Normal extensive air showers have lateral widths of at **over large distances** most 10 km, even at the highest energies. However, there are indications that *correlations between arrival times* of air showers over distances of more than 100 km exist. Such coincidences could be understood by assuming that energetic primary cosmic particles undergo interactions or fragmentations at large distances from Earth. The secondary particles produced in these interactions would initiate separate air showers in the atmosphere (Fig. 7.32).

Even moderate distances of only one parsec (3×10^{16} m) are sufficient to produce separations of air showers at Earth on the order of 100 km (primary energy 10^{20} eV, transverse

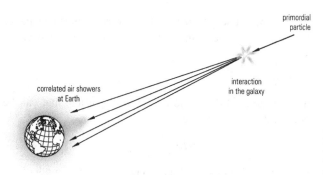

primordial
particle

correlated air showers
at Earth

interaction
in the galaxy

Fig. 7.32
Possible explanation for
correlations between distant
extensive air showers

momenta $\approx 0.3\,\mathrm{GeV}/c$). Variations in arrival times of these
showers could be explained by unequal energies of the frag-
ments which could cause different propagation times. Galac-   **correlated showers**
tic or extragalactic magnetic fields could also affect the trajectories of the fragments in a different way thus also influencing the arrival times.

7.5 Nature and Origin of the Highest-Energy Cosmic Rays

> *"The universe is full of magical things
> patiently waiting for our wits to grow
> sharper."*
>
> *Eden Phillpotts*

As already explained in Sect. 7.4, the highest-energy particles of cosmic rays appear to be of extragalactic origin.
The problem of the sources of these particles is closely related to the identity of these particles. Up to the present time   **identity**
one had always assumed that the chemical composition of   **of high-energy primaries?**
primary cosmic rays might change with energy. However,
one always anticipated that the highest-energy particles were
either protons, light, or possibly medium heavy nuclei (up
to iron). For particles with energies exceeding $10^{20}\,\mathrm{eV}$ this
problem is completely open. In the following the candidates
which might be responsible for cosmic-ray events with energies $> 10^{20}\,\mathrm{eV}$ will be critically reviewed.

Up to now only a handful of events with energies ex-   **events with $E > 10^{20}\,\mathrm{eV}$**
ceeding $10^{20}\,\mathrm{eV}$ have been observed. Due to the measure-
ment technique via extensive-air-shower experiments the
energy assignments are connected with an experimental error of typically $\pm 30\%$. For the accelerated parent particles

of these high-energy particles the gyroradii must be smaller than the size of the source. Therefore, one can derive from

$$\frac{mv^2}{R} \leq evB$$

galactic containment   a maximum value for the energy of a particle that can be accelerated in the source,

$$E_{max} \approx p_{max} \leq eBR \tag{7.19}$$

(v is a particle velocity, B is a magnetic field strength of the source, R is the size of the source, m is the relativistic mass of the particle). In units appropriate for astroparticle physics the maximum energy, which can be obtained by acceleration in the source, can be expressed in the following way:

$$E_{max} = 10^5 \, \text{TeV} \, \frac{B}{3 \times 10^{-6} \, \text{G}} \, \frac{R}{50 \, \text{pc}} \, . \tag{7.20}$$

With a typical value of $B = 3 \, \mu\text{G}$ for our Milky Way and the very generous gyroradius of $R = 5 \, \text{kpc}$ one obtains

$$E_{max} = 10^7 \, \text{TeV} = 10^{19} \, \text{eV} \, . \tag{7.21}$$

This equation implies that our Milky Way can hardly accelerate or store particles of these energies, so that for particles with energies exceeding $10^{20} \, \text{eV}$ one has to assume that they are of extragalactic origin.

protons   For protons the Greisen–Zatsepin–Kuzmin cutoff (GZK) of photoproduction of pions off blackbody photons through the Δ resonance takes an important influence on the propagation,

$$\gamma + p \rightarrow p + \pi^0 \, . \tag{7.22}$$

The energy threshold for this process is at $6 \times 10^{19} \, \text{eV}$ (see Sect. 6.1). Protons exceeding this energy lose rapidly their
mean free path of protons   energy by such photoproduction processes. The mean free path for photoproduction is calculated to be

$$\lambda_{\gamma p} = \frac{1}{N \, \sigma} \, , \tag{7.23}$$

where N is the number density of blackbody photons and $\sigma(\gamma p \rightarrow \pi^0 p) \approx 100 \, \mu\text{b}$ the cross section at threshold. This leads to

$$\lambda_{\gamma p} \approx 10 \, \text{Mpc} \, . \tag{7.24}$$

The Markarian galaxies Mrk 421 and Mrk 501, which have been shown to be sources of photons of the highest energies, would be candidates for the production of high-energy protons. Since they are residing at distances of approximately 100 Mpc, the arrival probability of protons from these distances with energies exceeding 10^{20} eV, however, is only

Markarian galaxies as cosmic-ray source?

$$\approx e^{-x/\lambda} \approx 4 \times 10^{-5} \; . \qquad (7.25)$$

Therefore protons can initiate the high-energy air-shower events only if they come from relatively nearby sources (i.e., from a local GZK sphere defined by distances < 30 Mpc, i.e., several mean free paths). The giant elliptical galaxy M87 lying in the heart of the Virgo cluster (distance ≈ 20 Mpc) is one of the most remarkable objects in the sky. It meets all of the conditions for being an excellent candidate for a high-energy cosmic-ray source.

M87 as particle accelerator?

It is, however, possible to shift the effect of the Greisen–Zatsepin–Kuzmin cutoff to higher energies by assuming that primary particles are nuclei. Since the threshold energy must be available per nucleon, the corresponding threshold energy, for example, for carbon nuclei ($Z = 6$, $A = 12$) would be correspondingly higher,

GZK sphere

$$E = E_{\text{cutoff}}^{p} A = 7.2 \times 10^{20} \, \text{eV} \; , \qquad (7.26)$$

so that the observed events would not be in conflict with the Greisen–Zatsepin–Kuzmin cutoff. It is, however, difficult to understand, how atomic nuclei can be accelerated to such high energies, without being disintegrated by photon interactions or by fragmentation or spallation processes.

heavy nuclei

photo disintegration

One remote and rather drastic assumption to explain the trans-GZK events would be a possible violation of Lorentz invariance. If Lorentz transformations would not only depend on the relative velocity difference of inertial frames, but also on the absolute velocities, the threshold energy for γp collisions for interactions of blackbody photons with high-energy protons would be washed out and different from γp collisions when photon and proton had comparable energies, thus evading the GZK cutoff.

Photons as possible candidates for the observed high-energy cascades are even more problematic. Because of the process of pair production of electrons and positrons off blackbody photons (see Sect. 6.3.3), photons have a relatively short mean free path of

photons

mean free path of photons

$$\lambda_{\gamma\gamma} \approx 10\,\text{kpc}\ . \tag{7.27}$$

The γ-ray sources have to be relatively near to explain the high-energy showers. This would mean that they must be of galactic origin, which appears rather unlikely, because of the limited possibility for their parent particles to be accelerated in our Milky Way up to the highest energies required. High-energy photons, furthermore, would initiate air showers at high altitudes above sea level ($\approx$ 3000 km) due to interactions with the Earth's magnetic field. Therefore one would theoretically expect that they would reach a shower maximum at $\approx$ 1075 g/cm^2 (calculated from sea level). The event observed by the Fly's Eye experiment has a shower maximum at (815 ± 40) g/cm^2, which is typical for a hadron-induced cascade. Photons as candidates for the highest-energy events can therefore be firmly excluded.

neutrinos   Recently, neutrinos were discussed as possible candidates for the high-energy events. But neutrinos also encounter severe problems in explaining such events. The ratio of the interaction cross section for neutrino–air and proton–air interactions at 10^{20} eV is

$$\left.\frac{\sigma(\nu\text{–air})}{\sigma(p\text{–air})}\right|_{E\approx10^{20}\,\text{eV}} \approx 10^{-6}\ . \tag{7.28}$$

Quite enormous neutrino fluxes are required to explain the events with energies $> 10^{20}$ eV. It has been argued that the measurements of the structure function of the protons at HERA[1] have shown that protons have a rich structure of partons at low x ($x = E_{\text{parton}}/E_{\text{proton}}$). Even in view of these results showing evidence for a large number of gluons in the proton, one believes that the neutrino interaction cross section with nuclei of air cannot exceed 0.3 µb. This makes interactions of extragalactic neutrinos in the atmosphere very improbable, compare (3.56):

$$\phi = \sigma(\nu\text{–air})\,\frac{N_A}{A}\,d$$
$$\leq 0.3\,\mu\text{b}\,\frac{6\times10^{23}}{14}\,\text{g}^{-1}\times1000\,\text{g/cm}^2$$
$$\approx 1.3\times10^{-5} \tag{7.29}$$

(N_A is the Avogadro number, d is a column density of the atmosphere).

photonic origin? (margin)

rising neutrino cross section? (margin)

---

[1] HERA – Hadron Elektron Ring Anlage at the Deutsches Elektronensynchrotron (DESY) in Hamburg

To obtain a reasonable interaction rate only neutrino interactions for inclined directions of incidence or in the Earth can be considered. The resulting expected distribution of primary vertices due to neutrino interactions is in contrast to observation. Therefore, neutrinos as well can very likely be excluded as candidates for the highest-energy cosmic air-shower events.

vertex distribution for neutrinos

It has been demonstrated that a large fraction of matter is in the form of dark matter. A possible way out concerning the question of high-energy particles in cosmic rays would be to assume that weakly interacting massive particles (WIMPs) could also be responsible for the observed showers with energies $> 10^{20}$ eV. It has to be considered that all these particles have only weak or even superweak interactions so that their interaction rate can only be on the order of magnitude of neutrino interactions.

WIMPs

The events with energies exceeding 10^{20} eV therefore represent a particle physics dilemma. One tends to assume that protons are the favoured candidates. They must come from relative nearby distances (< 30 Mpc), because otherwise they would lose energy by photoproduction processes and fall below the energy of 6×10^{19} eV. It is, however, true that up to these distances there are quite a number of galaxies (e.g., M87). The fact that the observed events do not clearly point back to a nearby source can be explained by the fact that the extragalactic magnetic fields are so strong that the directional information can be lost, even if the protons are coming from comparably close distances. Actually, there are hints showing that these fields are more in the μGauss rather than in the nGauss region [6].

extragalactic magnetic fields

Recent measurements, however, appear to indicate that the GZK cutoff might have been seen at least in the data of the HiRes experiment (see Fig. 6.5). On the other hand, this finding is in conflict with results from the large AGASA air-shower array (see also the comment on page 83).

Presently one assumes that in supernova explosions particles can only be accelerated to energies of 10^{15} eV by shock-wave mechanisms. At these energies the primary spectrum gets steeper ('knee of the primary spectrum'). As already shown, our Milky Way is too small to accelerate and store particles with energies exceeding 10^{20} eV. Furthermore, the arrival directions of the high-energy particles show practically no correlation to the galactic plane. Therefore, one has to assume that they are of extragalactic origin.

acceleration mechanisms

active galactic nuclei
blazars

BL-Lacertae objects, quasars

energy conversion efficiency

particle jets from blazars

supergalactic origin?

Active galactic nuclei (AGN's) are frequently discussed as possible sources for the highest cosmic-ray energies. In this group of galaxies *blazars* play an outstanding rôle. Blazar is a short for sources belonging to the class of BL-Lacertae objects and quasars. BL-Lacertae objects, equally as quasars, are Milky Way-like sources, whose nuclei outshine the whole galaxy making them to appear like stars. While the optical spectra of quasars exhibit emission and absorption lines, the spectra of BL-Lacertae objects show no structures at all. This is interpreted in such a way that the galactic nuclei of quasars are surrounded by dense gas, while BL-Lacertae objects reside in low-gas-density elliptical galaxies.

A characteristic feature of blazars is their high variability. Considerable brightness excursions have been observed on time scales as short as a few days. Therefore, these objects must be extremely compact, because the size of the sources can hardly be larger than the time required for light to travel across the diameter of the source. It is generally assumed that blazars are powered by black holes at their center. The matter falling into a black hole liberates enormous amounts of energy. While in nuclear fission only 1‰ and in nuclear fusion still only 0.7% of the mass is transformed into energy, an object of mass m can practically liberate all its rest energy mc^2 if it is swallowed by a black hole.

Many high-energy γ-ray sources which were found by the CGRO (Compton Gamma Ray Observatory) satellite, could be correlated with blazars. This led to the conjecture that these blazars could also be responsible for the acceleration of the highest-energy particles. The *particle jets* produced by blazars exhibit magnetic fields of more than 10 Gauss and extend over 10^{-2} pc and more. Therefore, according to (7.20), particles could be accelerated to energies exceeding 10^{20} eV. If protons are accelerated in such sources they could easily escape from these galaxies, because their interaction strength is smaller than that of the electrons which must certainly be accelerated as well. If these arguments are correct, blazars should also be a rich source of high-energy neutrinos. This prediction can be tested with the large water (or ice) Cherenkov counters.

It has already been mentioned before that for protons to arrive at Earth the sources must not be at too large distances. The best candidates for sources should therefore lie in the supergalactic plane. The local supergalaxy is a kind of

'Milky Way' of galaxies whose center lies in the direction of the Virgo cluster. The local group of galaxies, of which our Milky Way is a member, has a distance of about 20 Mpc from the center of this local supergalaxy and the members of this supergalaxy scatter around the supergalactic center only by about 20 Mpc.

Virgo cluster

Even though the origin of the highest-energy cosmic rays is still unknown, there are some hints that the sources for these high-energy events really lie in the supergalactic plane. Certainly more events are required to confirm in detail that such a correlation really exists. The Auger experiment under construction in Argentina should be able to solve the question of the origin of high-energy cosmic rays.

Finally, ideas have also been put forward that the extreme-energy cosmic rays are not the result of the acceleration of protons or nuclei but rather decay products of unstable primordial objects. Candidates discussed as possible sources are decays of massive GUT particles spread through the galactic halo, topological defects produced in the early stages of the universe like domain walls, 'necklaces' of magnetic monopoles connected by cosmic strings, closed cosmic loops containing a superconducting circulating current, or cryptons – relic massive metastable particles born during cosmic inflation.

exotic candidates

7.6 Problems

1. The pressure at sea level is 1013 hPa. Convert this pressure into a column density in kg/cm^2!

2. The barometric pressure varies with altitude h in the atmosphere (assumed to be isothermal) like

$$p = p_0 \, e^{-h/7.99 \, km} \, .$$

What is the residual pressure at 20 km altitude and what column density of residual gas does this correspond to?

3. For not too large zenith angles the angular distribution of cosmic-ray muons at sea level can be parameterized as $I(\theta) = I(0) \cos^2 \theta$. Motivate the $\cos^2 \theta$ dependence!

4. Figure 7.22 shows the rate of stopping muons underground. Work out the rate of stopping atmospheric muons (curve labeled 1) as a function of depth underground for shallow depths!

5. Figure 7.23 shows a muon shower in the ALEPH experiment. Typical energies of muons in this shower are 100 GeV. What is the r.m.s. scattering angle of muons in rock for such muons (overburden 320 m w.e., radiation length in rock $X_0 = 25\,\mathrm{g/cm^2} \cong 10\,\mathrm{cm}$)?

6. Narrow muon bundles with muons of typically 100 GeV originate in interactions of primary cosmic rays in the atmosphere. Estimate the typical lateral separation of cosmic-ray muons in a bundle at a depth of 320 m w.e. underground.

7. Due to the dipole character of the Earth's magnetic field the geomagnetic cutoff varies with geomagnetic latitude. The minimum energy for cosmic rays to penetrate the Earth's magnetic field and to reach sea level can be worked out to be

$$E_{\min} = \frac{ZeM}{4R^2} \cos^4 \lambda \, ,$$

where Z is the charge number of the incident particle, M is the moment of the Earth's magnetic dipole, R is the Earth radius, λ is the geomagnetic latitude ($0°$ at the equator). For protons one gets $E_{\min} = 15\,\mathrm{GeV} \cos^4 \lambda$.

The Earth's magnetic field has reversed several times over the history of our planet. In those periods when the dipole changed polarity, the magnetic field went through zero. In these times when the magnetic shield decayed, more cosmic-ray particles could reach the surface of the Earth causing a higher level of radiation for life developing on our planet. Whether this had a positive effect on the biological evolution or not is the object of much debate. The estimation of the increased radiation level for periods of zero field can proceed along the following lines:

- the differential energy spectrum of primary cosmic rays can be represented by a power law $N(E) \sim E^{-\gamma}$ with $\gamma = 2.7$.
- in addition to the geomagnetic cutoff there is also an atmospheric cutoff due to the energy loss of charged particles in the atmosphere of $\approx 2\,\mathrm{GeV}$.

Work out the increase in the radiation level using the above limits!

8. Neutrons as candidates for the highest-energy cosmic rays have not been discussed so far. What are the problems with neutrons?

8 Cosmology

"As far as the laws of mathematics refer to reality, they are not certain; as far as they are certain, they do not refer to reality."

Albert Einstein

In the following chapters the application of our knowledge of particle physics to the very early universe in the context of the Hot Big Bang model of cosmology will be explored. The basic picture is that the universe emerged from an extremely hot, dense phase about 14 billion years ago. The earliest time about which one can meaningfully speculate is about 10^{-43} seconds after the Big Bang (the Planck time). To go earlier requires a quantum-mechanical theory of gravity, and this is not yet available.

Hot Big Bang

quantum gravity

At early times the particle densities and typical energies were extremely high, and particles of all types were continually being created and destroyed. For the first 10^{-38} seconds or so, it appears that all of the particle interactions could have been 'unified' in a theory containing only a single coupling strength. It was not until after this, when typical particle energies dropped below around 10^{16} GeV, that the strong and electroweak interactions became distinct. At this time, from perhaps 10^{-38} to 10^{-36} seconds after the Big Bang, the universe may have undergone a period of *inflation*, a tremendous expansion where the distances between any two elements of the primordial plasma increased by a factor of perhaps e^{100}. When the temperature of the universe dropped below 100 GeV, the electroweak unification broke apart into separate electromagnetic and weak interactions.

inflation

Until around 1 microsecond after the Big Bang, quarks and gluons could exist as essentially free particles. After this point, energies dropped below around 1 GeV and the partons became bound into hadrons, namely, protons, neutrons, and their antiparticles. Had the universe contained at this point equal amounts of matter and antimatter, almost all of it would have annihilated, leaving us with photons, neutrinos, and little else. For whatever reason, nature apparently made one a bit more abundant than the other, so there was some matter left over after the annihilation phase to make

annihilation of matter and antimatter

annihilation phase

the universe as it is now. Essentially all of the positrons had annihilated with electrons within the first couple of seconds.

Around three minutes after the Big Bang, the temperature had dropped to the point where protons and neutrons could fuse to form deuterons. In the course of the next few minutes these combined to form helium, which makes up a quarter of the universe's nuclear matter by mass, and smaller quantities of a few light elements such as deuterium, lithium, **Big Bang Nucleosynthesis** and beryllium. The model of Big Bang Nucleosynthesis (BBN) is able to correctly predict the relative abundances of these light nuclei and this is one of the cornerstones of the Hot Big Bang model.

As the universe continued to expand over the next several hundred thousand years, the temperature finally dropped to the point where electrons and protons could join to form neutral atoms. After this the universe became essentially transparent to photons, and those which existed at that time have been drifting along unimpeded ever since. They can be **cosmic microwave** detected today as the cosmic microwave background radia-**background radiation** tion. Only small variations in the temperature of the radiation, depending on the direction, at a level of one part in 10^5 are observed. These are thought to be related to small density variations in the universe left from a much earlier period, perhaps as early as the inflationary epoch only 10^{-36} seconds after the Big Bang.

Studies of the cosmic microwave background radiation (CMB) also lead to a determination of the total density of **critical density** the universe, and one finds a value very close to the so-called critical density, above which the universe should recollapse **'Big Crunch'** in a 'Big Crunch'. The same CMB data and also observations of distant supernovae, however, show that about 70% of this is not what one would call matter at all, but rather a sort of energy density associated with empty space – a vac-**vacuum energy density** uum energy density.

The remaining 30% appears to be gravitating matter, but of what sort? One of the indirect consequences of Big Bang Nucleosynthesis is that only a small fraction of the matter in the universe appears to be composed of known particles. The **dark matter** remainder of the *dark matter* may consist of *neutralinos*, **neutralinos** particles predicted by a theory called supersymmetry.

The framework in which the early universe will be studied is based on the 'Standard Cosmological Model' or 'Hot Big Bang'. The basic ingredients are Einstein's theory of general relativity and the hypothesis that the universe is isotropic and homogeneous when viewed over sufficiently large distances. It is in the context of this model that the laws of particle physics will be applied in an attempt to trace the evolution of the universe at very early times. In this chapter the important aspects of cosmology that one needs will be reviewed.

Standard Cosmological Model

8.1 The Hubble Expansion

"The history of astronomy is a history of receding horizons."

Edwin Powell Hubble

The first important observation that leads to the Standard Cosmological Model is Hubble's discovery that all but the nearest galaxies are receding away from us (i.e., from the Milky Way) with a speed proportional to their distance. The speeds are determined from the Doppler shift of spectral lines. Suppose a galaxy receding from us (i.e., from the Milky Way) with a speed $v = \beta c$ emits a photon of wavelength λ_{em}. When the photon is observed, its wavelength will be shifted to λ_{obs}. To quantify this, the *redshift z* is defined as

redshift

$$z = \frac{\lambda_{obs} - \lambda_{em}}{\lambda_{em}} . \tag{8.1}$$

From relativity one obtains the relation between the redshift and the speed,

$$z = \sqrt{\frac{1 + \beta}{1 - \beta}} - 1 , \tag{8.2}$$

which can be approximated by

$$z \approx \beta \tag{8.3}$$

for $\beta \ll 1$.

To measure the distance of a galaxy one needs in it a light source of a calibrated brightness (a 'standard candle'). The light flux from the source falls off inversely as the square of the distance r, so if the absolute luminosity L is known and the source radiates isotropically, then the measured light flux is $F = L/4\pi r^2$. The *luminosity distance* can therefore be determined from

'standard candle'

luminosity distance

$$r = \sqrt{\frac{L}{4\pi F}} \,. \tag{8.4}$$

Various standard candles can be used, such as Cepheid variable stars, used by Hubble. A plot of speed versus distance determined from type-Ia supernovae[1] is shown in Fig. 8.1 [7].

type-Ia supernovae

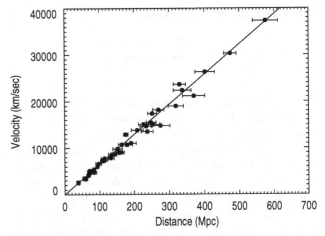

Fig. 8.1
The speed versus distance for a sample of type-Ia supernovae (from [7])

---

[1] The spectra of SN Ia are hydrogen poor. The absence of planetary nebulae allows to reconstruct the genesis of these events. It is generally believed that the progenitor of a SN Ia is a binary consisting of a white dwarf and a red giant companion. Both members are gravitationally bound. In white dwarves the electron degeneracy pressure compensates the inward bound gravitational pressure. The strong gravitational potential of the white dwarf overcomes the weaker gravity of the red giant. At the periphery of the red giant the gravitational force of the white dwarf is stronger than that of the red giant causing mass from its outer envelope to be accreted onto the white dwarf. Since for white dwarves the product of mass times volume is constant, it decreases in size during accretion. When the white dwarf reaches the Chandrasekhar limit ($1.44\,M_\odot$), the electron degeneracy pressure can no longer withstand the gravitational pressure. It will collapse under its own weight. This goes along with an increase in temperature causing hydrogen to fuse to helium and heavier elements. This sudden burst of energy leads to a thermonuclear explosion which destroys the star.

Since the Chandrasekhar limit is a universal quantity, all SN Ia explode in the same way. Therefore, they can be considered as standard candles.

For the distances covered in the plot, the data are clearly in good agreement with a linear relation,

$$v = H_0\, r \ , \tag{8.5}$$

which is Hubble's Law. The parameter H_0 is Hubble's constant, which from the data in Fig. 8.1 is determined to be 64 (km/s)/Mpc. Here the subscript 0 is used to indicate the value of the parameter today. The relation between speed and distance is not, however, constant in time.   **Hubble's constant**

Determinations of the Hubble constant based on different observations have yielded inconsistent results, although less so now than a decade ago. A compilation of data from 2004 concludes a value of

$$H_0 = 71 \pm 5 \,(\text{km/s})/\text{Mpc} \ , \tag{8.6}$$

with further systematic uncertainties of 5 to 10% [2]. In addition, one usually defines the parameter h by   **h parameter**

$$H_0 = h \times 100 \,(\text{km/s})/\text{Mpc} \ . \tag{8.7}$$

Quantities that depend on H_0 are then written with the corresponding dependence on h. To obtain a numerical value one substitutes for h the most accurate estimate available at the time ($h = 0.71^{+0.04}_{-0.03}$ in 2004). For purposes of this course one can use $h \approx 0.7 \pm 0.1$.

8.2 The Isotropic and Homogeneous Universe

> *"The center of the universe is everywhere, and the circumference is nowhere."*
>
> *Giordano Bruno*

The assumption of an isotropic and homogeneous universe, sometimes called the *cosmological principle*, was initially made by Einstein and others because it simplified the mathematics of general relativity. Today there is ample observational evidence in favour of this hypothesis. The cosmic microwave background radiation, for example, is found to be isotropic to a level of around one part in 10^5.   **cosmological principle**

isotropy and homogeneity

The isotropy and homogeneity appear today to hold at sufficiently large distances, say, greater than around 100 megaparsecs. At smaller distances, galaxies are found to

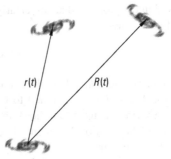

any galaxy
(e.g., Milky Way)

Fig. 8.2
Two galaxies at distances $r(t)$ and
$R(t)$ from our own

clump together forming clusters and voids. A typical inter-galactic distance is on the order of 1 Mpc, so a cube with sides of 100 Mpc could have a million galaxies. Therefore, one should think of the galaxies as the 'molecules' of a gas, which is isotropic and homogeneous when a volume large enough to contain large numbers of them is considered.

If the universe is assumed to be isotropic and homogeneous, then the only possible motion is an overall expansion or contraction. Consider, for example, two randomly chosen galaxies at distances $r(t)$ and $R(t)$ from ours, as shown in Fig. 8.2.

distance scale An isotropic and homogeneous expansion (or contraction) means that the ratio

$$\chi = r(t)/R(t) \tag{8.8}$$

is constant in time. Therefore, $r(t) = \chi R(t)$ and

$$\dot{r} = \chi \dot{R} = \frac{\dot{R}}{R} r \equiv H(t)r \, , \tag{8.9}$$

where dots indicate derivatives with respect to time. The ratio

$$H(t) = \dot{R}/R \tag{8.10}$$

Hubble parameter is called the *Hubble parameter*. It is the fractional change in the distance between any pair of galaxies per unit time. H is **expansion rate** often called the *expansion rate* of the universe.

Equation (8.9) is exactly Hubble's law, where $H(t)$ at the present time is identified as the Hubble constant H_0. So the hypothesis of an isotropic and homogeneous expansion explains why the speed with which a galaxy moves away from us is proportional to its distance.

8.3 The Friedmann Equation from Newtonian Gravity

"No theory is sacred."

Edwin Powell Hubble

The evolution of an isotropic and homogeneous expansion is completely determined by giving the time dependence of the distance between any representative pair of galaxies. One can denote this distance by $R(t)$, which is called the *scale factor*. An actual numerical value for R is not important. For example, one can define $R = 1$ at a particular time (e.g., now). It is the time dependence of R that gives information about how the universe as a whole evolves.

Friedmann equation

scale factor

The rigorous approach would now be to assume an isotropic and homogeneous matter distribution and to apply the laws of general relativity to determine $R(t)$. By fortunate coincidence, in this particular problem Newton's theory of gravity leads to the same answer, namely, to the Friedmann equation for the scale factor R. This approach will now be briefly reviewed.

Consider a spherical volume of the universe with a radius R sufficiently large to be considered homogeneous, as shown in Fig. 8.3. In today's universe this would mean taking R at least 100 Mpc. If one assumes that the universe is electrically neutral, then the only force that is significant over these distances is gravity. As a test mass, consider a galaxy of mass m at the edge of the volume. It feels the gravitational attraction from all of the other galaxies inside. As a consequence of the inverse-square nature of gravity, this

cosmological model

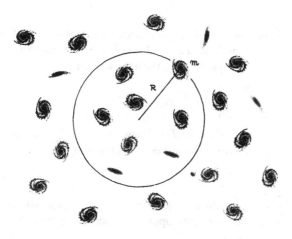

Fig. 8.3
A sphere of radius R containing many galaxies, with a test galaxy of mass m at its edge

force is the same as what one would obtain if all of the mass inside the sphere were placed at the center.

Another non-trivial consequence of the $1/r^2$ force is that the galaxies outside the sphere do not matter. Their total gravitational force on the test galaxy is zero. In Newtonian gravity these properties of isotropically distributed matter inside and outside a sphere follow from Gauss's law for a $1/r^2$ force. The corresponding law holds in *general relativity* as well, where it is known as Birkhoff's theorem.

Newtonian gravity

Birkhoff's theorem

If one assumes that the mass of the galaxies is distributed in space with an average density ϱ, then the mass inside the sphere is

$$M = \frac{4}{3}\pi R^3 \varrho \ . \tag{8.11}$$

The gravitational potential energy V of the test galaxy is therefore

$$V = -\frac{GmM}{R} = -\frac{4\pi}{3}GmR^2\varrho \ . \tag{8.12}$$

The sum of the kinetic energy T and potential energy V of the test galaxy gives its total energy E,

$$E = \frac{1}{2}m\dot{R}^2 - \frac{4\pi}{3}GmR^2\varrho = \frac{1}{2}mR^2\left(\frac{\dot{R}^2}{R^2} - \frac{8\pi}{3}G\varrho\right) \ . \tag{8.13}$$

curvature parameter

The *curvature parameter* k is now defined by

$$k = \frac{-2E}{m} = R^2\left(\frac{8\pi}{3}G\varrho - \frac{\dot{R}^2}{R^2}\right) \ . \tag{8.14}$$

If one were still dragging along the factors of c, k would have been defined as $-2E/mc^2$; in either case k is dimensionless. Equation (8.14) can be written as

$$\frac{\dot{R}^2}{R^2} + \frac{k}{R^2} = \frac{8\pi}{3}G\varrho \ , \tag{8.15}$$

Friedmann equation

which is called the *Friedmann equation*. The terms in this equation can be identified as representing

$$T - E = -V \ , \tag{8.16}$$

energy conservation

i.e., the Friedmann equation is simply an expression of conservation of energy applied to our test galaxy. Since the sphere and test galaxy could be anywhere in the universe, the equation for R applies to any pair of galaxies sufficiently far apart that one can regard the intervening matter as being homogeneously distributed.

The Friedmann equation can also be applied to the early **early universe** universe, before the formation of galaxies. It will hold even for an ionized plasma as long as the universe is electrically neutral overall and one averages over large enough distances. The scale factor R in that case represents the distance between any two elements of matter sufficiently far apart such that gravity is the only force that does not cancel out.

8.4 The Friedmann Equation from General Relativity[2]

> *"Since relativity is a piece of mathematics, popular accounts that try to explain it without mathematics are almost certain to fail."*
>
> Eric Rogers

As mentioned above, the Friedmann equation can also be obtained from general relativity. Even though this approach is mathematically demanding, this elegant method will be sketched in the following. More detailed presentations are given in the literature [8, 9].

In general relativity the scale factor $R(t)$ enters as a factor in the *metric tensor* $g_{\mu\nu}$. This relates the space-time in-   **metric tensor** terval ds^2 to changes in coordinates x^μ by

$$ds^2 = g_{\mu\nu}\, dx^\mu\, dx^\nu\,, \tag{8.17}$$

where $\mu, \nu = 0, 1, 2, 3$ and summation over repeated indices is implied. The most general form of the metric tensor for an isotropic and homogeneous universe is given by the   **Robertson–Walker metric** *Robertson–Walker* metric,

$$ds^2 = dt^2$$
$$- R^2(t)\left[\frac{dr^2}{1 - kr^2} + r^2\left(d\theta^2 + \sin^2\theta\, d\phi^2\right)\right]\,, \tag{8.18}$$

---

[2] This section is mathematically demanding and should be skipped by those readers, who are not familiar with tensors.

where k is the curvature parameter and R the scale factor. The metric tensor is obtained as a solution to the Einstein field equations,

$$\mathcal{R}_{\mu\nu} - \frac{1}{2}\mathcal{R}g_{\mu\nu} = 8\pi G T_{\mu\nu} + \Lambda g_{\mu\nu} , \qquad (8.19)$$

where $\mathcal{R}_{\mu\nu}$ is the Ricci tensor, $\mathcal{R}$ is the Ricci scalar (not to be confused with the scale factor R), both of which are spe-

cosmological constant   cific functions of $g_{\mu\nu}$. On the right-hand side, Λ is the cosmological constant and $T_{\mu\nu}$ is the energy–momentum tensor.

An isotropic and homogeneous universe implies that $T_{\mu\nu}$ is of the form

$$T_{\mu\nu} = \mathrm{diag}(\varrho, -P, -P, -P) , \qquad (8.20)$$

energy–momentum tensor   where ϱ is the energy density, P is the pressure, and 'diag' means a square matrix with diagonal elements given by (8.20) and zeros everywhere else.

Combining the Robertson–Walker metric and the $T_{\mu\nu}$ from (8.20) together with the field equations results in a differential equation for the scale factor R,

$$\frac{\dot{R}^2}{R^2} + \frac{k}{R^2} = \frac{8\pi}{3}G\varrho + \frac{\Lambda}{3} . \qquad (8.21)$$

This is essentially the Friedmann equation that was found from Newtonian gravity, but with three differences. First, ϱ here represents the energy density, not just the mass density. It includes all forms of energy, including, for example, photons.

Second, the curvature parameter k really represents the curvature of space, which is why it has this name. In the Newtonian case k was simply a measure of the total energy.

Third, there is an additional term in the equation from the cosmological constant Λ. It can be absorbed into ϱ by

vacuum energy density   defining the *vacuum energy density* ϱ_v as

$$\varrho_v = \frac{\Lambda}{8\pi G} \qquad (8.22)$$

and then regarding the complete energy density to include ϱ_v. It is called the vacuum energy because such a term is predicted by quantum mechanics to arise from virtual particles that 'fluctuate' in and out of existence from the vacuum.

This phenomenon is present in quantum field theories such as the Standard Model of particle physics. In certain circumstances it can even be observed experimentally, such as in the *Casimir effect*, where virtual particles from the vacuum lead to an attractive force between two metal plates with a very small separation, see Fig. 8.4.[3] Nevertheless, naïve estimates of the magnitude of Λ from quantum field theories lead to values that are too large by over 120 orders of magnitude!

A more classical approach to obtain the generalized differential equation for the scale factor is to start from (8.15) and interpret the density ϱ as the sum of the classical density and a vacuum energy density ϱ_v defined by (8.22).

There is clearly much that is not understood about the cosmological constant, and Einstein famously regretted proposing it. The more modern view of physical laws, however, leads one to believe that when a term is absent from an equation, it is usually because its presence would violate some symmetry principle. The cosmological constant is consistent with the symmetries on which general relativity is based. And in the last several years clear evidence from the redshifts of distant supernovae has been presented that the cosmological constant is indeed non-zero and that vacuum energy makes a large contribution to the total energy density of the universe. In Chaps. 12 and 13 this will be discussed in more detail.

Casimir effect

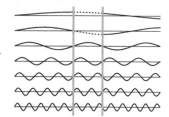

Fig. 8.4
Illustration of the Casimir effect: Only certain wavelengths fit into the space between the plates. The outside of the plates does not limit the number of possible frequencies

symmetry principle

---

[3] In the framework of quantum mechanics it is usually demonstrated that the harmonic oscillator has a non-zero zero-point energy. Also in quantum field theories the vacuum is not empty. Heisenberg's uncertainty relation suggests that the vacuum contains infinitely many virtual particle–antiparticle pairs. The evidence that such quantum fluctuations exist was demonstrated by the Casimir effect. Consider two parallel metal plates at a very small separation in vacuum. Because the distance is so small, not every possible wavelength can exist in the space between the two plates, quite in contrast to the surrounding vacuum. The effect of this limited choice of quanta between the plates leads to a small attractive force of the plates. This pressure of the surrounding vacuum was experimentally confirmed. This force can be imagined in such a way that the reduced number of field quanta in the space between the plates cannot resist the pressure of the unlimited number of field quanta in the surrounding vacuum.

8.5 The Fluid Equation

*"The universe is like a safe to which
there is a combination – but the combination is locked in the safe."*

Peter de Vries

The Friedmann equation cannot be solved yet because one
does not know how the energy density ϱ varies with time. Instead, in the following a relation between ϱ, its time derivative $\dot{\varrho}$, and the pressure P will be derived. This relation,
fluid equation   called *fluid equation*, follows from the *first law of thermodynamics* for a system with energy U, temperature T, entropy
S, and volume V,

$$dU = T \, dS - P \, dV \ . \tag{8.23}$$

The first law of thermodynamics will now be applied to a
volume R^3 in our expanding universe. Since by symmetry
there is no net heat flow across the boundary of the volume,
adiabatic expansion   one has $dQ = T \, dS = 0$, i.e., the expansion is adiabatic.
Dividing (8.23) by the time interval dt then gives

$$\frac{dU}{dt} + P \frac{dV}{dt} = 0 \ . \tag{8.24}$$

The total energy U is

$$U = R^3 \varrho \ . \tag{8.25}$$

The derivative dU/dt is therefore

$$\frac{dU}{dt} = \frac{\partial U}{\partial R} \dot{R} + \frac{\partial U}{\partial \varrho} \dot{\varrho} = 3R^2 \varrho \dot{R} + R^3 \dot{\varrho} \ . \tag{8.26}$$

For the second term in (8.24) one gets

$$\frac{dV}{dt} = \frac{d}{dt} R^3 = 3R^2 \dot{R} \ . \tag{8.27}$$

Putting (8.26) and (8.27) into (8.24) and rearranging terms
gives

$$\dot{\varrho} + \frac{3\dot{R}}{R}(\varrho + P) = 0 \ , \tag{8.28}$$

which is the *fluid equation*. Unfortunately, this is still not
equation of state   enough to solve the problem, since an *equation of state* relating ϱ and P is needed. This can be obtained from the laws
of statistical mechanics as will be shown in Sect. 9.2. With
these ingredients the Friedmann equation can then be used
to find $R(t)$.

8.6 The Acceleration Equation

> *"Observations always involve theory."*
> *Edwin Powell Hubble*

In a number of cases it can be useful to combine the Friedmann and fluid equations to obtain a third equation involving the second derivative $\ddot{R}$. Here only the relevant steps will be outlined and the results will be given; the derivation is straightforward (see Problem 4 in this chapter).

If the Friedmann equation (8.15) is multiplied by R^2 and then differentiated with respect to time, one will obtain an equation involving $\ddot{R}$, $\dot{R}$, R, and $\dot{\varrho}$. The fluid equation (8.28) can then be solved for $\dot{\varrho}$ and substituted into the derivative of the Friedmann equation. This gives

$$\frac{\ddot{R}}{R} = -\frac{4\pi G}{3}(\varrho + 3P) , \qquad (8.29)$$

which is called the *acceleration equation*. It does not add any new information beyond the Friedmann and fluid equations from which it was derived, but in a number of problems it will provide a more convenient path to a solution.

acceleration equation

8.7 Nature of Solutions to the Friedmann Equation

> *"In every department of physical science there is only so much science, properly so-called, as there is mathematics."*
> *Immanuel Kant*

Without explicitly solving the Friedmann equation one can already make some general statements about the nature of possible solutions. The Friedmann equation (8.15) can be written as

$$H^2 = \frac{8\pi G}{3}\varrho - \frac{k}{R^2} , \qquad (8.30)$$

where, as always, $H = \dot{R}/R$. From the observed redshifts of galaxies, it is known that the current expansion rate H is positive. One expects, however, that the galaxies should be slowed by their gravitational attraction. One can therefore

decelerated expansion?

ask whether this attraction will be sufficient to slow the expansion to a halt, i.e., whether H will ever decrease to zero.

If the curvature parameter k is negative, then this cannot happen, since everything on the right-hand side of (8.30) is positive. Recall that $k = -2E/m$ basically gives the total energy of the test galaxy on the edge of the sphere of galaxies. Having $k < 0$ means that the total energy of the test galaxy is positive, i.e., it is not gravitationally bound. In this **open universe** case the universe is said to be *open*; it will continue to expand forever.

If, for example, the energy density of the universe is dominated by non-relativistic matter, then one will find that ϱ decreases as $1/R^3$. So, eventually, the term with the curvature parameter will dominate on the right-hand side of (8.30) **eternal expansion** leading to

$$\frac{\dot{R}^2}{R^2} = -\frac{k}{R^2} , \tag{8.31}$$

which means $\dot{R}$ is constant or $R \sim t$.

If, on the other hand, one has $k > 0$, then as ϱ decreases, the first and second terms on the right-hand side of (8.30) will eventually cancel with the results of $H = 0$. That is, the **contraction of the universe** expansion will stop. At this point, all of the kinetic energy of the galaxies is converted to gravitational potential energy, just as when an object thrown vertically into the air reaches its highest point. And just as with the thrown object, the motion then reverses and the universe begins to contract. In this **closed universe** case the universe is said to be *closed*.

One can also ask what happens if the curvature parameter k and hence also the energy E of the test galaxy are zero, i.e., if the universe is just on the borderline between being open and closed. In this case the expansion will be decelerated but H will approach zero asymptotically. The **flat universe** universe is then said to be *flat*. This is analogous to throwing a projectile upwards with a speed exactly equal to the escape velocity.

Which of the three scenarios one obtains – open, closed, or flat – depends on what is in the universe to slow or otherwise affect the expansion. To see how this corresponds to the energy density, the Friedmann equation (8.15) can be solved for the curvature parameter k, which gives

$$k = R^2 \left(\frac{8\pi}{3} G\varrho - H^2 \right) . \tag{8.32}$$

One can then define the *critical density* ϱ_c by

$$\varrho_c = \frac{3H^2}{8\pi G} \qquad (8.33)$$

and express the energy density ϱ by giving the ratio Ω,   **Ω parameter**

$$\Omega = \frac{\varrho}{\varrho_c} . \qquad (8.34)$$

Using this with $H = \dot{R}/R$, (8.32) becomes

$$k = R^2(\Omega - 1)H^2 . \qquad (8.35)$$

So, if $\Omega < 1$, i.e., if the density is less than the critical density, then $k < 0$ and the universe is open; the expansion continues forever. Similarly, if $\Omega > 1$, the universe is closed; H will decrease to zero and then become negative. If the density ϱ is exactly equal to the critical density ϱ_c, then one has $k = 0$, corresponding to $\Omega = 1$, and the universe is flat.

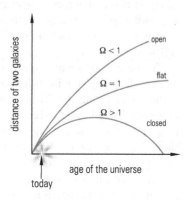

Fig. 8.5
The scale factor R as a function of time for $\Omega < 1$, $\Omega > 1$, and $\Omega = 1$

Figure 8.5 illustrates schematically how the scale factor R depends on time for the three scenarios. The names closed, open, and flat refer to the geometrical properties of space-time that one finds in the corresponding solutions using general relativity.

A completely different type of solution will be found if the total energy density is dominated by vacuum energy. In that case the expansion increases exponentially. This scenario will be dealt with in Chap. 12 when inflation will be   **inflation** discussed.

8.8 Experimental Evidence
for the Vacuum Energy

> *"It happened five billion years ago. That was when the Universe stopped slowing down and began to accelerate, experiencing a cosmic jerk."*
>
> *Adam Riess*

deceleration parameter

acceleration parameter

negative pressure

dark energy

For a given set of contributions to the energy density of the universe, the Friedmann equation predicts the scale factor, $R(t)$, as a function of time. From an observational standpoint, one would like to turn this around: from measurements of $R(t)$ one can make inferences about the contents of the universe. Naïvely one would expect the attractive force of gravity to slow the Hubble expansion, leading to a deceleration, i.e., $\ddot{R} < 0$. One of the most surprising developments of recent years has been the discovery that the expansion is, in fact, accelerating, and apparently has been for several billion years. This can be predicted by the Friedmann equation if one assumes a contribution to the energy density with *negative pressure*, such as the vacuum energy previously mentioned. Such a contribution to ϱ is sometimes called *dark energy*.

Because of the finite speed of light, observations of galaxies far away provide information about the conditions of the universe long ago. By accurately observing the motion of very distant galaxies and comparing to that of those closer to us, one can try to discern whether the universe's expansion is slowing down or speeding up. From each observed galaxy, two pieces of information are required: its speed of recession and its distance. As type-Ia supernovae are extremely bright and therefore can be found over cosmological distances, they are well suited for this type of study. To obtain the speed, the redshift z of spectral lines can be used.

type-Ia supernovae

As type-Ia supernovae have to good approximation a constant absolute luminosity, the apparent brightness provides information on their distance. This is essentially the same type of analysis that was carried out to determine the Hubble parameter H_0 as described in Sect. 8.1. Here, however, one is interested in pushing the measurement to distances so far that the rate of expansion itself may have changed in the time that it took for the light to arrive at Earth.

relation between brightness and redshift

The relation between the apparent brightness of a supernova and its redshift can be obtained once one knows the scale factor as a function of time. This in turn is determined by the Friedmann equation once the contributions to the energy density are specified. Suppose the universe contains (non-relativistic) matter and vacuum energy; the latter could be described by a cosmological constant Λ. Their current energy densities divided by the critical density ϱ_c can be written as $\Omega_{m,0}$ and $\Omega_{\Lambda,0}$, respectively. Here as usual the subscripts 0 denote present-day values. One should also consider the energy density of radiation, $\Omega_{r,0}$, from photons and neutrinos. Such a contribution is well determined from measurements of the cosmic microwave background and is very small compared to the other terms for the time period relevant to the observations. So neglecting the relativistic particles, one can show that the *luminosity distance*, d_L, is given by

cosmological constant

luminosity distance

$$d_L(z) = \frac{1+z}{H_0} \int_0^z \left[\Omega_{\Lambda,0} + (1+z')^3 \Omega_{m,0} \right.$$
$$\left. + (1+z')^2 (1 - \Omega_0) \right]^{-1/2} dz' . \quad (8.36)$$

It is convenient to carry out the integral numerically, yielding $d_L(z)$ for any hypothesized values of $\Omega_{m,0}$ and $\Omega_{\Lambda,0}$.

Recall the luminosity distance is defined by $F = L/4\pi d_L^2$, where F is the flux one measures at Earth and L is the intrinsic luminosity of the source. Astronomers usually replace F and L by the apparent and absolute *magnitudes*, m and M, which are related to the base-10 logarithm of F and L, respectively (precise definitions can be found in standard astronomy texts such as [10] and in the glossary). The observation of a supernova allows one to determine its apparent magnitude, m. The absolute magnitude, M, is unknown *a priori*, but is assumed to be the same (after some corrections and adjustments) for all type-Ia supernovae. These quantities are related to the luminosity distance by

magnitudes

$$m = 5 \log_{10} \left(\frac{d_L}{1\,\text{Mpc}} \right) + 25 + M . \quad (8.37)$$

Notice that a higher apparent magnitude m corresponds to a fainter supernova, i.e., one further away.

A plot of the apparent magnitude m of a sample of distant supernovae versus the redshift is shown in Fig. 8.6 [11]. The data points at low z determine the Hubble constant H_0. The relation at higher z, however, depends on the matter and dark-energy content of the universe. The various curves on the plot show the predictions from (8.36) and (8.37) for different values of $\Omega_{m,0}$ and $\Omega_{\Lambda,0}$. The curve with no dark

Hubble diagram

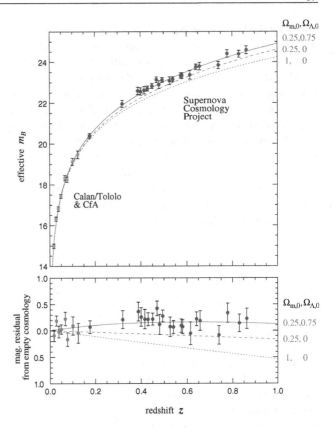

Fig. 8.6
Magnitudes and residuals of
supernovae of type Ia as a function
of redshift of their host galaxies in
comparison to the expectation of
various models. The data are
consistent with a flat universe with
a fraction of about 75% of dark
energy. Shown are data from the
Supernova Cosmology Project, the
Calan/Tololo group, and the
Harvard–Smithsonian Center for
Astrophysics (CfA) {26}

energy, $\Omega_{m,0} = 1$, i.e., $\Omega_{\Lambda,0} = 0$, is in clear disagreement. The data are well described by $\Omega_{m,0} \approx 0.25$ and
$\Omega_{\Lambda,0} \approx 0.75$. Further measurements of the cosmic microwave background support this view. This picture will be
completed in Chaps. 11 and 13.

Qualitatively the behaviour shown in Fig. 8.6 can be understood in the following way. The data points at high red-

accelerating universe   shift lie above the curve for zero vacuum energy, i.e., at
higher magnitudes, which means that they are dimmer than
expected. Thus the supernovae one sees at a given z are farther away than one would expect, therefore the expansion
must be speeding up.

The exact physical origin of the vacuum energy remains
a mystery. On the one hand, vacuum energy is expected in
a quantum field theory such as the Standard Model of elementary particles. Naïvely, one would expect its value to be

vacuum energy   of the order

$$\varrho_v \approx E_{max}^4 , \tag{8.38}$$

where $E_{\max}$ is the maximum energy at which the field theory is valid. One expects the Standard Model of particle physics to be incomplete, for example, at energies higher than the Planck energy, $E_{Pl} \approx 10^{19}$ GeV, where quantum-gravitational effects come into play. So the vacuum energy might be roughly

$$\varrho_v \approx E_{Pl}^4 \approx 10^{76}\,\mathrm{GeV}^4 \,. \tag{8.39}$$

But from the observed present-day value $\Omega_{\Lambda,0} \approx 0.7$, the vacuum energy density is

discrepancy between expectation and observation

$$\varrho_{\Lambda,0} = \Omega_{\Lambda,0}\varrho_{c,0} = \Omega_{\Lambda,0}\frac{3H_0^2}{8\pi G} \approx 10^{-46}\,\mathrm{GeV}^4 \,. \tag{8.40}$$

The discrepancy between the naïve prediction and the observed value is 122 orders of magnitude.

So something has clearly gone wrong with the prediction. One could argue that not much is understood about the physics at the Planck scale, which is surely true, and so a lower energy cutoff should be tried. Suppose one takes the maximum energy at the electroweak scale, $E_{EW} \approx 100$ GeV, roughly equal to the masses of the W and Z bosons. At these energies the Standard Model has been tested to high accuracy. The prediction for the vacuum energy density becomes

$$\varrho_v \approx E_{EW}^4 \approx 10^8\,\mathrm{GeV}^4 \,. \tag{8.41}$$

Now the discrepancy with the observational limit is 'only' 53 orders of magnitude. Perhaps an improvement but clearly not enough.

Finally one might argue that the entire line of reasoning which predicts vacuum energy might be incorrect. But phenomena such as the *Casimir effect* discussed in Sect. 8.4 have been observed experimentally and provide an important confirmation of this picture. So one cannot dismiss the vacuum energy as fiction; there must be some other reason why its contribution to Ω is much less than expected. This is currently one of the most gaping holes in our understanding of the universe.

Casimir effect

8.9 Problems

1. Derive the relativistic relation (8.2) between redshift and velocity of a receding galaxy.
2. A gas cloud gets unstable if the gravitational energy exceeds the thermal energy of the molecules constituting the cloud, i.e.,

$$\frac{GM^2}{R} > \frac{3}{2}kT\frac{M}{\mu},$$

where $\frac{3}{2}kT$ is the thermal energy of a molecule, $\frac{M}{\mu}$ is the number of molecules, and μ is the molecule mass. Derive from this condition the stability limit of a gas cloud (Jeans criterion).

3. Let us assume that a large astrophysical object of constant density not stabilized by internal pressure is about to contract due to gravitation. Estimate the minimum rotational velocity so that this object is stabilized against gravitational collapse! How does the rotational velocity depend on the distance from the galactic center?

4. Derive the acceleration equation (8.29) from the Friedmann equation (8.15) with the help of the fluid equation (8.28).

5. Show that the gravitational redshift of light emitted from a massive star (mass M) of radius R is

$$\frac{\Delta \nu}{\nu} = \frac{GM}{c^2 R}.$$

6. Estimate the classical value for the deflection of starlight passing near the Sun.

7. Clocks in a gravitational potential run slow relative to clocks in empty space. Estimate the slowing-down rate for a clock on a pulsar ($R = 10\,\text{km}$, $M = 10^{30}\,\text{kg}$)!

8. Estimate the gravitational pressure at the center of the Sun (average density $\varrho = 1.4\,\text{g/cm}^3$) and the Earth ($\varrho = 5.5\,\text{g/cm}^3$).

9. Estimate the average density of
   a) a large black hole residing at the center of a galaxy of 10^{11} solar masses,
   b) a solar-mass black hole,
   c) a mini black hole ($m = 10^{15}\,\text{kg}$).

10. The orbital velocity v of stars in our galaxy varies up to distances of 20 000 light-years as if the density were homogeneous and constant ($\varrho = 6 \times 10^{-21}\,\text{kg/m}^3$). For larger distances the velocities of stars follow the expectation from Keplerian motion.
    a) Work out the dependence of $v(R)$ for $R < 20\,000$ light-years.
    b) Estimate the mass of the Milky Way.
    c) The energy density of photons is on the order of $0.3\,\text{eV/cm}^3$. Compare the critical density to this number!

9 The Early Universe

> *"Who cares about half a second after the*
> *Big Bang; what about the half second be-*
> *fore?"*
>
> *Fay Weldon quoted by Paul Davies*

In this chapter the history of the universe through the first ten microseconds of its existence will be described. First, in Sect. 9.1 the *Planck scale*, where quantum-mechanical and gravitational effects both become important, will be defined. This sets the starting point for the theory to be described. In Sect. 9.2 some formulae from statistical and thermal physics will be assembled which are needed to describe the hot dense phase out of which the universe then evolved. Then, in Sects. 9.3 and 9.4 these formulae will be used to solve the Friedmann equation and to investigate the properties of the universe at very early times. Finally, in Sect. 9.5 one of the outstanding puzzles of the Hot Big Bang model will be presented, namely, why the universe appears to consist almost entirely of matter rather than a mixture of matter and antimatter.

Planck scale

matter-dominated universe

9.1 The Planck Scale

> *"A new scientific truth does not triumph*
> *by convincing its opponents and mak-*
> *ing them see the light, but rather be-*
> *cause its opponents eventually die, and*
> *a new generation grows up that is famil-*
> *iar with it."*
>
> *Max Planck*

The earliest time about which one can meaningfully speculate with our current theories is called the *Planck era*, around 10^{-43} seconds after the Big Bang. Before then, the quantum-mechanical aspects of gravity are expected to be important, so one would need a quantum theory of gravity to describe this period. Although superstrings could perhaps provide such a theory, it is not yet in such a shape that one can use it to make specific predictions.

Planck time

quantum theory of gravity

Schwarzschild radius   To see how the Planck scale arises, consider the Schwarzschild radius for a mass m (see Problem 5 in this chapter),

$$R_S = \frac{2mG}{c^2}, \tag{9.1}$$

event horizon   where for the moment factors of c and $\hbar$ will be explicitly inserted. The distance R_S gives the event horizon of a black hole. It represents the distance at which the effects of space-time curvature due to the mass m become significant.

Now consider the Compton wavelength of a particle of mass m,

$$\lambda_C = \frac{h}{mc}. \tag{9.2}$$

This represents the distance at which quantum effects become important. The Planck scale is thus defined by the condition $\lambda_C/2\pi = R_S/2$, i.e.,

$$\frac{\hbar}{mc} = \frac{mG}{c^2}. \tag{9.3}$$

Planck mass   Solving for the Planck mass gives

$$m_{Pl} = \sqrt{\frac{\hbar c}{G}} \approx 2.2 \times 10^{-5}\,\mathrm{g}, \tag{9.4}$$

or the mass of a water droplet about $1/3$ mm in diameter.

Planck energy   The rest-mass energy of m_{Pl} is the Planck energy,

$$E_{Pl} = \sqrt{\frac{\hbar c^5}{G}} \approx 1.22 \times 10^{19}\,\mathrm{GeV}, \tag{9.5}$$

which is about $2\,\mathrm{GJ}$ or $650\,\mathrm{kg}$ TNT equivalent. Using the Planck mass in the reduced Compton wavelength, $\hbar/mc$,

Planck length   gives the Planck length,

$$l_{Pl} = \sqrt{\frac{\hbar G}{c^3}} \approx 1.6 \times 10^{-35}\,\mathrm{m}. \tag{9.6}$$

Planck time   The time that it takes light to travel l_{Pl} is the Planck time,

$$t_{Pl} = \frac{l_{Pl}}{c} = \sqrt{\frac{\hbar G}{c^5}} \approx 5.4 \times 10^{-44}\,\mathrm{s}. \tag{9.7}$$

The Planck mass, length, time, etc. are the unique quantities with the appropriate dimension that can be constructed from the fundamental constants linking quantum mechanics and relativity: $\hbar$, c, and G. Since henceforth $\hbar$ and c will be set equal to one, and one has

$$m_{\mathrm{Pl}} = E_{\mathrm{Pl}} = 1/\sqrt{G} \,, \tag{9.8}$$

$$t_{\mathrm{Pl}} = l_{\mathrm{Pl}} = \sqrt{G} \,. \tag{9.9}$$

So the Planck scale basically characterizes the strength of gravity. As the Planck mass (or energy), 1.2×10^{19} GeV, is a number people tend to memorize, often $1/m_{\mathrm{Pl}}^2$ will be used as a convenient replacement for G.

strength of gravity

9.2 Thermodynamics of the Early Universe

> *"We are startled to find a universe we did not expect."*
>
> *Walter Bagehot*

In this section some results from statistical and thermal physics will be collected that will be needed to describe the early universe. Some of the relations presented may differ from those covered in a typical course in statistical mechanics. This is for two main reasons. First, the particles in the very hot early universe typically have speeds comparable to the speed of light, so the relativistic equation $E^2 = p^2 + m^2$ must be used to relate energy and momentum. Second, the temperatures will be so high that particles are continually being created and destroyed, e.g., through reactions such as $\gamma\gamma \leftrightarrow e^+e^-$. This is in contrast to the physics of low-temperature systems, where the number of particles in a system is usually constrained to be constant. The familiar exception is blackbody radiation, since massless photons can be created and destroyed at any non-zero temperature. For a gas of relativistic particles expressions for ϱ, n, and P will be found that are similar to those for blackbody radiation. Here ϱ, n, and P are the density or – more generally – the energy density, the number density, and the pressure.

creation and annihilation of particles

gas of relativistic particles

The formulae in this section are derived in standard texts on statistical mechanics. Some partial derivations are given in Appendix B; here merely the results in a form appropriate for the early universe will be quoted. In a more rigorous treatment one would need to consider conservation of various quantum numbers such as charge, baryon number, and lepton number. For each conserved quantity one has a chemical potential μ, which enters into the expressions for the energy and number densities. For most of the treatment of the very early universe one can neglect the *chemical potentials*, and thus they will not appear in the formulae which are given here.

statistical mechanics

chemical potential

9.2.1 Energy and Number Densities

In the limit where the particles are relativistic, i.e., $T \gg m$,[1] the energy density for a given particle type is

$$
\varrho = \begin{cases}
\dfrac{\pi^2}{30} g T^4 & \text{for bosons,} \\[2ex]
\dfrac{7}{8} \dfrac{\pi^2}{30} g T^4 & \text{for fermions.}
\end{cases}
\tag{9.10}
$$

internal degrees of freedom Here g is the number of internal degrees of freedom for the particle. For a particle of spin J, for example, one has $2J + 1$ spin degrees of freedom. In addition, the factor g includes different colour states for quarks and gluons. For example, for electrons one has $g_{e^-} = 2$ or, for electrons and positrons considered together, $g_e = 4$. The spin-1/2 u quark together with the $\bar{u}$ has $g_u = 12$, i.e., 2 from spin, 3 from colour, and 2 for considering particle and antiparticle together. Note that the photon has $J = 1$ but only two

photon spin states spin states, which correspond to the two transverse polarization states. The longitudinal polarization is absent as a consequence of these particles having zero mass. For photons this gives $\varrho_\gamma = (\pi^2/15) T^4$, the well-known formula for the energy density of blackbody radiation (Stefan–Boltzmann

Stefan–Boltzmann law law). Indeed, one sees that all relativistic particles have a similar behaviour with $\varrho \sim T^4$, the only differences arising from the number of degrees of freedom and from the factor 7/8 if one considers fermions.

In a similar manner one can show that the number den-

number densities sity n is given by

$$
n = \begin{cases}
\dfrac{\zeta(3)}{\pi^2} g T^3 & \text{for bosons,} \\[2ex]
\dfrac{3}{4} \dfrac{\zeta(3)}{\pi^2} g T^3 & \text{for fermions.}
\end{cases}
\tag{9.11}
$$

Here ζ is the Riemann zeta function and $\zeta(3) \approx 1.202\,06\ldots$. Notice that in particle physics units the number density has dimension of energy cubed. To convert this to a normal number per unit volume, one has to divide by $(\hbar c)^3 \approx (0.2\,\text{GeV}\,\text{fm})^3$.

From the number and energy densities one can obtain the average energy per particle, $\langle E \rangle = \varrho/n$. For $T \gg m$ one finds

---

[1] Note that Boltzmann's constant k has also been set equal to one.

$$\langle E \rangle = \begin{cases} \dfrac{\pi^4}{30\,\zeta(3)}T \approx 2.701\,T & \text{for bosons,} \\[2mm] \dfrac{7\pi^4}{180\,\zeta(3)}T \approx 3.151\,T & \text{for fermions.} \end{cases} \tag{9.12}$$

In the non-relativistic limit one finds for the energy density

energy densities

$$\varrho = mn\,, \tag{9.13}$$

where n is the number density. This is given in the non-relativistic limit by

$$n = g\left(\frac{mT}{2\pi}\right)^{3/2} \mathrm{e}^{-m/T}\,, \tag{9.14}$$

where the same result is obtained for both the Fermi–Dirac distribution and Bose–Einstein distribution and, as above, it is assumed that chemical potentials can be neglected. One therefore finds that for a non-relativistic particle species, the number density is exponentially suppressed by the factor $\mathrm{e}^{-m/T}$. In the non-relativistic limit, the average energy is the sum of mass and kinetic terms,

Fermi–Dirac distribution
Bose–Einstein distribution

$$\langle E \rangle = m + \frac{3}{2}T \approx m\,. \tag{9.15}$$

The final approximation holds in the non-relativistic limit where $T \ll m$.

9.2.2 The Total Energy Density

The formulae above give ϱ and n for a single particle type. What is needed for the Friedmann equation, however, is the total energy density from all particles. This is simply the sum of the ϱ values for all particle types, where at a given temperature some types will be relativistic and others not.

Boltzmann factor

The expression that one will obtain for the total energy density is greatly simplified by recalling that the number density of a non-relativistic particle species is exponentially suppressed by the factor $\mathrm{e}^{-m/T}$. This will hold as long as the reduction in density is not prevented by a conserved quantum number, which would imply a non-zero chemical potential. Assuming this is not the case, to a good approximation the non-relativistic matter contributes very little to the total energy density. At early times in the universe ($t < 10^5$

years), it will be seen that the total energy density was dominated by relativistic particles. This is called the 'radiation-dominated era', and in this period one can ignore the contribution to ϱ from non-relativistic particles. Then, using the appropriate relativistic formulae from (9.10) for bosons and fermions, one obtains for the total energy density

radiation-dominated era

$$\varrho = \sum_{i=\text{bosons}} \frac{\pi^2}{30} g_i T^4 + \sum_{j=\text{fermions}} \frac{7}{8} \frac{\pi^2}{30} g_j T^4 , \quad (9.16)$$

where the sums include only those particle types that are relativistic, i.e., having $m \ll T$. This can be equivalently written as

$$\varrho = \frac{\pi^2}{30} g_* T^4 , \quad (9.17)$$

where the *effective number of degrees of freedom* g_* is defined as

effective number of degrees of freedom

$$g_* = \sum_{i=\text{bosons}} g_i + \frac{7}{8} \sum_{j=\text{fermions}} g_j . \quad (9.18)$$

Here as well, the sums only include particles with $m \ll T$.

T is always assumed to mean the photon temperature, since its value for the present era is very accurately measured from the cosmic microwave background radiation to be $T \approx 2.73$ K. Some particle types may have a different temperature, however, since they may no longer be in thermal contact with photons. Neutrinos, for example, effectively decoupled from other particles at a time before most electrons and positrons annihilated into photons. As a result, the neutrino temperature today is around 1.95 K. In general, one can modify (9.18) to account for different temperatures by using

decoupling of neutrinos

neutrino temperature

$$g_* = \sum_{i=\text{bosons}} g_i \left(\frac{T_i}{T}\right)^4 + \frac{7}{8} \sum_{j=\text{fermions}} g_j \left(\frac{T_j}{T}\right)^4 .$$
$$(9.19)$$

This more general form of g_* will be rarely needed except for the example of neutrinos mentioned above.

In order to compute g_* at a given temperature T, one needs to know what particle types have $m \ll T$, and also the number of degrees of freedom for these types is required.

particle	mass	spin states	colour states	g (particle and antiparticle)
photon (γ)	0	2	1	2
W^+, W^-	80.4 GeV	3	1	6
Z	91.2 GeV	3	1	3
gluon (g)	0	2	8	16
Higgs	> 114 GeV	1	1	1
bosons				28
$u, \bar{u}$	3 MeV	2	3	12
$d, \bar{d}$	6 MeV	2	3	12
$s, \bar{s}$	100 MeV	2	3	12
$c, \bar{c}$	1.2 GeV	2	3	12
$b, \bar{b}$	4.2 GeV	2	3	12
$t, \bar{t}$	175 GeV	2	3	12
e^+, e^-	0.511 MeV	2	1	4
μ^+, μ^-	105.7 MeV	2	1	4
τ^+, τ^-	1.777 GeV	2	1	4
$\nu_e, \bar{\nu}_e$	< 3 eV	1	1	2
$\nu_\mu, \bar{\nu}_\mu$	< 0.19 MeV	1	1	2
$\nu_\tau, \bar{\nu}_\tau$	< 18.2 MeV	1	1	2
fermions				90

Table 9.1
Particles of the Standard Model and their properties [2]

Table 9.1 shows the masses and g values of the particles of the Standard Model ('color states' 1 means a colour-neutral particle).

Here the neutrinos will be treated as only being left-handed and antineutrinos as only right-handed, i.e., they only have one spin state each, which again is related to their being considered massless. In fact, recent evidence (see Sect. 6.2 on neutrino astronomy) indicates that neutrinos do have non-zero mass, but the coupling to the additional spin states is so small that their effect on g_* can be ignored.

neutrino mass

In addition to the particles listed, there are possible X and Y bosons of a Grand Unified Theory and perhaps super-symmetric partners for all particle types. It will be the task of hadron colliders to find out whether such particles can be produced with present-day accelerator technology. Possibly such new particles – if they existed in the early universe – have left imprints on astrophysical data, so that investigations on *cosmoarcheology* might find evidence for them. If one restricts oneself here to the Standard Model and considers a temperature much greater than any of the masses, e.g., $T > 1$ TeV, one has

supersymmetric particles

cosmoarcheology

$$g_* = 28 + \frac{7}{8} \times 90 = 106.75 \,. \tag{9.20}$$

For the accuracy required it will usually be sufficient to take $g_* \approx 10^2$ for $T > 1\,\text{TeV}$.

A few further subtleties come into play at lower temperatures. For example, at around $T \approx 0.2\,\text{GeV}$, quarks and gluons become confined into colour-neutral hadrons, namely, protons and neutrons. So, to obtain g_* at, say, $T = 100\,\text{MeV}$, one would not have any quarks or gluons. Instead one gets a contribution of $8 \times 7/8 = 7$ from hadrons, i.e., two spin states each for protons and neutrons and their antiparticles.

9.2.3 Equations of State

energy density

Next the required equations of state will be recalled, that is, relations between energy density ϱ and pressure P. These are derived in Appendix B. They are needed in conjunction with the acceleration and fluid equations in order to solve the Friedmann equation for $R(t)$. For the pressure for a gas

pressure of relativistic matter

of relativistic particles one finds

$$P = \frac{\varrho}{3} \,. \tag{9.21}$$

This is the well-known result from blackbody radiation, but in fact it applies for any particle type in the relativistic limit

pressure of non-relativistic matter

$T \gg m$. In the non-relativistic limit, the pressure is given by the ideal-gas law, $P = nT$. In this case, however, one has an energy density $\varrho = mn$, so for $T \ll m$ one has $P \ll \varrho$ and in the acceleration and fluid equations one can approximate

$$P \approx 0 \,. \tag{9.22}$$

In addition, one can show that for the case of vacuum energy

pressure of the vacuum

density from a cosmological constant,

$$P = -\varrho_{\text{v}} \,. \tag{9.23}$$

That is, a vacuum energy density leads to a negative pressure. In general, the equation of state can be expressed as

$$P = w\varrho \,, \tag{9.24}$$

w parameter

where the parameter w is $1/3$ for relativistic particles, 0 for (non-relativistic) matter, and -1 for vacuum energy.

9.2.4 Relation between Temperature and Scale Factor

Finally, in this section a general relation between the temperature T and the scale factor R will be noted. All lengths, when considered over distance scales of at least 100 Mpc, increase with R. Since the de Broglie wavelength of a particle, $\lambda = h/p$, is inversely proportional to the momentum, one sees that particle momenta decrease as $1/R$. For photons, one has $E = p$, and so their energy decreases as $1/R$. Furthermore, the temperature of photons in thermal equilibrium is simply a measure of the photons' average energy, so one gets the important relation

momentum and scale factor

$$T \sim R^{-1} \,. \tag{9.25}$$

This relation holds as long as T is interpreted as the photon temperature and as long as the Hubble expansion is what provides the change in T. In fact this is not exact, because there are other processes that affect the temperature as well. For example, as electrons and positrons become non-relativistic and annihilate into photons, the photon temperature receives an extra contribution. These effects can be taken into account by thermodynamic arguments using conservation of entropy. The details of this are not critical for the present treatment, and one will usually be able to assume (9.25) to hold.

photon temperature

9.3 Solving the Friedmann Equation

> *"No one will be able to read the great book of the universe if he does not understand its language which is that of mathematics."*
>
> *Galileo Galilei*

Now enough information is available to solve the Friedmann equation. This will allow to derive the time dependence of the scale factor R, temperature T, and energy density ϱ. If, for the start, very early times will be considered, one can simplify the problem by seeing that the term in the Friedmann equation (8.15) with the curvature parameter, k/R^2, can be neglected. To show this, recall from Sect. 8.5 the fluid equation (8.28),

curvature parameter

$$\dot{\varrho} + \frac{3\dot{R}}{R}(\varrho + P) = 0 \,, \tag{9.26}$$

which relates the time derivative of the energy density ϱ and the pressure P. One can suppose that ϱ is dominated by radiation, so that the equation of state (9.21) can be used,

$$P = \frac{\varrho}{3} .$$ (9.27)

fluid equation Substituting this into the fluid equation (9.26) gives

$$\dot{\varrho} + \frac{4\varrho\dot{R}}{R} = 0 .$$ (9.28)

The left-hand side is proportional to a total derivative, so one can write

$$\frac{1}{R^4} \frac{\mathrm{d}}{\mathrm{d}t} \left(\varrho R^4 \right) = 0 .$$ (9.29)

radial dependence of ϱ This implies that ϱR^4 is constant in time, and therefore

$$\varrho \sim \frac{1}{R^4} .$$ (9.30)

If, instead, one would have assumed that ϱ was dominated by non-relativistic matter, one would have used the equation of state $P = 0$, and in a similar way one would have found (see Problem 1 in this chapter)

$$\varrho \sim \frac{1}{R^3} .$$ (9.31)

In either case the dependence of ϱ on R is such that for very early times, that is, for sufficiently small R, the term $8\pi G\varrho/3$ on the right-hand side of the Friedmann equation will be much larger than k/R^2. One can then ignore the **curvature parameter** curvature parameter and effectively set $k = 0$; this is definitely valid at the very early times that will be considered in this chapter and it is still a good approximation today, 14 **modified Friedmann** billion years later. The Friedmann equation then becomes **equation**

$$\frac{\dot{R}^2}{R^2} = \frac{8\pi}{3}G\varrho .$$ (9.32)

In the following (9.32) will be solved for the case where ϱ is radiation dominated. One can write (9.30) as

$$\varrho = \varrho_0 \left(\frac{R_0}{R} \right)^4 ,$$ (9.33)

where here ϱ_0 and R_0 represent the values of ϱ and R at some particular (early) time. One can guess a solution of the form $R = At^p$ and substitute this along with (9.33) for ϱ into the Friedmann equation (9.32). With this ansatz the Friedmann equation can only be satisfied if $p = 1/2$, i.e., **radiation-dominated era**

$$R \sim t^{1/2} . \tag{9.34}$$

The expansion rate H is therefore **expansion rate**

$$H = \frac{\dot{R}}{R} = \frac{1}{2t} . \tag{9.35}$$

If, instead, one assumes that ϱ is dominated by non-relativistic matter and if one uses the equation of state $P = 0$, then one finds in a similar way (see Problem 2 in this chapter) **matter-dominated era**

$$R \sim t^{2/3} \tag{9.36}$$

and

$$H = \frac{2}{3t} . \tag{9.37}$$

First, one can combine the Friedmann equation (9.32) with the energy density (9.17) and use $G = 1/m_{\mathrm{Pl}}^2$ to replace the gravitational constant. Taking the square root then gives the expansion rate H as a function of the temperature (see Problem 3 in this chapter),

$$H = \sqrt{\frac{8\pi^3 g_*}{90}} \frac{T^2}{m_{\mathrm{Pl}}} \approx 1.66\sqrt{g_*} \frac{T^2}{m_{\mathrm{Pl}}} . \tag{9.38}$$

This can be combined with (9.35) to give a relation between the temperature and the time, **relation between temperature and time**

$$t = \frac{1}{2}\sqrt{\frac{90}{8\pi^3 g_*}} \frac{m_{\mathrm{Pl}}}{T^2} \approx \frac{0.301}{\sqrt{g_*}} \frac{m_{\mathrm{Pl}}}{T^2} . \tag{9.39}$$

Remember that both of these equations use the particle physics system of units, i.e., the expansion rate in (9.38) is in GeV and the time in (9.39) is in GeV^{-1}. One can also combine the Friedmann equation with the solution $H = 1/2t$ to give the energy density as a function of time, **time dependence of the energy density**

$$\varrho = \frac{3m_{\mathrm{Pl}}^2}{32\pi} \frac{1}{t^2} . \tag{9.40}$$

9.3.1 Digression on Thermal Equilibrium

Having derived the relations for quantities such as number and energy density as a function of temperature, it is worth asking when one expects them to apply. In order for a system to be characterized by a temperature, there must exist inter-actions between the particles that allow their numbers and momentum distribution to adjust to those of thermal equi-

thermal equilibrium librium. Furthermore, one has to wait long enough for equi-
conditions librium to be attained, namely, much longer than the time scale of the individual microscopic interactions.

Now in any change in the temperature, the microscopic interactions must take place quickly enough for the thermal distribution to adjust. One can express this condition by re-quiring that the rate Γ of the reaction needed to maintain equilibrium must be much greater than the fractional change in the temperature per unit time, i.e.,

$$\Gamma \gg |\dot{T}/T| \ . \tag{9.41}$$

But for a system of relativistic particles one has from (9.25) that $T \sim 1/R$, so $\dot{T}/T = -\dot{R}/R = -H$. Therefore, (9.41) is equivalent to the requirement

$$\Gamma \gg H \ , \tag{9.42}$$

where the absolute value has been dropped since the expan-sion rate is always assumed to be positive. For a given tem-perature it is straightforward to use (9.38) to determine H. It is basically proportional to T^2, ignoring the small temper-ature dependence of g_*.

interaction rates The reaction rate Γ is the number of interactions of a specified type per unit time per particle. It is the reciprocal of the mean time that it will take for the particle to undergo the interaction in question. It can be calculated as a function of the particle's speed v, the number density of target particles

cross section n, and the interaction cross section σ by

$$\Gamma = n \langle \sigma v \rangle \ , \tag{9.43}$$

where the brackets denote an average of σv over a thermal distribution of velocities.

If thermal equilibrium has been attained, one can find the number density n using the appropriate formulae from Sect. 9.2. To find the cross section, one needs to bring in the knowledge of particle physics.

9.4 Thermal History of the First Ten Microseconds

> *"An elementary particle that does not exist in particle theory should also not exist in cosmology."*
>
> *Anonymous*

The relations from the previous sections can now be used to work out the energy density and temperature of the universe as a function of time. As a start, one can use (9.40) to give the energy density at the Planck time, although one needs to keep in mind from the previous section that the assumption of thermal equilibrium may not be valid. In any case the formula gives

$$\varrho(t_{\mathrm{Pl}}) = \frac{3m_{\mathrm{Pl}}^2}{32\pi}\frac{1}{t_{\mathrm{Pl}}^2} = \frac{3}{32\pi}m_{\mathrm{Pl}}^4 \approx 6\times 10^{74}\,\mathrm{GeV}^4\;, \quad (9.44)$$

where $m_{\mathrm{Pl}} = 1/t_{\mathrm{Pl}} \approx 1.2\times 10^{19}\,\mathrm{GeV}$ has been used. One can convert this to normal units by dividing by $(\hbar c)^3$, **Planckian energy densities**

$$\varrho(t_{\mathrm{Pl}}) \approx 6\times 10^{74}\,\mathrm{GeV}^4 \times \frac{1}{(0.2\,\mathrm{GeV\,fm})^3}$$

$$\approx 8\times 10^{76}\,\mathrm{GeV/fm}^3\;. \quad (9.45)$$

This density corresponds to about 10^{77} proton masses in the volume of a single proton!

Proceeding now more systematically, one can find the times and energy densities at which different temperatures were reached. By combining this with the knowledge of particle physics, one will see what types of particle interactions were taking place at what time.

To relate the temperature to the time, one needs the number of degrees of freedom, g_*. Assuming that nature only contains the known particles of the Standard Model, then for T greater than several hundred GeV all of them can be **characteristic temperatures in the early universe** treated as relativistic. From (9.20) one has $g_* = 106.75$. If, say, GUT bosons or supersymmetric particles also exist, then one would have a higher value. For the order-of-magnitude values that one is interested in here this uncertainty in g_* will not be critical.

Table 9.2 shows values for the temperature and energy density at several points within the first 10 microseconds after the Big Bang, where most of the values have been rounded to the nearest order of magnitude.

Table 9.2
Thermal history of the first 10
microseconds

'scale'	T [GeV]	ϱ [GeV^4]	t [s]
Planck	10^{19}	10^{78}	10^{-45}
GUT	10^{16}	10^{66}	10^{-39}
Electroweak	10^{2}	10^{10}	10^{-11}
QCD	0.2	0.01	10^{-5}

cosmoparticle physics

Higgs field

**vacuum expectation value
of the Higgs field**

**spontaneous symmetry
breaking**

electroweak scale

QCD scale

At the Planck scale, i.e., with energies on the order of
10^{19} GeV, the limit of our ability to speculate about cosmology and cosmoparticle physics has been reached. Notice
from (9.39) that the time when the temperature is equal to
the Planck energy is not the Planck time but is in fact somewhat earlier, so the limit has even been overstepped somewhat.

After 10^{-39} seconds, when the universe had cooled to
temperatures around 10^{16} GeV (the 'GUT scale'), the strong,
electromagnetic, and weak interactions start to become distinct, each with a different coupling strength. One expects
that around this temperature a phase transition related to the
Higgs field of the Grand Unified Theory took place. At temperatures above the phase transition, the vacuum expectation value of the Higgs should be zero, and therefore all elementary particles would be massless, including the X and Y
bosons. During the transition, the GUT Higgs field acquires
a vacuum expectation value different from zero; this phenomenon is called *spontaneous symmetry breaking* (SSB).
As a result, the X and Y bosons go from being massless to
having very high masses on the order of the GUT scale. So,
at lower temperatures, baryon-number-violating processes
mediated by exchange of X and Y bosons are highly suppressed.

After around 10^{-11} seconds, the temperature is on the
order of 100 GeV; this is called the 'electroweak scale'. Here
another SSB phase transition is expected to occur whereby
the electroweak Higgs field acquires a non-zero vacuum expectation value. As a result, W and Z bosons as well as the
quarks and leptons acquire their masses. At temperatures
significantly lower than the electroweak scale, the masses
$M_W \approx 80$ GeV and $M_Z \approx 91$ GeV are large compared to
the kinetic energies of other colliding particles, and the W
and Z propagators effectively suppress the strength of the
weak interaction.

At temperatures around 0.2 GeV (the 'QCD scale'), the
effective coupling strength of the strong interaction, α_s, becomes very large. At this point quarks and gluons become

confined into colour-neutral hadrons: protons, neutrons, and their antiparticles. This process, called *hadronization*, occurs around $t \approx 10^{-5}\,\text{s}$, where to obtain this time from (9.39) one should use a value of g_* somewhat smaller than before, since not all of the particles are relativistic. If one takes photons, gluons, u, d, s, e, μ, and all families of ν as relativistic just before hadronization then one would get $g_* = 61.75$.

hadronization

It is interesting to convert the energy density at the QCD scale to normal units, which gives about $1\,\text{GeV/fm}^3$. This is about seven times the density of ordinary nuclear matter. Experiments at the Relativistic Heavy Ion Collider (RHIC) at the Brookhaven National Laboratory near New York are currently underway to recreate these conditions by colliding together heavy ions at very high energies [12]. This will allow more detailed studies of the 'quark–gluon plasma' and its transition to colour-neutral hadrons.

energy density

quark–gluon plasma

This short sketch of the early universe has ignored at least two major issues. First, it has not yet been explained why the universe appears to be composed of matter, rather than a mixture of matter and antimatter. There is no definitive answer to this question but there are plausible scenarios whereby the so-called baryon asymmetry of the universe could have arisen at very early times, perhaps at the GUT scale or later at the electroweak scale. This question will be looked at in greater detail in the next section.

baryon asymmetry

Second, one will see that the model that has been developed thus far fails to explain several observational facts, the most important of which are related to the cosmic microwave background radiation. A possible remedy to these problems will be to suppose that the energy density at some very early time was dominated by vacuum energy. In this case one will not find $R \sim t^{1/2}$ but rather an exponential increase, known as *inflation*.

vacuum energy

inflation

9.5 The Baryon Asymmetry of the Universe

> *"Astronomy, the oldest and one of the most juvenile of the sciences, may still have some surprises in store. May antimatter be commended to its case."*
>
> *Arthur Schuster*

For several decades after the discovery of the positron, it appeared that the laws of nature were completely symmetric between matter and antimatter. The universe known to us, however, seems to consist of matter only. The relative abundance of baryons will now be looked at in detail and therefore this topic is called the *baryon asymmetry of the universe*. One has to examine how this asymmetry could evolve from a state which initially contained equal amounts of matter and antimatter, a process called *baryogenesis*. For every proton there is an electron, so the universe also seems to have a non-zero lepton number. This is a bit more difficult to pin down, however, since the lepton number could in principle be compensated by unseen antineutrinos. In any case, models of baryogenesis generally incorporate in some way lepton production (*leptogenesis*) as well.

baryogenesis

leptogenesis

Baryogenesis provides a nice example of the interplay between particle physics and cosmology. In the final analysis it will be seen that the Standard Model as it stands cannot explain the observed baryon asymmetry of the universe. This is a compelling indication that the Standard Model is incomplete.

9.5.1 Experimental Evidence of Baryon Asymmetry

If antiparticles were to exist in significant numbers locally, one would see evidence of this from proton–antiproton or electron–positron annihilation. $p\bar{p}$ annihilation produces typically several mesons including neutral pions, which decay into two photons. So, one would see γ rays in an energy range up to around 100 MeV. No such gamma rays resulting from asteroid impacts on other planets are seen, man-made space probes landing on Mars survived, and Neil Armstrong landing on the Moon did not annihilate, so one can conclude that the entire solar system is made of matter.

$p\bar{p}$ annihilation

One actually finds some antiprotons bombarding the Earth as cosmic rays at a level of around 10^{-4} compared to cosmic-ray protons (see Sect. 6.1), so one may want to leave open the possibility that more distant regions are made of antimatter. But the observed antiproton rate is compatible with production in collisions of ordinary high-energy protons with interstellar gas or dust through reactions of the type

$$p + p \rightarrow 3p + \bar{p} \,. \tag{9.46}$$

antinuclei in cosmic rays?

There is currently no evidence of antinuclei in cosmic rays, although the Alpha Magnetic Spectrometer based on the

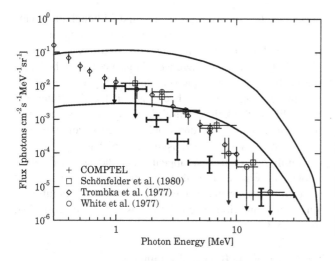

Fig. 9.1
The measured gamma-ray flux (*data points*) along with the levels predicted to arise from interaction between domains of matter and antimatter. The *upper curve* corresponds to domain sizes of 20 Mpc, the *lower* for 1000 Mpc [14, 15]

International Space Station will carry out more sensitive searches for this in the next several years [13].

If there were to exist antimatter domains of the universe, then their separation from matter regions would have to be very complete or else one would see the γ-ray flux from proton–antiproton or electron–positron annihilation. The flux that one would expect depends on the size of the separated domains. Figure 9.1 shows the measured gamma-ray flux (data points) along with the predicted levels (curves) that would arise from collisions of matter and antimatter regions [14, 15]. The upper curve corresponds to domain sizes of 20 Mpc and is clearly excluded by the data. The lower curve is for domains of 1000 Mpc and it as well is incompatible with the measurements. So one can conclude that if antimatter regions of the universe exist, they must be separated by distances on the order of a gigaparsec, which is a significant fraction of the observable universe. Given that there is no plausible mechanism for separating matter from antimatter over such large distances, it is far more natural to assume that the universe is made of matter, i.e., that it has a net non-zero baryon number. Also the absence of a significant flux of 511 keV γ rays from electron–positron annihilation adds to this conclusion.

e^+e^- annihilation

absence of annihilation radiation

If one then takes as working hypothesis that the universe contains much more matter than antimatter, one needs to ask how this could have come about. One possibility is that the non-zero baryon number existed as an initial condition, and that this was preserved up to the present day. This is not an attractive idea for several reasons. First, although the asymmetry between baryons and antibaryons today appears to be large, i.e., lots of the former and none of the latter, at times closer to the Big Bang, there were large amounts of both and the relative imbalance was very small. This will be quantified in Sect. 9.5.2. Going back towards the Big Bang one would like to think that nature's laws become in some sense more fundamental, and one would prefer to avoid the need to impose any sort of small asymmetry by hand.

net baryon number   Furthermore, it now appears that the laws of nature allow, or even require, that a baryon asymmetry would arise from a state that began with a net baryon number of zero. The conditions needed for this will be discussed in Sect. 9.5.3.

9.5.2 Size of the Baryon Asymmetry

Although the universe today seems completely dominated by baryons and not antibaryons, the relative asymmetry was very much smaller at earlier times. This can be seen roughly by considering a time when quarks and antiquarks were all highly relativistic, at a temperature of, say, $T \approx 1\,\mathrm{TeV}$, and suppose that since that time there have been no baryon-**baryon-number-violating**   number-violating processes. The net baryon number in a co-
processes   moving volume R^3 is then constant, so one has

$$(n_{\mathrm{b}} - n_{\bar{\mathrm{b}}})R^3 = (n_{\mathrm{b},0} - n_{\bar{\mathrm{b}},0})R_0^3 \;, \tag{9.47}$$

where the subscript 0 on the right-hand side denotes present values. Today, however, there are essentially no antibaryons, so one can approximate $n_{\bar{\mathrm{b}},0} \approx 0$. The *baryon–antibaryon*
baryon–antibaryon   *asymmetry A* is therefore
asymmetry

$$A \equiv \frac{n_{\mathrm{b}} - n_{\bar{\mathrm{b}}}}{n_{\mathrm{b}}} = \frac{n_{\mathrm{b},0}}{n_{\mathrm{b}}} \frac{R_0^3}{R^3} \;. \tag{9.48}$$

One can now relate the ratio of scale factors to the ratio of temperatures, using the relation $R \sim 1/T$. Therefore, one gets

$$A \approx \frac{n_{\mathrm{b},0}}{n_{\mathrm{b}}} \frac{T^3}{T_0^3} \;. \tag{9.49}$$

Now one can use the fact that the number densities are related to the temperature. Equation (9.11) had shown

$$n_{\mathrm{b}} \approx T^3 \;, \tag{9.50}$$

$$n_{\gamma,0} \approx T_0^3 \;, \tag{9.51}$$

where these are rough approximations with the missing factors of order unity. Using these ingredients one can express the asymmetry as

$$A \approx \frac{n_{b,0}}{n_{\gamma,0}} . \tag{9.52}$$

Further, the baryon-number-to-photon ratio can be defined:

baryon-to-photon ratio

$$\eta = \frac{n_b - n_{\bar{b}}}{n_\gamma} . \tag{9.53}$$

One expects this ratio to remain constant as long as there are no further baryon-number-violating processes and there are no extra influences on the photon temperature beyond the Hubble expansion. So one can also assume that η refers to the current value, $(n_{b,0} - n_{\bar{b},0})/n_{\gamma,0} \approx n_{b,0}/n_{\gamma,0}$, although – strictly speaking – one should call this η_0. So one finally obtains that the baryon–antibaryon asymmetry A is roughly equal to the current baryon-to-photon ratio η. A more careful analysis which keeps track of all the missing factors gives $A \approx 6\eta$.

The current photon density $n_{\gamma,0}$ is well determined from the CMB temperature to be $410.4\,\mathrm{cm}^{-3}$. In principle one could determine $n_{b,0}$ by adding up all of the baryons that one finds in the universe. This in fact is expected to be an underestimate, since some matter such as gas and dust will not be visible and these will also obscure stars further away. A more accurate determination of η comes from the model of Big Bang Nucleosynthesis combined with measurements of the ratio of abundances of deuterium to hydrogen. From this one finds $\eta \approx 5 \times 10^{-10}$. So, finally, the baryon asymmetry can be expressed as

current photon density

nucleosynthesis

$$A \approx 6\eta \approx 3 \times 10^{-9} . \tag{9.54}$$

This means that at early times, for every billion antiquarks there were a billion and three quarks. The matter in the universe one sees today is just the tiny amount left over after essentially all of the antibaryons annihilated.

9.5.3 The Sakharov Conditions

In 1967 Andrei Sakharov pointed out that three conditions must exist in order for a universe with non-zero baryon number to evolve from an initially baryon-symmetric state [16]. Nature must provide:

baryon-number-violating
processes
violation
of C and CP symmetry
departure
from thermal equilibrium

1. baryon-number-violating processes;
2. violation of C and CP symmetry;
3. departure from thermal equilibrium.

The first condition must clearly hold, or else a universe with $B = 0$ will forever have $B = 0$. In the second condition, C refers to charge conjugation and P to parity. C and CP symmetry roughly means that a system of particles behaves the same as the corresponding system made of antiparticles. If all matter and antimatter reactions proceed at the same rate, then no net baryon number develops; thus, violation of C and CP symmetry is needed. The third condition on departure from equilibrium is necessary in order to obtain unequal occupation of particle and antiparticle states, which necessarily have the same energy levels.

Sakharov conditions

A given theory of the early universe that satisfies at some level the Sakharov conditions will in principle predict a baryon density or, equivalently, a baryon-to-photon ratio η. One wants the net baryon number to be consistent with the measured baryon-to-photon ratio $\eta = n_b/n_\gamma \approx 5 \times 10^{-10}$. It is not entirely clear how this can be satisfied, and a detailed discussion goes beyond the scope of this book. Here only some of the currently favoured ideas will be mentioned.

quantum anomalies

Baryon-number violation is predicted by Grand Unified Theories, but it is difficult there to understand how the resulting baryon density could be preserved when this is combined with other ingredients such as inflation. Surprisingly, a non-zero baryon number is also predicted by *quantum anomalies*[2] in the usual Standard Model, and this is currently a leading candidate for baryogenesis.

incompleteness
of the Standard Model

CP violation is observed in decays of K and B mesons and it is predicted by the Standard Model of particle physics but at a level far too small to be responsible for baryogenesis. If nature includes further CP-violating mechanisms from additional Higgs fields, as would be present in supersymmetric models, then the effect could be large enough to account for the observed baryon density. This is one of the clearest indications from cosmology that the Standard Model is incomplete and that other particles and interactions must exist. It has been an important motivating factor in the experimental

---

[2] Quantum anomalies can arise if a classical symmetry is broken in the process of quantization and renormalization. The perturbative treatment of quantum field theories requires a renormalization, and this adds non-invariant counter terms to the invariant Lagrange density that one gets at the classical level.

investigation of *CP*-violating decays of K and B mesons. These experiments have not, however, revealed any effect incompatible with Standard Model predictions.

The departure from thermal equilibrium could be achieved simply through the expansion of the universe, i.e., when the reaction rate needed to maintain equilibrium falls below the expansion rate: $\Gamma \ll H$. Alternatively, it could result from a phase transition such as those associated with spontaneous symmetry breaking.

deviation from thermal equilibrium

So, the present situation with the baryon asymmetry of the universe is a collection of incomplete experimental observations and partial theories which point towards the creation of a non-zero baryon density at some point in the early universe. The details of baryogenesis are still murky and remain an active topic of research. It provides one of the closest interfaces between particle physics and cosmology. Until the details are worked out one needs to take the baryon density of the universe or, equivalently, the baryon-to-photon ratio, as a free parameter that must be obtained from observation, see also Fig. 9.2.

interface between particle physics and cosmology

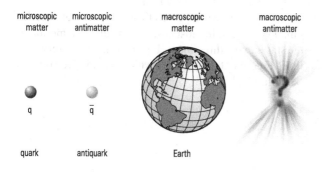

microscopic matter   microscopic antimatter   macroscopic matter   macroscopic antimatter

q $\bar{q}$

quark   antiquark   Earth

Fig. 9.2
The matter–antimatter symmetry observed at microscopic scales appears to be broken at the macroscopic level

9.6 Problems

1. Derive the relation between the scale factor R and the energy density ϱ for a universe dominated by non-relativistic matter!

2. Derive the relation between the scale factor R and time for an early universe dominated by non-relativistic matter!

3. The total energy density of the early universe varies with the temperature like

$$\varrho = \frac{\pi^2}{30} g_* T^4 \,. \tag{9.55}$$

Work out the expansion rate H as a function of the temperature and Planck mass using the Friedmann equation!

4. In Sect. 9.1 the Planck length has been derived using the argument that quantum and gravitational effects become equally important. One can also try to combine the relevant constants of nature $(G, \hbar, c)$ in such a way that a characteristic length results. What would be the answer?

5. Estimate a value for the Schwarzschild radius assuming weak gravitational fields (which do not really apply to the problem). What is the Schwarzschild radius of the Sun and the Earth?

   Hint: Consider the non-relativistic expression for the escape velocity and replace formally v by c.

6. A gas cloud becomes unstable and collapses under its own gravity if the gravitational energy exceeds the thermal energy of its molecules. Estimate the critical density of a hydrogen cloud at $T = 1000\,\mathrm{K}$ which would have collapsed into our Sun! Refer to Problem 8.2 for the stability condition (Jeans criterion).

7. Occasionally it has appeared that stars move at superluminal velocities. In most cases this is due to a motion of the star into the direction of Earth during the observation time. Figure out an example which would give rise to a 'superluminal' speed!

10 Big Bang Nucleosynthesis

> *"In fact, it seems that present day science, with one sweeping step back across millions of centuries, has succeeded in bearing witness to that primordial 'Fiat lux' [let there be light] uttered at the moment when, along with matter, there burst forth from nothing a sea of light and radiation, while the particles of the chemical elements split and formed into millions of galaxies. Hence, creation took place in time, therefore, there is a Creator, therefore, God exists!"*
>
> *Pope Pius XII*

At times from around 10^{-2} seconds through the first several minutes after the Big Bang, the temperature passed through the range from around 10 to below 10^{-1} MeV. During this period protons and neutrons combined to produce a significant amount of ^{4}He – one quarter of the universe's nuclei by mass – plus smaller amounts of deuterium (D, i.e., ^{2}H), tritium (^{3}H), ^{3}He, ^{6}Li, ^{7}Li, and ^{7}Be. Further synthesis of nuclei in stars accounts for all of the heavier elements plus only a relatively small additional amount of helium. The predictions of Big Bang Nucleosynthesis (BBN) are found to agree remarkably well with observations, and provide one of the most important pillars of the Big Bang model. **primordial elements**

Big Bang Nucleosynthesis

The two main ingredients of BBN are the equations of cosmology and thermal physics that have already been described, plus the rates of nuclear reactions. Although the nuclear cross sections are difficult to calculate theoretically, they have for the most part been well measured in laboratory experiments. Of crucial importance is the rate of the reaction $\nu_e n \leftrightarrow e^- p$, which allows transformation between neutrons and protons. The proton is lighter than the neutron by $\Delta m = m_n - m_p \approx 1.3$ MeV, and as long as this reaction proceeds sufficiently quickly, one finds that the neutron-to-proton ratio is suppressed by the Boltzmann factor $e^{-\Delta m/T}$.[1] At a temperature around 0.7 MeV the reaction is no longer fast enough to keep up and the neutron-to-proton ratio 'freezes out' at a value of around 1/6. To first approximation one can estimate the helium abundance simply by assuming that all of the available neutrons end up in ^{4}He. **transformation between neutrons and protons**

freeze-out temperature

---

[1] In the following the natural constants c, $\hbar$, and k are set to unity.

baryon density   The one free parameter of BBN is the baryon density Ω_b or, equivalently, the baryon-to-photon ratio η. By comparing the observed abundances of the light elements with those predicted by BBN, the value of η can be estimated. The result will turn out to be of fundamental importance for the dark-matter problem, which will be dealt with in Chap. 13.

10.1 Some Ingredients for BBN

"The point of view of a sinner is that the church promises him hell in the future, but cosmology proves that the glowing hell was in the past."

Ya. B. Zel'dovich

To model the synthesis of light nuclei one needs the equations of cosmology and thermal physics relevant for temperatures in the MeV range. At this point the total energy density of the universe is still dominated by radiation (i.e., relativistic particles), so the pressure and energy density are related by $P = \varrho/3$. In Chap. 9 it was shown that this led

expansion rate   to relations for the expansion rate and time as a function of temperature,

$$H = 1.66\sqrt{g_*}\,\frac{T^2}{m_{\text{Pl}}}\,,\tag{10.1}$$

$$t = \frac{0.301}{\sqrt{g_*}}\,\frac{m_{\text{Pl}}}{T^2}\,.\tag{10.2}$$

ingredients for BBN   To use these equations one needs to know the effective number of degrees of freedom g_*. Recall that quarks and gluons have already become bound into protons and neutrons at around $T = 200\,\text{MeV}$, and since the nucleon mass is around $m_N \approx 0.94\,\text{GeV}$, these are no longer relativistic. Nucleons and antinucleons can remain in thermal equilibrium down to temperatures of around $50\,\text{MeV}$, and during this period their number densities are exponentially suppressed by

disappearance of antimatter   the factor $e^{-m_N/T}$. At temperatures below several tens of MeV the antimatter has essentially disappeared and the resulting nucleon density therefore does not make a significant contribution to the energy density.

At temperatures in the MeV range, the relativistic particles are photons, e^-, ν_e, ν_μ, ν_τ, and their antiparticles. The photons contribute $g_\gamma = 2$ and e^+ and e^- together give

neutrino families $g_e = 4$. In general, if there are N_ν families of neutrinos, these contribute together $g_\nu = 2N_\nu$. From (9.18) one therefore obtains

$$g_* = 2 + \frac{7}{8}(4 + 2N_\nu) \; . \tag{10.3}$$

For $N_\nu = 3$ one has $g_* = 10.75$. Using this value, (10.2) can be written in a form convenient for description of the BBN era,

$$tT^2 \approx 0.74 \, \mathrm{s \, MeV}^2 \; . \tag{10.4}$$

10.2 Start of the BBN Era

> "By the word of the Lord were the heavens made. For he spoke, and it came to be; he commanded, and it stood firm."
> The Bible; Psalm 33:6,9

From (10.4) a temperature of $T = 10\,\mathrm{MeV}$ is reached at a time $t \approx 0.007\,\mathrm{s}$. At this temperature, all of the relativistic particles – γ, e^-, ν_e, ν_μ, ν_τ, and their antiparticles – are in thermal equilibrium through reactions of the type $e^+e^- \leftrightarrow \nu\bar{\nu}$, $e^+e^- \leftrightarrow \gamma\gamma$, etc. The number density of the neutrinos, for example, is given by the equilibrium formula appropriate for relativistic fermions, see (9.11), **equilibrium of relativistic particles**

$$n_\nu = \frac{3}{4}\frac{\zeta(3)}{\pi^2} g_\nu T^3 \; , \tag{10.5}$$

with a similar formula holding for the electron density.

Already at temperatures around $20\,\mathrm{MeV}$, essentially all of the antiprotons and antineutrons annihilated. The baryon-to-photon ratio is a number that one could, in principle, predict, if a complete theory of baryogenesis were available. **baryogenesis** Since this is not the case, however, the baryon density has to be treated as a free parameter. Since one does not expect any more baryon-number-violating processes at temperatures near the BBN era, the total number of protons and neutrons in a comoving volume remains constant. That is, even though protons and neutrons are no longer relativistic, baryon-number conservation requires that the sum of their **baryon-number conservation** number densities follows

$$n_n + n_p \sim \frac{1}{R^3} \sim T^3 \; . \tag{10.6}$$

At temperatures much greater than the neutron–proton mass difference, $\Delta m = m_n - m_p \approx 1.3\,\mathrm{MeV}$, one has $n_n \approx n_p$.

10.3 The Neutron-to-Proton Ratio

"The most serious uncertainty affecting the ultimate fate of the universe is the question whether the proton is absolutely stable against decay into lighter particles. If the proton is unstable, all matter is transitory and must dissolve into radiation."

Freeman J. Dyson

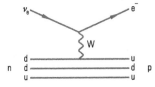

Fig. 10.1
Feynman diagram for the reaction
$n\nu_e \leftrightarrow pe^-$

neutron-to-proton ratio

Although the total baryon number is conserved, protons and neutrons can be transformed through reactions like $n\nu_e \leftrightarrow pe^-$ and $ne^+ \leftrightarrow p\bar{\nu}_e$. A typical Feynman diagram is shown in Fig. 10.1.

The crucial question is whether these reactions proceed faster than the expansion rate so that thermal equilibrium is maintained. If this is the case, then the ratio of neutron-to-proton number densities is given by

$$\frac{n_n}{n_p} = \left(\frac{m_n}{m_p}\right)^{3/2} e^{-(m_n - m_p)/T} \approx e^{-\Delta m/T} , \qquad (10.7)$$

where $m_p = 938.272\,\text{MeV}$, $m_n = 939.565\,\text{MeV}$, and $\Delta m = m_n - m_p = 1.293\,\text{MeV}$. To find out whether equilibrium is maintained, one needs to compare the expansion rate H from (10.1) to the reaction rate Γ. In Sect. 9.3 this rate was found to be given by (9.43)

$$\Gamma = n\langle \sigma v \rangle . \qquad (10.8)$$

reaction rates

Γ is the the reaction rate per neutron for $n\nu_e \leftrightarrow pe^-$, where the brackets denote an average of σv over a thermal distribution of velocities. The number density n in (10.8) refers to the target particles, i.e., neutrinos, which is therefore given by (10.5).

The cross section for the reaction $\nu_e n \leftrightarrow e^- p$ can be predicted using the Standard Model of electroweak interactions. An exact calculation is difficult but to a good approximation one finds for the thermally averaged speed times

weak cross section

cross section

$$\langle \sigma v \rangle \approx G_F^2 T^2 . \qquad (10.9)$$

Here $G_F = 1.166 \times 10^{-5}\,\text{GeV}^{-2}$ is the Fermi constant, which characterizes the strength of weak interactions.

To find the reaction rate one has to multiply (10.9) by the number density n from (10.5). But since the main interest here is to get a rough approximation, factors of order unity can be ignored and one can take $n \approx T^3$ to obtain

$$\Gamma(\nu_e n \to e^- p) \approx G_F^2 T^5 . \qquad (10.10)$$

A similar expression is found for the inverse reaction $e^- p \to \nu_e n$.

Now the question of whether this reaction proceeds quickly enough to maintain thermal equilibrium can be addressed. Figure 10.2 shows the expansion rate H from (10.1) and reaction rate Γ from (10.10) as a function of the temperature.

The point where $\Gamma = H$ determines the decoupling or *freeze-out temperature* T_f.

Equating the expressions for Γ and H,

$$G_F^2 T^5 = 1.66\sqrt{g_*}\frac{T^2}{m_{Pl}} , \qquad (10.11)$$

and solving for T gives

$$T_f = \left(\frac{1.66}{G_F^2 m_{Pl}}\right)^{1/3} g_*^{1/6} . \qquad (10.12)$$

Evaluating this numerically using $g_* = 10.75$ gives $T_f \approx$ 1.5 MeV. It is interesting to note that (10.12) contains nothing directly related to the neutron–proton mass difference, and yet it gives a value very close to $m_n - m_p \approx 1.3$ MeV. This coincidence is such that the actual freeze-out temperature is somewhat lower than that of the naïve calculation. A more careful analysis gives

$$T_f \approx 0.7 \, \text{MeV} . \qquad (10.13)$$

At temperatures below T_f the reaction $\nu_e n \leftrightarrow e^- p$ can no longer proceed quickly enough to maintain the equilibrium number densities. The neutron density is said to *freeze out*, i.e., the path by which neutrons could be converted to protons is effectively closed. If neutrons were stable, this would mean that the number of them in a comoving volume would be constant, i.e., their number density would follow $n_n \sim 1/R^3$. Actually, this is not quite true because free neutrons can still decay. But the neutron has a mean lifetime $\tau_n \approx 886$ s, which is relatively long, but not entirely negligible, compared to the time scale of nucleosynthesis.

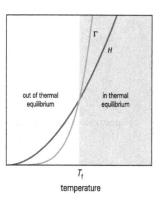

Fig. 10.2
The reaction rate $\Gamma(\nu_e n \leftrightarrow e^- p)$ and the expansion rate H as a function of temperature

freeze-out temperature T_f

freeze-out conditions

neutron decay

Ignoring for the moment the effect of neutron decay, the neutron-to-proton ratio at the freeze-out temperature is

$$\frac{n_n}{n_p} = e^{-(m_n - m_p)/T_f} \approx e^{-1.3/0.7} \approx 0.16 \ . \qquad (10.14)$$

neutron-to-proton ratio

According to (10.4), this temperature is reached at a time $t \approx 1.5\,\text{s}$. By the end of the next five minutes, essentially all of the neutrons become bound into ^{4}He, and thus the neutron-to-proton ratio at the freeze-out temperature is the dominant factor in determining the amount of helium produced. Before proceeding to predict the helium abundance, a few questions will be addressed in somewhat more detail, namely, neutrino decoupling, positron annihilation, and neutron decay.

10.4 Neutrino Decoupling, Positron Annihilation, and Neutron Decay

> *"Neutrinos, they are very small.*
> *They have no charge and have no mass*
> *and do not interact at all."*
>
> *John Updike*

effective number of degrees of freedom

For an estimate of the neutron freeze-out temperature of $T_f \approx 0.7\,\text{MeV}$ the value $g_* = 10.75$ had been used for the effective number of degrees of freedom. This corresponds to photons, e^+, e^-, and three families of ν and $\bar{\nu}$ as relativistic particles. The timeline of BBN is complicated slightly by the fact that g_* changes from 10.75 to 3.36 as the temperature drops below the electron mass, $m_e \approx 0.511\,\text{MeV}$, see (10.16). This means that somewhat more time elapses before the synthesis of deuterium can begin and, as a result, more neutrons have a chance to decay than would be the case with the higher value of g_*.

neutrino decoupling

Neutrinos are held in equilibrium with e^+ and e^- through the reaction $e^+ e^- \leftrightarrow \nu \bar{\nu}$. At a temperature around 1 MeV, however, the rate of this reaction drops below the expansion rate. Muon- and tau-type neutrinos decouple completely from the rest of the particles while ν_e continues to interact for a short while further through $\nu_e n \leftrightarrow p e^-$. After the neutron freeze-out at around 0.7 MeV, all neutrino flavours are decoupled from other particles. As they are stable, however, their number densities continue to decrease in proportion to $1/R^3$, just like other relativistic particles, and therefore they still contribute to the effective number of degrees of freedom.

For $T \gg m_e \approx 0.5\,\mathrm{MeV}$, the reaction $e^+ e^- \leftrightarrow \gamma\gamma$ proceeds at the same rate in both directions. As the temperature drops below the electron mass, however, the photons no longer have enough energy to allow $\gamma\gamma \to e^+ e^-$. The positrons as well as all but a small fraction of the initially present electrons annihilate through $e^+ e^- \to \gamma\gamma$. They no longer make a significant contribution to the total energy density, and therefore they do not contribute towards g_*.

annihilation and creation

The fact that the reaction $e^+ e^- \to \gamma\gamma$ produces photons means that the photon temperature decreases less quickly than it otherwise would. The neutrinos, however, are oblivious to this, and the neutrino temperature continues to scale as $T_\nu \sim 1/R$. Using thermodynamic arguments based on conservation of entropy, one can show that after positron annihilation the neutrino temperature is lower than that of the photons by a factor [17]

neutrino temperature

$$\frac{T_\nu}{T_\gamma} = \left(\frac{4}{11}\right)^{1/3} \approx 0.714 . \tag{10.15}$$

In the following T is always assumed to mean the photon temperature.

photon temperature

So, to obtain the total energy density, one needs to take into account that the neutrinos have a slightly lower temperature, and thus one has to use (9.19),

$$g_* = g_\gamma + \frac{7}{8} 2 N_\nu \frac{T_\nu^4}{T^4} = 2 + \frac{7}{8} 2 N_\nu \left(\frac{4}{11}\right)^{4/3} \approx 3.36 , \tag{10.16}$$

where for the final value $N_\nu = 3$ was used.

This new value of g_* alters the relation between expansion rate and temperature, and therefore also between time and temperature. Using (10.2) with $g_* = 3.36$ now gives

$$t T^2 = 1.32\,\mathrm{s\,MeV}^2 . \tag{10.17}$$

In the next section it will be shown that the neutrons, present at freeze-out, will be able to decay until deuterium production begins at around $T = 0.085\,\mathrm{MeV}$. From (10.17) this takes place at a time $t \approx 180\,\mathrm{s}$.

freeze-out temperature

To first approximation one can say that the neutrons can decay for around 3 minutes, after which time they are

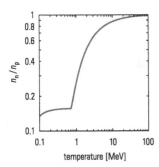

Fig. 10.3
The ratio n_n/n_p as a function of the temperature

neutron-to-proton ratio

quickly absorbed to form deuterium and then helium.[2] So, from freeze-out to the start of deuterium production, the neutron-to-proton ratio is

$$\frac{n_n}{n_p} = e^{-(m_n-m_p)/T_f}\, e^{-t/\tau_n} \, , \tag{10.18}$$

where the mean neutron lifetime is $\tau_n = 885.7\,\text{s}$. Figure 10.3 shows n_n/n_p as a function of the temperature. The freeze-out temperature is at $T_f = 0.7\,\text{MeV}$, below which the ratio is almost constant, falling slightly because of neutron decay. At a time $t = 180\,\text{s}$ ($T = 0.086\,\text{MeV}$), a value of

$$\frac{n_n}{n_p} \approx 0.13 \, . \tag{10.19}$$

is found.

10.5 Synthesis of Light Nuclei

> *"Give me matter and I will construct a world out of it."*
>
> *Immanuel Kant*

The synthesis of ^4He proceeds through a chain of reactions which includes, for example,

$$p\,n \to d\,\gamma \, , \tag{10.20}$$

$$d\,p \to {}^3\text{He}\,\gamma \, , \tag{10.21}$$

$$d\,{}^3\text{He} \to {}^4\text{He}\,p \, . \tag{10.22}$$

deuterium production

The binding energy of deuterium is $E_{\text{bind}} = 2.2\,\text{MeV}$, so if the temperature is so high that there are many photons with energies higher than this, then the deuterium will be broken apart as soon as it is produced. One might naïvely expect that the reaction (10.20) would begin to be effective as soon as the temperature drops to around 2.2 MeV. In fact this does not happen until a considerably lower temperature. This is because there are so many more photons than baryons, and the photon energy distribution, i.e., the Planck distribution, has a long tail towards high energies.

baryon-to-photon ratio

The nucleon-to-photon ratio is at this point essentially the same as the baryon-to-photon ratio $\eta = n_b/n_\gamma$, which

---

[2] The calculations are based on average values of thermodynamic distributions. The approximations therefore may show discontinuities which, however, would disappear if Maxwell–Boltzmann or Planck distributions, respectively, were used.

is around 10^{-9}. One can estimate roughly when deuterium production can begin to proceed by finding the temperature where the number of photons with energies greater than 2.2 MeV is equal to the number of nucleons. For a nucleon-to-photon ratio of 10^{-9}, this occurs at $T = 0.086$ MeV, which is reached at a time of $t \approx 3$ minutes.

Over the next several minutes, essentially all of the neutrons, except those that decay, are processed into ^{4}He. The abundance of ^{4}He is usually quoted by giving its mass fraction,

^{4}He mass fraction

$$Y_{\rm P} = \frac{\text{mass of }^4\text{He}}{\text{mass of all nuclei}} = \frac{m_{\rm He} n_{\rm He}}{m_{\rm N}(n_n + n_p)}, \qquad (10.23)$$

where the neutron and proton masses have both been approximated by the nucleon mass, $m_{\rm N} \approx m_n \approx m_p \approx 0.94$ GeV. There are four nucleons in ^{4}He, so, neglecting the binding energy, one has $m_{\rm He} \approx 4m_{\rm N}$. Furthermore, there are two neutrons per ^{4}He nucleus, so if one assumes that all of the neutrons end up in ^{4}He, one has $n_{\rm He} = n_n/2$. This will turn out to be a good approximation, as the next most common nucleus, deuterium, ends up with an abundance four to five orders of magnitude smaller than that of hydrogen. The ^{4}He mass fraction is therefore

$$Y_{\rm P} = \frac{4m_{\rm N}(n_n/2)}{m_{\rm N}(n_n + n_p)} = \frac{2(n_n/n_p)}{1 + n_n/n_p}$$
$$\approx \frac{2 \times 0.13}{1 + 0.13} \approx 0.23 . \qquad (10.24)$$

This rough estimate turns out to agree quite well with more detailed calculations. 23% ^{4}He mass fraction means that the number fraction of ^{4}He is about 6% with respect to hydrogen.

^{4}He number fraction

The fact that the universe finally ends up containing around one quarter ^{4}He by mass can be seen as the result of a number of rather remarkable coincidences. For example, mean lifetimes for weak decays can vary over many orders of magnitude. The fact that the mean neutron lifetime turns out to be 885.7 s is a complicated consequence of the rather close neutron and proton masses combined with strong and weak interaction physics. If it had turned out that τ_n were, say, only a few seconds or less, then essentially all of the neutrons would decay before they could be bound up into deuterium, and the chain of nuclear reactions could not have started.

tuning of parameters

The exact value of the decoupling temperature is also a complicated mixture of effects, being sensitive, for example, to g_*, which depends on the number of relativistic particle species in thermal equilibrium. If, for example, the decoupling temperature T_f had been not $0.7\,\text{MeV}$ but, say, $0.1\,\text{MeV}$, then the neutron-to-proton ratio would have been $e^{-1.3/0.1} \approx 2 \times 10^{-6}$, and essentially no helium would have formed. On the other hand, if it had been at a temperature

fine-tuning of parameters much higher than $m_n - m_p$, then there would have been equal amounts of protons and neutrons. Then the entire uni-

astrochemistry, astrobiology verse would have been made of helium. Usual hydrogen-burning stars would be impossible, and the universe would certainly be a very different place (see also Chap. 14: Astrobiology).

10.6 Detailed BBN

"Little by little, time brings out each several thing into view, and reason raises it up into the shores of light."

Lucretius

mass fractions A detailed modeling of nucleosynthesis uses a system of differential equations involving all of the abundances and reaction rates. The rates of nuclear reactions are parameterizations of experimental data. Several computer programs that numerically solve the system of rate equations are publicly available [18]. An example of predicted mass fractions versus temperature and time is shown in Fig. 10.4.

Around $t \approx 1000\,\text{s}$, the temperature has dropped to $T \approx 0.03\,\text{MeV}$. At this point the kinetic energies of nuclei are too low to overcome the Coulomb barriers and the fusion processes stop.

In order to compare the predictions of Big Bang Nucleosynthesis with observations, one needs to measure the *primordial* abundances of the light elements, i.e., as they were just after the BBN era. This is complicated by the fact that the abundances change as a result of *stellar* nucleosynthe-

stellar nucleosynthesis sis. For example, helium is produced in stars and deuterium is broken apart.

primordial ^{4}He abundance To obtain the most accurate measurement of the ^{4}He mass fraction, for example, one tries to find regions of hot ionized gas from 'metal-poor' galaxies, i.e., those where rel-

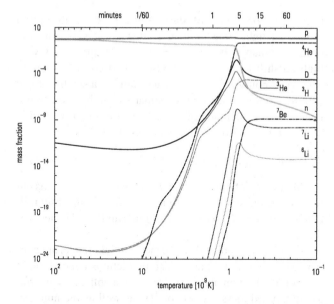

Fig. 10.4
Evolution of the mass and number
fractions of primordial elements.
^{4}He is given as a mass fraction
while the other elements are
presented as number fractions

atively small amounts of heavier elements have been produced through stellar burning of hydrogen. A recent survey
of data [2] concludes for the primordial ^{4}He mass fraction

$$Y_{\mathrm{P}} = 0.238 \pm 0.002 \pm 0.005 , \qquad (10.25)$$

where the first error is statistical and the second reflects systematic uncertainties. In contrast to the helium-4 content of
the universe which is traditionally given as a mass fraction,
the abundances of the other primordial elements are pre-
sented as number fractions, e.g., $n_{^7\mathrm{Li}}/n_p \equiv n_{^7\mathrm{Li}}/n_{\mathrm{H}}$ for ^{7}Li.

The best determinations of the ^{7}Li abundance come from
hot metal-poor stars from the galactic halo. As with ^{4}He, one
extrapolates to zero metallicity to find the primordial value.
Recent data [2] give a lithium-to-hydrogen ratio of

^{7}Li abundance
metal-poor stars

$$n_{^7\mathrm{Li}}/n_{\mathrm{H}} = 1.23 \times 10^{-10} . \qquad (10.26)$$

The systematic uncertainty on this value is quite large, however, corresponding to the range from about 1 to 2×10^{-10}.

Although deuterium is produced by the first reaction in
hydrogen-burning stars through $pp \rightarrow de^+\nu_e$, it is quickly
processed further into heavier nuclei. Essentially, no net
deuterium production takes place in stars and any present
would be quickly fused into helium. So to measure the primordial deuterium abundance, one needs to find gas clouds
at high redshift, hence far away and far back in time, that

primordial deuterium

have never been part of stars. These produce absorption
spectra in light from even more distant quasars. The hydrogen Lyman-α line at $\lambda = 121.6\,\text{nm}$ appears at very
high redshift ($z \geq 3$) in the visible part of the spectrum.

deuterium spectroscopy   The corresponding line from deuterium has a small isotopic
shift to shorter wavelengths. Comparison of the two compo-
nents gives an estimate of the deuterium-to-hydrogen ratio
D/H $\,\hat{=}\, n_d/n_p$. A recent measurement finds [19]

$$n_d/n_p = (3.40 \pm 0.25) \times 10^{-5} \,. \tag{10.27}$$

primordial element   Measuring of the primordial abundance of ^{3}He turns out
abundances   to be more difficult. There do not yet exist sufficiently reli-
able values for ^{3}He to test or constrain Big Bang Nucleosynthesis.

Finally, the predicted abundances of light nuclei are con-
fronted with measurements. The predictions depend, how-
ever, on the baryon density n_b or, equivalently, $\eta = n_b/n_\gamma$.
The predicted mass fraction of ^{4}He as well as the numbers
relative to hydrogen for D, ^{3}He, and ^{7}Li are shown as a function of η in Fig. 10.5 [2].

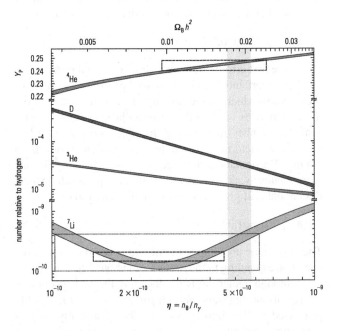

Fig. 10.5
Predictions for the abundances of
^{4}He, D, and ^{7}Li as a function of
the baryon-to-photon ratio η. Y_P is
the primordial ^{4}He mass fraction.
Traditionally, the ^{4}He content of
the universe is given as mass
fraction, while the other primordial
elements are presented as number
fraction (see also the *broken
vertical scale*). The *larger box* for
^{7}Li/H includes the systematical
error added in quadrature to the
statistical error

The deuterium fraction D/H decreases for increasing η
because a higher baryon density means that deuterium is
processed more completely into helium. Since the resulting

prediction for D/H depends quite sensitively on the baryon density, the measured D/H provides the most accurate determination of η. The measured value from (10.27), namely, $n_d/n_p = (3.40 \pm 0.25) \times 10^{-5}$, gives the range of allowed η values, which are shown by the vertical band in Fig. 10.5. This corresponds to **sensitive D/H ratio**

$$\eta = (5.1 \pm 0.5) \times 10^{-10} . \tag{10.28}$$

The boxes in Fig. 10.5 indicate the measured abundances of ^{4}He and ^{7}Li (from slightly different analyses than those mentioned above). The size of the boxes shows the measurement uncertainty. These measurements agree remarkably well with the predictions, especially when one considers that the values span almost 10 orders of magnitude.

The value of $\eta = n_b/n_\gamma$ determines the baryon density, since the photon density is well-known from the measured CMB temperature, $T = 2.725$ K, through (9.11) to be $n_\gamma = 2\zeta(3)T^3/\pi^2$. Therefore, the value of η can be converted into **baryon-to-photon ratio** a prediction for the energy density of baryons divided by the critical density,

$$\Omega_b = \frac{\varrho_b}{\varrho_c} . \tag{10.29}$$

The critical density is given by (8.33) as $\varrho_c = 3H_0^2/8\pi G$. The baryons today are non-relativistic, so their energy density is simply the number of nucleons per unit volume times the mass of a nucleon, i.e., $\varrho_b = n_b m_N$, where $m_N \approx 0.94$ GeV. Putting these ingredients together gives **Ω_b determination**

$$\Omega_b = 3.67 \times 10^7 \times \eta h^{-2} , \tag{10.30}$$

where h is defined by $H_0 = 100\, h\, \mathrm{km\, s}^{-1}\, \mathrm{Mpc}^{-1}$. Using $h = 0.71^{+0.04}_{-0.03}$ and η from (10.28) gives **baryon fraction of the universe**

$$\Omega_b = 0.038 \pm 0.005 , \tag{10.31}$$

where the uncertainty originates both from that of the Hubble constant and also from that of η. Recently η and Ω_b have been measured to higher acuracy using the temperature variations in the cosmic microwave background radiation, leading to $\Omega_b = 0.044 \pm 0.004$; this will be followed up in Chap. 11 (see also Table 11.1). The values from the BBN and CMB studies are consistent with each other and, taken together, provide a convincing confirmation of the Big Bang model.

10.7 Constraints on the Number of Neutrino Families

*"If there were many neutrino genera-
tions, we would not be here to count
them, because the whole universe would
be made of helium and no life could develop."*

Anonymous

N_ν **from cosmology**

In this section it will be shown how the comparison of the
measured and predicted ^{4}He mass fractions can result in
constraints on the particle content of the universe at BBN
temperatures. For example, the Standard Model has $N_\nu = 3$,
but one can ask whether additional families exist. It will be
seen that BBN was able to constrain N_ν to be quite close
to three – a number of years earlier than accelerator experiments were able to determine the same quantity to high precision using electron–positron collisions at energies near the
Z resonance.

how many neutrino families?

Once the parameter η has been determined, the predicted
^{4}He mass fraction is fixed to a narrow range of values close
to $Y_P = 0.24$. As was noted earlier, this prediction is in
good agreement with the measured abundance. The prediction depended, however, on the effective number of degrees
of freedom,

$$g_* = 2 + \frac{7}{8}(4 + 2N_\nu) , \qquad (10.32)$$

where N_ν is the number of neutrino families. Earlier the
Standard Model value $N_\nu = 3$ was used, which gave $g_* =
10.75$. This then determines the expansion rate through

$$H = 1.66\sqrt{g_*}\frac{T^2}{m_{Pl}} . \qquad (10.33)$$

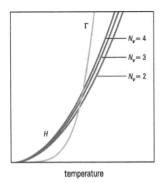

temperature

Fig. 10.6
The reaction rate $\Gamma(\nu_e n \leftrightarrow e^- p)$
and the expansion rate H for
$N_\nu = 2, 3,$ and 4 as a function of
temperature

The effective number of degrees of freedom therefore has an
impact on the freeze-out temperature, where one has $H =
\Gamma(n\nu_e \leftrightarrow pe^-)$, see (10.11). This can be seen in Fig. 10.6,
which shows the reaction rate $\Gamma(n\nu_e \leftrightarrow pe^+) \approx G_F^2 T^5$ and
the expansion rate H versus temperature. The expansion rate
is shown using three different values of g_*, corresponding to
$N_\nu = 2, 3,$ and 4.

From Fig. 10.6 one can see that the freeze-out tempera-
ture T_f is higher for larger values of g_*, i.e., for higher N_ν.
At T_f the neutron-to-proton ratio freezes out to $n_n/n_p =$

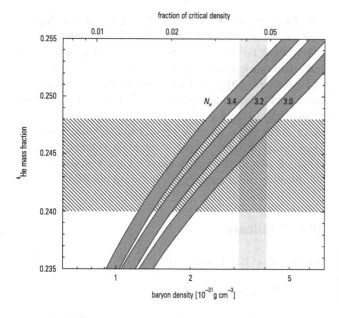

Fig. 10.7
The predicted ^{4}He mass fraction as a function of η for different values of N_ν

$e^{-(m_n-m_p)/T_f}$. If this occurs at a higher temperature, then the ratio is higher, i.e., there are more neutrons available to make helium, and the helium abundance will come out higher.

This can be seen in Fig. 10.7, which shows the predicted helium abundance as a function of the baryon density. The three diagonal bands show the predicted Y_P for different values of an equivalent number of neutrino families $N_\nu = 3.0$, 3.2, and 3.4. Of course, this no longer represents the (integer) number of neutrino flavours but rather an effective parameter that simply gives g_*. The data are consistent with $N_\nu = 3$ and are clearly incompatible with values much higher than this [19].

N_ν and helium abundance

The equivalent number of light (i.e., with masses $\leq m_Z/2$) neutrino families has also been determined at the Large Electron–Positron (LEP) collider from the total width of the Z resonance, as was shown in Fig. 2.1 (see also Problem 3 in this chapter). From a combination of data from the LEP experiments one finds [20]

accelerator data on N_ν

$$N_\nu = 2.9835 \pm 0.0083 . \tag{10.34}$$

Although this is 1.7 times the quoted error bar below 3, it is clear that $N_\nu = 3$ fits reasonably well and that any other integer value is excluded.

Before around 1990, when N_ν was determined to high precision in accelerator experiments, BBN measurements had already found that there could be at most one additional neutrino family [21] (see also Chap. 2). The interplay between these two different determinations of N_ν played an important rôle in alerting particle physicists to the relevance of cosmology. Although the example that has just been discussed is for the number of neutrino families, the same arguments apply to any particles that would contribute to g_* such as to affect the neutron freeze-out temperature. Thus the abundances of light elements provide important constraints for any theory involving new particles that would contribute significantly to the energy density during the BBN era.

impact of cosmology on particle physics

10.8 Problems

1. Equation (10.4) shows that

   $$tT^2 \approx 0.74 \,\text{s}\,\text{MeV}^2 \quad.$$

   How can this numerical result be derived from the previous equations?

2. In the determination of the neutron-to-proton ratio the cross section for the $(n \leftrightarrow p)$ conversion reaction

   $$\nu_e + n \rightarrow e^- + p$$

   was estimated from

   $$\langle \sigma\, v \rangle \approx G_F^2 T^2 \quad.$$

   How can the T^2 dependence of the cross section be motivated?

3. The measurement of the total width Γ_Z of the Z resonance at LEP led to a precise determination of the number of light neutrinos N_ν. How can N_ν be obtained from the Z width?

   The solution of this problem requires some intricate details of elementary particle physics.

11 The Cosmic Microwave Background

> "God created two acts of folly. First, He created the universe in a Big Bang. Second, He was negligent enough to leave behind evidence for this act, in the form of the microwave radiation."
>
> *Paul Erdös*

In this chapter the description of the early universe to cover the first several hundred thousand years of its existence will be presented. This leads to one of the most important pillars of the Big Bang model: the cosmic microwave background radiation or CMB. It will be seen how and when the CMB was formed and what its properties are. Most important among these are its blackbody energy spectrum characterized by an average temperature of $T = 2.725$ K, and the fact that one sees very nearly the same temperature independent of direction. Recent measurements of the slight dependence of the temperature on direction have been used to determine a number of cosmological parameters to a precision of several percent.

overview of cosmic microwave and blackbody radiation

precision determination of cosmological parameters

11.1 Prelude: Transition to a Matter-Dominated Universe

> "We know too much about matter today to be materialists any longer."
>
> *Harides Chaudhuri*

Picking up the timeline from the last chapter, Big Bang Nucleosynthesis was completed by around $t \approx 10^3$ s, i.e., when the temperature had fallen to several hundredths of an MeV. Any neutrons that by this time had not become bound into heavier nuclei soon decayed.

nucleosynthesis

The break-neck pace of the early universe now shifts gears noticeably. The next interesting event takes place when the energy density of radiation, i.e., relativistic particles (photons and neutrinos), drops below that of the matter (non-relativistic nuclei and electrons), called the time of 'matter–radiation equality'. In order to trace the time evolution of the

matter–radiation equality

universe, one needs to determine when this occurred, since the composition of the energy density influences the time dependence of the scale factor $R(t)$.

present-day energy densities

Estimates of the time of matter–radiation equality depend on what is assumed for the contents of the universe. For matter it will be seen that estimates based on the CMB properties as well as on the motion of galaxies in clusters give $\Omega_{m,0} = \varrho_{m,0}/\varrho_{c,0} \approx 0.3$, where as usual the subscript 0 indicates a present-day value. For photons it will be found from the CMB temperature: $\Omega_{\gamma,0} = 5.0 \times 10^{-5}$. Taking into account neutrinos brings the total for radiation to $\Omega_{r,0} = 8.4 \times 10^{-5}$, so currently, matter contributes some 3 600 times more to the total energy density than does radiation.

R dependence of energy densities

In Chap. 8 it has been derived by solving the Friedmann equation how to predict the time dependence of the different components of the energy density. For radiation $\varrho_r \sim 1/R^4$ was obtained, whereas for matter $\varrho_m \sim 1/R^3$ was found. Therefore, the ratio follows $\varrho_m/\varrho_r \sim R$, and it was thus equal to unity when the scale factor R was 3 600 times smaller than its current value.

time dependence of R

In order to pin down when this occurred, one needs to know how the scale factor varies in time. If it is assumed that the universe has been matter dominated from the time of matter–radiation equality up to now, then one has $R \sim t^{2/3}$, and this leads to a time of matter–radiation equality, t_{mr}, of around 66 000 years. In fact, there is now overwhelming evidence that vacuum energy makes up a significant portion of the universe, with $\Omega_\Lambda \approx 0.7$. Taking this into account leads to a somewhat earlier time of matter–radiation equality of around $t_{mr} \approx 50\,000$ years.

transition to a matter-dominated energy density

When the dominant component of the energy density changes from radiation to matter, this alters the relation between the temperature and time. For non-relativistic particle types with mass m_i and number density n_i, one now gets for the energy density

$$\varrho \approx \sum_i m_i n_i \, , \tag{11.1}$$

where the sum includes at least baryons and electrons, and perhaps also 'dark-matter' particles. The Friedmann equation (neglecting the curvature term) then reads

$$H^2 = \frac{8\pi G}{3} \sum_i m_i n_i \, . \tag{11.2}$$

Assuming the particles that contribute to the energy density are stable, one obtains

$$n_i \sim 1/R^3 \sim T^3 . \tag{11.3}$$

Furthermore, since $R \sim t^{2/3}$, the expansion rate is now given by

time dependence of R and T under matter domination

$$H = \frac{\dot{R}}{R} = \frac{2}{3t} . \tag{11.4}$$

Combining (11.3) with (11.4) or $R \sim t^{2/3}$ then leads to

$$T^3 \sim \frac{1}{t^2} . \tag{11.5}$$

This is to be contrasted with the relation $T^2 \sim t^{-1}$ valid for the era when the energy density was dominated by relativistic particles.

11.2 Discovery and Basic Properties of the CMB

> *"What we have found is evidence for the birth of the universe. It's like looking at God."*
>
> *George Smoot*

The existence of the CMB was predicted by Gamow [22] in connection with Big Bang Nucleosynthesis. It was shown in Chap. 10 that BBN requires temperatures around $T \approx 0.08\,\text{MeV}$, which are reached at a time $t \approx 200\,\text{s}$. By knowing the cross section for the first reaction, $p + n \rightarrow d + \gamma$, and the number density n of neutrons and protons, one can predict the reaction rate $\Gamma = n\langle\sigma v\rangle$.

In order for BBN to produce the observed amount of helium, one needs a sufficiently high rate for the deuterium fusion reaction over the relevant time scale. This corresponds to requiring Γt to be at least on the order of unity at $t \approx 200\,\text{s}$, when the temperature passes through the relevant range. This assumption determines the nucleon density during the BBN phase.

deuterium fusion

Since the BBN era, the nucleon and photon densities have both followed $n \sim 1/R^3 \sim T^3$. So, by comparing the nucleon density in the BBN era to what one finds today, the current temperature of the photons can be predicted. Reasoning along these lines, Alpher and Herman [23] estimated

photon temperature estimation

**photon contribution
to the cosmic energy density**

**prediction and observation
of the CMB**

**earth-based vs.
satellite-supported
CMB measurements**

a CMB temperature of around 5 K, which turned out to be not far off.

Even without invoking BBN one can argue that the contribution of photons to the current energy density cannot exceed by much the critical density. If one assumes, say, $\Omega_\gamma \leq 1$, then this implies $T \leq 32$ K.

Gamow's prediction of the CMB was not pursued for a number of years. In the 1960s, a team at Princeton (Dicke, Peebles, Roll, and Wilkinson) did take the prediction seriously and set about building an experiment to look for the CMB. Unknown to them, a pair of radio astronomers, A. Penzias and R. Wilson at Bell Labs in New Jersey, were calibrating a radio antenna in preparation for studies unrelated to the CMB. They reported finding an "effective zenith noise temperature ... about 3.5°K higher than expected. This excess temperature is, within the limits of our observations, isotropic, unpolarized, and free from seasonal variations ..." [24]. The Princeton team soon found out about Penzias' and Wilson's observation and immediately supplied the accepted interpretation [25].

Although the initial observations of the CMB were consistent with a blackbody spectrum, the earth-based observations were only able to measure accurately the radiation at wavelengths of several cm; shorter wavelengths are strongly absorbed by the water in the atmosphere. The peak of the blackbody spectrum for a temperature of 3 K, however, is around 2 mm. It was not until 1992 that the COBE satellite made accurate measurements of the CMB from space. This showed that the form of the energy distribution is extremely close to that of blackbody radiation, i.e., to a Planck distribution, as shown in Fig. 11.1.

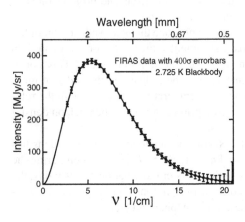

Fig. 11.1
The spectrum of the CMB measured by the COBE satellite together with the blackbody curve for $T = 2.725$ K. The *error bars* have been enlarged by a factor of 400; any deviations from the Planck curve are less than 0.005% (from [26])

11.3 Formation of the CMB

> "The old dream of wireless communica-
> tion through space has now been real-
> ized in an entirely different manner than
> many had expected. The cosmos' short
> waves bring us neither the stock market
> nor jazz from different worlds. With soft
> noises they rather tell the physicists of
> the endless love play between electrons
> and protons."
>
> *Albrecht Unsöld*

At very early times, any protons and electrons that managed
to bind together into neutral hydrogen would be dissociated
very quickly by collision with a high-energy photon. As the
temperature decreased, the formation of hydrogen eventually became possible and the universe transformed from an
ionized plasma to a gas of neutral atoms; this process is
called *recombination*. The reduction of the free electron den- **recombination: creation**
sity to almost zero meant that the mean free path of a photon **of hydrogen atoms**
soon became so long that most photons have not scattered
since. This is called the *decoupling* of photons from mat- **decoupling of photons**
ter. By means of some simple calculations one can estimate **from matter**
when recombination and decoupling took place.

Neutral hydrogen has an electron binding energy of
13.6 eV and is formed through the reaction

$$p + e^- \rightarrow H + \gamma \ . \tag{11.6}$$

Naïvely one would expect the fraction of neutral hydrogen to become significant when the temperature drops below 13.6 eV. But because the baryon-to-photon ratio $\eta \approx$
5×10^{-10} is very small, the temperature must be significantly lower than this before the number of photons with
$E > 13.6$ eV is comparable to the number of baryons.
(This is the same basic argument for why deuterium production began not around $T = 2.2$ MeV, the binding energy of deuterium, but rather much lower.) One finds that
the numbers of neutral and ionized atoms become equal at
a *recombination temperature* of $T_{rec} \approx 0.3$ eV (3500 K). At **recombination temperature**
this point the universe transforms from an ionized plasma to
an essentially neutral gas of hydrogen and helium.

It can be estimated when recombination took place by **estimation**
comparing the temperature of the CMB one observes today, **of the recombination era**
$T_0 \approx 2.73$ K, to the value of the recombination temperature
$T_{rec} \approx 0.3$ eV. Recall from Sect. 9.2.4 that the wavelength

of a photon follows $\lambda \sim R$. Therefore, the ratio of the scale factor R at a previous time to its value R_0 now is related to the redshift z through

$$\frac{R_0}{R} = \frac{\lambda_0}{\lambda} = 1 + z \, . \tag{11.7}$$

Furthermore, today's CMB temperature is measured to be $T_0 \approx 2.73\,\text{K}$ and it is known that $T \sim 1/R$. Therefore one gets

$$1+z = \frac{T}{T_0} \approx \frac{0.3\,\text{eV}}{2.73\,\text{K}} \times \frac{1}{8.617 \times 10^{-5}\,\text{eV K}^{-1}} \approx 1300 \, , \tag{11.8}$$

where Boltzmann's constant was inserted to convert temperature from units of eV to K. If one assumes that the scale factor follows $R \sim t^{2/3}$ from this point up to the present, then it is found that recombination was occurring at

recombination time $\approx 300\,000$ years

$$t_{\text{rec}} = t_0 \left(\frac{R}{R_0}\right)^{3/2} = t_0 \left(\frac{T_0}{T_{\text{rec}}}\right)^{3/2} = \frac{t_0}{(1 + z_{\text{rec}})^{3/2}}$$

$$\approx \frac{1.4 \times 10^{10}\,\text{years}}{(1300)^{3/2}} \approx 300\,000\,\text{years} \, . \tag{11.9}$$

Shortly after recombination, the mean free path for a photon became so long that photons effectively decoupled from matter. While the universe was an ionized plasma, the photon scattering cross section was dominated by Thomson scattering, i.e., elastic scattering of a photon by an electron. The mean free path of a photon is determined by the number density of electrons, which can be predicted as a function of time, and by the Thomson scattering cross section, which can be calculated. As the universe expands, the electron density decreases leading to a longer mean free path for the photons. This path length becomes longer than the horizon distance (the size of the observable universe at a given time) at a *decoupling temperature* of $T_{\text{dec}} \approx 0.26\,\text{eV}$ (3000 K) corresponding to a redshift of $1 + z \approx 1100$. This condition defines *decoupling* of photons from matter. The decoupling time is

decoupling of photons from matter

decoupling temperature

$$t_{\text{dec}} = t_0 \left(\frac{T_0}{T_{\text{dec}}}\right)^{3/2} = \frac{t_0}{(1 + z_{\text{dec}})^{3/2}} \approx 380\,000\,\text{years} \, . \tag{11.10}$$

Once the photons and matter decoupled, the photons simply continue unimpeded to the present day. One can define a *surface of last scattering* as the sphere centered about us with a radius equal to the mean distance to the last place where the CMB photons scattered. To a good approximation this is equal to the distance to where decoupling took place and the time of last scattering is essentially the same as t_{dec}. So, when the CMB is detected, one is probing the conditions in the universe at a time of approximately 380 000 years after the Big Bang.

surface of last scattering

transparent universe

11.4 CMB Anisotropies

> *"As physics advances farther and farther every day and develops new axioms, it will require fresh assistance from mathematics."*
>
> Francis Bacon

The initial measurements of the CMB temperature by Penzias and Wilson indicated that the temperature was independent of direction, i.e., that the radiation was isotropic, to within an accuracy of around 10%. More precise measurements eventually revealed that the temperature is about one part in one thousand hotter in one particular direction of the sky than in the opposite. This is called the dipole anisotropy and is interpreted as being caused by the motion of the Earth through the CMB. Then in 1992 the COBE satellite found anisotropies at smaller angular separations at a level of one part in 10^5. These small variations in temperature have recently been measured down to angles of several tenths of a degree by several groups, including the WMAP[1] satellite, from which one can extract a wealth of information about the early universe.

dipole anisotropy

**discovery
of small-angle anisotropies**

In order to study the CMB anisotropies one begins with a measurement of the CMB temperature as a function of direction, i.e., $T(\theta, \phi)$, where θ and ϕ are spherical coordinates, i.e., polar and azimuthal angle, respectively. As with any function of direction, it can be expanded in spherical harmonic functions $Y_{lm}(\theta, \phi)$ (a *Laplace series*),

Laplace series

$$T(\theta, \phi) = \sum_{l=0}^{\infty} \sum_{m=-l}^{l} a_{lm} Y_{lm}(\theta, \phi) . \qquad (11.11)$$

---

[1] WMAP – Wilkinson Microwave Anisotropy Probe

Some of the mathematical formalism of the Laplace series is given in Appendix A.2. This expansion is analogous to a Fourier series, where the higher-order terms correspond to higher frequencies. Here, terms at higher l correspond to structures at smaller angular scales. The same mathematical **multipole expansion** technique is used in the *multipole expansion* of the potential from an electric charge distribution. The terminology is borrowed from this example and the terms in the series are referred to as multipole moments. The $l = 0$ term is the monopole, $l = 1$ the dipole, etc.

Once one has estimates for the coefficients a_{lm}, the amplitude of regular variation with angle can be summarized by defining

$$C_l = \frac{1}{2l + 1} \sum_{m=-l}^{l} |a_{lm}|^2 \ . \tag{11.12}$$

angular power spectrum The set of numbers C_l is called the *angular power spectrum*. The value of C_l represents the level of structure found at an angular separation

$$\Delta\theta \approx \frac{180°}{l} \ . \tag{11.13}$$

The measuring device will in general only be able to resolve angles down to some minimum value; this determines the maximum measurable l.

11.5 The Monopole and Dipole Terms

"The universe contains the record of its past the way that sedimentary layers of rock contain the geological record of the Earth's past."

Heinz R. Pagels

The $l = 0$ term in the expansion of $T(\theta, \phi)$ gives the tem- **monopole term** perature averaged over all directions. The most accurate determination of this value comes from the COBE satellite,

$$\langle T \rangle = 2.725 \pm 0.001 \text{ K} \ . \tag{11.14}$$

In the 1970s it was discovered that the temperature of the CMB in a particular direction was around 0.1% hotter than in the opposite direction. This corresponds to a non-**dipole term** zero value of the $l = 1$ or *dipole* term in the Laplace expansion. The dipole anisotropy has recently been remeasured by the WMAP experiment, which finds a temperature difference of

$$\frac{\Delta T}{T} = 1.23 \times 10^{-3} . \tag{11.15}$$

The temperature appears highest in the direction (right ascension and declination, see Appendix C for the definition of these coordinates) $(\alpha, \delta) = (11.20^{h}, -7.22°)$. This temperature variation has a simple interpretation, namely, the movement of the Earth through the (local) unique reference frame in which the CMB has no dipole anisotropy. This frame is in some sense the (local) 'rest frame' of the universe. The solar system and with it the Earth are moving through it with a speed of $v = 371$ km/s towards the constellation Crater (between Virgo and Hydra). The CMB is blueshifted to slightly higher temperature in the direction of motion and redshifted in the opposite direction. This dipole pattern in the map of the CMB temperature is shown in Fig. 11.2. The map is an equal-area projection in galactic coordinates with the plane of the Milky Way running horizontally through the plot.

local rest frame

Fig. 11.2
Map of the CMB temperature measured by the COBE satellite. The dipole pattern is due to the motion of the Earth through the CMB (from [27]) {27}

11.5.1 Small-Angle Anisotropy

If small density fluctuations were to exist in the early universe, then one would expect these to be amplified by gravity, with more dense regions attracting even more matter, until the matter of the universe was separated into clumps. This is how galaxy formation is expected to have taken place. Given that one sees a certain amount of clumpiness today, one can predict what density variations must have existed at the time of last scattering. These variations would correspond to regions of different temperature, and so from the

**fluctuations
and galaxy formation**

observed large-scale structure of the universe one expected to see anisotropies in the CMB temperature at a level of around one part in 10^5.

angular resolution
of COBE and WMAP

These small-angle anisotropies were finally observed by the COBE satellite in 1992. COBE had an angular resolution of around 7° and could therefore determine the power spectrum up to a multipole number of around $l = 20$. In the following years, balloon experiments were able to resolve much smaller angles but with limited sensitivity. Finally, in 2003 the WMAP project made very accurate measurements of the CMB temperature variations with an angular resolution of 0.2°. One of the maps (with the dipole term subtracted) is shown in Fig. 11.3.

Fig. 11.3
Cosmographic map of the CMB temperature measured by the WMAP satellite with the dipole component subtracted (from [28]) {28}

Because of the better angular resolution of WMAP compared to COBE, it was possible to make an accurate measurement of the angular power spectrum up to $l \approx 1000$, as shown in Fig. 11.4.

11.6 Determination of Cosmological Parameters

"There are probably few features of theoretical cosmology that could not be completely upset and rendered useless by new observational discoveries."

Sir Hermann Bondi

**conclusions from the CMB
angular power spectrum**

The angular power spectrum of the CMB can be used to make accurate determinations of many of the most important cosmological parameters, including the Hubble constant H, the baryon-to-photon ratio η, the total energy density over

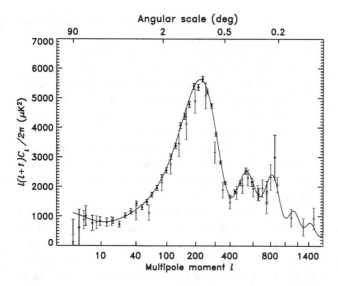

Fig. 11.4
CMB power spectrum. The set of measurements with the smaller error bars is from WMAP; those with the larger errors represent an average of measurements prior to WMAP (from [29])

the critical density, Ω, as well as the components of the energy density from baryons, Ω_b, and from all non-relativistic matter Ω_m.

Ω determination

As an example, in the following a rough idea will be given of how the angular power spectrum is sensitive to Ω. Consider the largest region that could be in causal contact at the time of last scattering $t_{ls} \approx t_{dec} \approx 380\,000$ years. This distance is called the *particle horizon* d_H. Naïvely one would expect this to be $d_H = t$ (i.e., ct, but $c = 1$ has been assumed). This is not quite right because the universe is expanding. The correct formula for the particle-horizon distance at a time t in an isotropic and homogeneous universe is (see, e.g., [30]),

particle horizon

distances in an expanding universe

$$d_H(t) = R(t) \int_0^t \frac{dt'}{R(t')} .\qquad(11.16)$$

If the time before matter–radiation equality is considered ($t_{mr} \approx 50\,000$ years), then $R \sim t^{1/2}$ and $d_H(t) = 2t$. For the matter-dominated era one has $R \sim t^{2/3}$ and $d_H = 3t$. If a sudden switch from $R \sim t^{1/2}$ to $R \sim t^{2/3}$ is assumed at t_{mr}, then one finds from integrating (11.16) a particle horizon at t_{ls} of

$$d_H(t_{ls}) = 3t_{ls} - t_{ls}^{2/3} t_{mr}^{1/3} \approx 950\,000 \text{ (light-)years} .$$

$$(11.17)$$

As most of the time up to t_{ls} is matter dominated, the result is in fact close to $3t_{ls}$.

density fluctuations as sound waves   A detailed modeling of the density fluctuations in the early universe predicts a large level of structure on distance scales roughly up to the horizon distance. These fluctuations are essentially sound waves in the primordial plasma, i.e., regular pressure variations resulting from the infalling of matter into small initial density perturbations. These initial perturbations may have been created at a much earlier time, e.g., at the end of the inflationary epoch.

By looking at the angular separation of the temperature fluctuations, in effect one measures the distance between the density perturbations at the time when the photons were emitted.

proper distance   To relate the angles to distances, one needs to review

angular diameter distance   briefly the *proper distance* and *angular diameter distance*. The *proper distance* d_p at a time t is the length one would measure if one could somehow stop the Hubble expansion and lay meter sticks end to end between two points. In an expanding universe one finds that the current proper distance (i.e., at t_0) to the surface of last scattering is given by

$$d_p(t_{ls}) = R(t_0) \int_{t_{ls}}^{t_0} \frac{dt}{R(t)} \, , \tag{11.18}$$

Note that this is the current proper distance to the position of the photon's emission, assuming that that place has been carried along with the Hubble expansion. (The particle-horizon distance used above is simply the proper distance to the source of a photon emitted at $t = 0$.) If matter domination is assumed, i.e., $R \sim t^{2/3}$ since t_{ls}, then one gets $d_p(t_{ls}) = 3(t_0 - t_{ls})$, which can be approximated by $d_p(t_{ls}) \approx 3t_0$.

angular variations   Now, what one wants to know is the angle subtended by

of temperature variations   a temperature variation which was separated by a distance perpendicular to our line of sight of $\delta = 3t_{ls}$ when the photons were emitted. To obtain this one needs to divide δ not by the current proper distance from us to the surface of last scattering, but rather by the distance that it *was* to us at the time when the photons started their journey. This location has been carried along with the Hubble expansion and is now further away by a factor equal to the ratio of the scale factors, $R(t_0)/R(t_{ls})$. Using (11.7), therefore, one finds

$$\Delta\theta = \frac{\delta}{d_p(t_{ls})} \frac{R(t_0)}{R(t_{ls})} = \frac{\delta}{d_p(t_{ls})} (1 + z) \, , \tag{11.19}$$

where $z \approx 1100$ is the redshift of the surface of last scattering. Thus, if a region is considered whose size was equal to the particle-horizon distance at the time of last scattering as given by (11.17) as viewed from today, then it will subtend an angle

$$\Delta\theta \approx \frac{3t_{ls} - t_{ls}^{2/3} t_{mr}^{1/3}}{3t_0}(1+z) \tag{11.20}$$

$$\approx \frac{950\,000\,a}{3 \times 1.4 \times 10^{10}\,a} \times 1100 \times \frac{180°}{\pi} \approx 1.4° \ .$$

angular separation of the particle-horizon distance at last scattering as seen today

The structure at or just below this angular scale corresponds to the 'acoustic peaks' visible in the power spectrum starting at around $l \approx 200$.

The naming of the structures in the power spectrum as 'acoustic peaks' comes about for the following reason: as already mentioned, the density fluctuations in the early universe caused gravitational instabilities. When the matter fell into these gravitational potential wells, this matter was compressed, thereby getting heated up. This hot matter radiated photons causing the plasma of baryons to expand, thereby cooling down and producing less radiation as a consequence. With decreasing radiation pressure the irregularities reach a point where gravity again took over initiating another compression phase. The competition between gravitational accretion and radiation pressure caused longitudinal acoustic oscillations in the baryon fluid. After decoupling of matter from radiation the pattern of acoustic oscillations became frozen into the CMB. CMB anisotropies therefore are a consequence of sound waves in the primordial proton fluid.

'acoustic peaks'

sound waves in the proton fluid

The angle subtended by the horizon distance at the time of last scattering depends, however, on the geometry of the universe, and this is determined by Ω, the ratio of the energy density to the critical density. This is illustrated schematically in Fig. 11.5. In Fig. 11.5(a), $\Omega = 1$ is assumed and therefore the universe is described by a flat geometry. The angles of the triangle sum to 180° and the angle subtended by the acoustic horizon has a value a bit less than 1°. If, however, one has $\Omega < 1$, then the universe is open and is described by a geometry with negative curvature. The photons then follow the trajectories shown in Fig. 11.5(b), and the angles of the triangle sum to less than 180°. In this case, the angle that one observes for the acoustic-horizon distance is reduced, and therefore one would see the first acoustic peak

dependence of measured angles on the Ω parameter

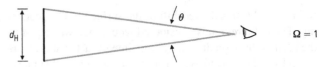

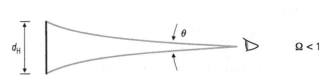

Fig. 11.5
The horizon distance at the time of last scattering as viewed by us today in **(a)** a flat universe ($\Omega = 1$) and **(b)** an open universe ($\Omega < 1$)

at a higher value of the multipole number l. It can be shown [31] that the position of the first acoustic peak in the angular power spectrum is related to Ω by

$$l_{\text{peak}} \approx \frac{220}{\sqrt{\Omega}} \,. \tag{11.21}$$

This behaviour can be seen in Fig. 11.6, which shows the predicted power spectra for several values of Ω.

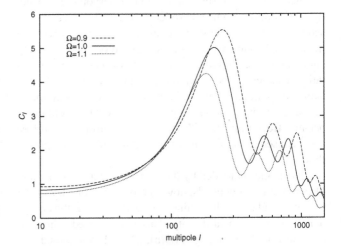

Fig. 11.6
Predicted CMB power spectra for different values of the current total energy density (values computed with the program CMBFAST [32])

The detailed structure of the peaks in the angular power spectrum depends not only on the total energy density but on many other cosmological parameters as well, such as the Hubble constant H_0, the baryon-to-photon ratio η, the energy-density contributions from matter, baryons, vacuum energy, etc. Using the power-spectrum measurement shown in Fig. 11.4, the WMAP team has determined many of these parameters to a precision of several percent or better. Some are shown in Table 11.1. Among the most important of these, one sees that the Hubble constant is now finally known to an accuracy of several percent, and the value obtained is in good agreement with the previous average ($h = 0.7 \pm 0.1$). Furthermore, the universe is flat, i.e., Ω is so close to unity that surely this cannot be a coincidence.

parameter	value and experimental error
Hubble constant h	$0.71^{+0.04}_{-0.03}$
ratio of total energy density to ϱ_c, Ω	1.02 ± 0.02
baryon-to-photon ratio η	$6.1^{+0.3}_{-0.2} \times 10^{-10}$
baryon energy density over ϱ_c, Ω_b	0.044 ± 0.004
matter energy density over ϱ_c, Ω_m	0.27 ± 0.04
vacuum energy density over ϱ_c, Ω_Λ	0.73 ± 0.04

Table 11.1
Some of the cosmological parameters determined by the WMAP experiment through measurement of the CMB angular power spectrum

Data from the Cosmic Background Imager (CBI) [33] – an array of antennae operating at frequencies from 26 to 36 GHz in the Atacama Desert – confirm the picture of a flat universe, giving $\Omega = 0.99 \pm 0.12$ in agreement with results from the Boomerang[2] and Maxima[3] experiments, and, of course, with WMAP.

In addition, WMAP has determined the baryon-to-photon ratio η to an accuracy of several percent. The value found is a bit higher than that obtained from the deuterium abundance, $\eta = (5.1 \pm 0.5) \times 10^{-10}$, but the latter measurement's error did not represent the entire systematic uncertainty and, in fact, the agreement between the two is quite reasonable. The fact that two completely independent measurements yield such close values is surely a good sign. The densities from non-relativistic matter Ω_m and from vacuum energy Ω_Λ are similarly well determined, and these measurements confirm earlier values based on completely different observables.

baryon-to-photon ratio measured by WMAP

Further improvement in the determination of these and other parameters is expected from the PLANCK project (formerly called COBRAS/SAMBA), scheduled to be launched by the European Space Agency in 2007 [34].

improved measurements in future

[2] Boomerang – Balloon Observations of Millimetric Extragalactic Radiation and Geophysics
[3] Maxima – Millimeter Anisotropy Experiment Imaging Array

particle physics
in the early universe These experiments provide another close bridge between cosmology and the particle physics of the early universe. In addition to helping pin down important cosmological parameters, the CMB may provide the most important clues needed to understand particle interactions at ultrahigh energy scales – energies that will never be attained by manmade particle accelerators. This theme will be further explored in the next chapter on *inflation*.

11.7 Problems

1. What is the probability that a photon from the Big Bang undergoes a scattering after having passed the *surface of last scattering*?
2. Estimate a limit for the cosmological constant!
3. The Friedmann equation extended by the cosmological constant is nothing but a relation between different forms of energy:

$$\underbrace{\frac{m}{2}\dot{R}^2}_{\text{kinetic}} + \underbrace{\left(-\frac{GmM}{R} - \frac{1}{6}m\Lambda c^2 R^2 \right)}_{\text{potential}} = \underbrace{-kc^2\frac{m}{2}}_{\text{total energy}}$$

   (m is the mass of a galaxy at the edge of a galactic cluster of mass M).
   Work out the pressure as a function of R related to the classical potential energy in comparison to the pressure caused by the term containing the cosmological constant.
4. Work out the average energy of blackbody microwave photons! For the integration of the Planck distribution refer to Chap. 27 in [35] or Formulae 3.411 in [36] or check with the web page:
   http://jove.prohosting.com/
      ~skripty/page_998.htm.
   For the Riemann zeta function look at
   http://mathworld.wolfram.com/
      RiemannZetaFunction.html.
5. Estimate the energy density of the cosmic microwave background radiation at present and at the time of last scattering.
6. Why is the decoupling temperature for photons at last scattering (0.3 eV) much lower compared to the ionization energy of hydrogen (13.6 eV)?

12 Inflation

"Inflation hasn't won the race, but so far it's the only horse."

Andrei Linde

The Standard Cosmological Model appears to be very successful in describing observational data, such as the abundances of light nuclei, the isotropic and homogeneous expansion of the universe, and the existence of the cosmic microwave background. The CMB's very high degree of isotropy and the fact that the total energy density is close to the critical density, however, pose problems in that they require a very specific and seemingly arbitrary choice of initial conditions for the universe. Furthermore, it turns out that Grand Unified Theories (GUTs) as well as other possible particle physics theories predict the existence of stable particles such as magnetic monopoles, which no one has yet succeeded in observing. These problems can be solved by assuming that at some very early time, the total energy density of the universe was dominated by vacuum energy. This leads to a rapid, accelerating increase in the scale factor called *inflation*.

This chapter takes a closer look at the problems mentioned above, how inflation solves them, what else inflation predicts, and how these predictions stand in comparison to observations. In doing this it will be necessary to make predictions for the expansion of the universe from very early times up to the present. For these purposes it will be sufficient to treat the universe as being matter dominated after a time of around 50 000 years, preceded by an era of radiation domination (except for the period of inflation itself). In fact it is now believed this picture is not quite true, and the current energy density is dominated again by a sort of vacuum energy. This fact will not alter the arguments relevant for the present chapter and it will be ignored here; the topic of vacuum energy in the present universe will be taken up again in the next chapter.

initial conditions for the universe

monopoles?

predictions of inflation

eras of the universe

vacuum energy

12.1 The Horizon Problem

"The existing universe is bounded in none of its dimensions; for then it must have an outside."

Lucretius

cosmic microwave background

In the previous chapter it was shown that, after correcting for the dipole anisotropy, the CMB has the same temperature to within one part in 10^5 coming from all directions. This radiation was emitted from the surface of last scattering at a time of around $t_{ls} \approx 380\,000$ years. In Sect. 11.6 it was calculated that two places separated by the particle-horizon distance at t_{ls} are separated by an angle of around $1.4°$ as viewed today. This calculation assumed $R \sim t^{1/2}$ for the first $50\,000$ years during the radiation-dominated era, followed by a matter-dominated phase with $R \sim t^{2/3}$ from then up to t_{ls}. For any mixture of radiation and matter domination one would find values in the range from one to two degrees.

regions in causal contact

So, if regions of the sky separated in angle by more than around $2°$ are considered, one would expect them not to have been in causal contact at the time of last scattering. Furthermore, the projected entire sky can be divided into more than 10^4 patches which should not have been in causal contact, and yet, they are all at almost the same temperature.

horizon problem

The unexplained uniform temperature in regions that appear to be causally disconnected is called the 'horizon problem'. It is not a problem in the sense that this model makes a prediction that is in contradiction with observation. The different temperatures could have perhaps all had the same temperature 'by chance'. This option is not taken seriously. The way that systems come to the same temperature is by interacting, and it is hard to believe that any mechanism other than this was responsible.

12.2 The Flatness Problem

"This type of universe, however, seems to require a degree of fine-tuning of the initial conditions that is in apparent conflict with 'common wisdom'."

Idit Zehavi and Avishai Dekel

$\Omega \approx 1$

It was found in Sect. 11.6 that the total energy density of the universe is very close to the critical density or $\Omega \approx 1$.

Although this is now known to hold to within around 2%, it has been clear for many years that Ω is at least constrained to roughly $0.2 < \Omega < 2$. (A lower limit can be obtained from, e.g., the motions of galaxies in clusters, and an upper bound comes from the requirement that the universe is at least as old as the oldest observed stars.) The condition $\Omega = 1$ gives a flat universe, i.e., one with zero spatial curvature.

flatness of the universe

The problem with having an almost flat universe today, i.e., $\Omega \approx 1$, is that it then must have been very much closer to unity at earlier times. To see this, recall first from Chap. 8 the Friedmann equation,

Friedmann equation

$$H^2 + \frac{k}{R^2} = \frac{8\pi G}{3}\varrho , \tag{12.1}$$

where as usual $H = \dot{R}/R$ and k is the curvature parameter. One obtains $k = 0$, i.e., a flat universe, if the density ϱ is equal to the critical density

critical density

$$\varrho_c = \frac{3H^2}{8\pi G} . \tag{12.2}$$

By dividing both sides of the Friedmann equation by H^2, using (12.2) and $H = \dot{R}/R$, and then rearranging terms, one finds

$$\Omega - 1 = \frac{k}{\dot{R}^2} . \tag{12.3}$$

Now, for the matter-dominated era one has $R \sim t^{2/3}$, which gives $\dot{R} \sim t^{-1/3}$ and thus $R\dot{R}^2$ is constant. Therefore, the difference between Ω and unity is found to be as follows:

matter-dominated era

$$\Omega - 1 \sim kR \sim kt^{2/3} . \tag{12.4}$$

If matter domination from the time of matter–radiation equality, $t_{mr} \approx 50\,000$ years, to the present, $t_0 \approx 1.4 \times 10^{10}$ years, is assumed, then (12.4) implies

early values of Ω

$$\frac{\Omega(t_{mr}) - 1}{\Omega(t_0) - 1} = \frac{R(t_{mr})}{R(t_0)} = \left(\frac{t_{mr}}{t_0}\right)^{2/3}$$

$$= \left(\frac{50\,000\,\text{a}}{1.4 \times 10^{10}\,\text{a}}\right)^{2/3} \approx 2 \times 10^{-4} . \tag{12.5}$$

From the recent CMB data $\Omega(t_0) - 1$ is currently measured to be less than around 0.04. Using this with (12.5) implies that $\Omega - 1$ was less than 10^{-5} at $t = 50\,000$ years after the Big Bang.

Going further back in time makes the problem more acute. Suppose that the universe was radiation dominated for times less than t_{mr} going all the way back to the Planck time, $t_{Pl} \approx 10^{-43}$ s. Then one should take $R \sim t^{1/2}$, which

Ω at the Planck time means $\dot{R} \sim t^{-1/2}$ and therefore[1]

$$\Omega - 1 \sim kR^2 \sim kt . \tag{12.6}$$

Using this dependence for times earlier than t_{mr}, one finds for $\Omega - 1$ at the Planck time relative to the value now:

$$\frac{\Omega(t_{Pl}) - 1}{\Omega(t_0) - 1} = \left(\frac{R(t_{Pl})}{R(t_{mr})} \right)^2 \frac{R(t_{mr})}{R(t_0)} = \frac{t_{Pl}}{t_{mr}} \left(\frac{t_{mr}}{t_0} \right)^{2/3}$$

$$\approx \frac{10^{-43} \text{ s}}{50\,000 \text{ a} \times 3.2 \times 10^7 \text{ s/a}} \left(\frac{50\,000 \text{ a}}{1.4 \times 10^{10} \text{ a}} \right)^{2/3}$$

$$\approx 10^{-59} . \tag{12.7}$$

That is, to be able to find Ω of order unity today, the model requires it to be within 10^{-59} of unity at the Planck time.

flatness problem As with the horizon problem, the issue is not one of a prediction that stands in contradiction with observation. There is nothing to prevent Ω from being arbitrarily close to unity at early times, but within the context of the cosmo-

fine-tuning of Ω logical model that has been described so far, it could have just as easily had some other value. And if other values are *a priori* just as likely, then it seems ridiculous to believe that nature would pick Ω 'by chance' to begin so close to unity. One feels that there must be some reason why Ω came out the way it did.

12.3 The Monopole Problem

> *"From the theoretical point of view one would think that monopoles should exist, because of the prettiness of the mathematics."*
>
> *Paul Adrian Maurice Dirac*

In order to understand the final 'problem' with the model that has been presented up to now, one needs to recall

phase transitions something about phase transitions in the early universe. In
in the early universe Chap. 9 it was remarked that a sort of phase transition took

---

[1] The Boltzmann constant and the curvature parameter are traditionally denoted with the same letter k. From the context it should always be clear which parameter is meant.

place at a critical temperature around the GUT energy scale, $T_c \approx E_{GUT} \approx 10^{16}$ GeV.[2] As the temperature dropped below T_c, the Higgs field, which is responsible for the masses of the X and Y bosons, acquired a non-zero vacuum expectation value. There is an analogy between this process and the cooling of a ferromagnet, whereby the magnetic dipoles suddenly line up parallel to their neighbours. Any direction is equally likely, but once a few of the dipoles have randomly chosen a particular direction, their neighbours follow along. The same is true for the more common transition of water when it freezes to snowflakes. In water the molecules move randomly in all directions. When the water is cooled below the freezing point, ice crystals or snowflakes form where the water molecules are arranged in a regular pattern which breaks the symmetry of the phase existing at temperatures above zero degree Celsius. In the case of the Higgs, the analogue of the dipole's direction or crystal orientation is not a direction in physical space, but rather in an abstract space where the axes correspond to components of the Higgs field. As with the ferromagnet or the orientation axis of the ice crystals, the components of the field tend to give the same configuration the same way in a given local region.

Higgs field

ferromagnet

symmetry-breaking mechanisms

If one considers two regions which are far enough apart so they are not in causal contact, then the configuration acquired by the Higgs field in each will not in general be the same. At the boundary between the regions there will be what is called a 'topological defect', analogous to a dislocation in a ferromagnetic crystal. The simplest type of defect is the analogue of a point dislocation, and in typical Grand Unified Theories, these carry a magnetic charge: they are magnetic monopoles. Magnetic monopoles behave as particles with masses of roughly

topological defect

magnetic monopoles

$$m_{mon} \approx \frac{M_X}{\alpha_U} \approx 10^{17} \text{ GeV} , \qquad (12.8)$$

where the X boson's mass, $M_X \approx 10^{16}$ GeV, is roughly the same as the GUT scale and the effective coupling strength is around $\alpha_U \approx 1/40$.

A further crucial prediction is that these monopoles are stable. Owing to their high mass they contribute essentially from the moment of their creation to the non-relativistic matter component of the universe's energy density. One expects monopoles to be produced with a number density of

stable monopoles

[2] As usual, here and in the following the standard notation $c = 1$, $\hbar = 1$, and $k = 1$ (Boltzmann constant) is used.

causally isolated region   roughly one in every causally isolated region. The size of such a region is determined by the distance which light can travel from the beginning of the Big Bang up to the time of the phase transition at t_c. This distance is simply the particle horizon from (11.16) at the time t_c. If a radiation-dominated phase for times earlier than t_c is assumed, then one gets $R \sim t^{1/2}$ and therefore a particle horizon of $2t_c$.

monopole number density   The monopole number density is therefore predicted to be

$$n_{\mathrm{mon}} \approx \frac{1}{(2t_c)^3} . \tag{12.9}$$

The critical temperature T_c is expected to be on the order of the GUT scale, $M_X \approx 10^{16}\,\mathrm{GeV}$, from which one finds the time of the phase transition to be $t_c \approx 10^{-39}\,\mathrm{s}$. As the monopoles are non-relativistic, their energy density is given

monopole energy density   by

$$\varrho_{\mathrm{mon}} = n_{\mathrm{mon}} m_{\mathrm{mon}} \approx \frac{M_X}{\alpha_U} \frac{1}{(2t_c)^3} \approx 2 \times 10^{57}\,\mathrm{GeV}^4 . \tag{12.10}$$

This can compared with the energy density of photons at the

photon energy density   same time, which is

$$\varrho_\gamma = \frac{\pi^2}{15} T_{\mathrm{GUT}}^4 \approx 2 \times 10^{63}\,\mathrm{GeV}^4 . \tag{12.11}$$

So, initially, the energy of photons still dominates over monopoles by a factor $\varrho_\gamma / \varrho_{\mathrm{mon}} \approx 10^6$. But as photons are always relativistic, one has $\varrho_\gamma \sim 1/R^4$, whereas for the monopoles, which are non-relativistic, $\varrho_{\mathrm{mon}} \sim 1/R^3$ holds. The two energy densities would become equal after

plethora of monopoles R increased by 10^6, which is to say, after the temperature dropped by a factor of 10^6, because of $R \sim 1/T$. Since the time follows $t \sim T^{-2}$, equality of ϱ_γ and ϱ_{mon} occurs after the time increases by a factor of 10^{12}. So, starting at the GUT scale around $T_{\mathrm{GUT}} \approx 10^{16}\,\mathrm{GeV}$ or at a time $t_{\mathrm{GUT}} \approx 10^{-39}\,\mathrm{s}$, one would predict $\varrho_\gamma = \varrho_{\mathrm{mon}}$ at a temperature $T_{\gamma\mathrm{mon}} \approx 10^{10}\,\mathrm{GeV}$ or at a time $t_{\gamma\mathrm{mon}} \approx 10^{-27}\,\mathrm{s}$.

This is clearly incompatible with what one observes today. Searches for magnetic monopoles have been carried out, and in a controversial experiment by Cabrera in 1982, evidence for a single magnetic monopole was reported [37]. This appears, however, to have been a one-time glitch, as no further monopoles were found in more sensitive experiments. Indeed, far more serious problems would arise from

the predicted monopoles. Foremost among these is that the
energy density would be so high that the universe would
have recollapsed long ago. Given the currently observed expansion rate and photon density, the predicted contribution
from monopoles would lead to a recollapse in a matter of
days.

recollapse
due to magnetic monopoles

So, here is not merely a failure to explain the universe's
initial conditions, but a real contradiction between a prediction and what one observes. One could always argue, of
course, that Grand Unified Theories are not correct, and indeed there is no direct evidence that requires such a picture
to be true. One can also try to arrange for types of GUTs
which do not produce monopoles, but these sorts of the theory are usually disfavoured for other reasons. Historically,
the monopole problem was one of the factors that motivated
Alan Guth to propose the mechanism of *inflation*, an accelerating phase in the early universe. In the next sections it
will first be defined more formally what inflation is and how
it could come about, and then it will be shown how it solves
not only the monopole problem but also the horizon and flatness problems as well.

monopole problem

inflation solves the puzzles

12.4 How Inflation Works

> *"No point is more central than this, that*
> *empty space is not empty. It is the seat*
> *of the most violent physics."*
>
> John Archibald Wheeler

Inflation is defined as meaning a period of accelerating expansion, i.e., where $\ddot{R} > 0$. In this section it will be investigated how this can arise. Recall first the Friedmann equation,
which can be written as

$$\frac{\dot{R}^2}{R^2} + \frac{k}{R^2} = \frac{8\pi G}{3}\varrho \,, \tag{12.12}$$

where here the term ϱ is understood to include all forms
of energy including that of the vacuum, ϱ_v. It was shown
in Chap. 8 that this arises if one considers a cosmological
constant Λ. This gives rise to a constant contribution to the
vacuum which can be interpreted as vacuum energy density
of the form

vacuum energy

cosmological constant

$$\varrho_v = \frac{\Lambda}{8\pi G} \,. \tag{12.13}$$

One can now ask what will happen if the vacuum energy or, in general, if any constant term dominates the total energy density. Suppose this is the case, i.e., $\varrho \approx \varrho_v$, and assume as well that one can neglect the k/R^2 term; this should always be a good approximation for the early universe. The Friedmann equation then reads

$$\frac{\dot{R}^2}{R^2} = \frac{8\pi G}{3}\varrho_v \,. \tag{12.14}$$

Thus, one finds that the expansion rate H is a constant,

$$H = \frac{\dot{R}}{R} = \sqrt{\frac{8\pi G}{3}\varrho_v} \,. \tag{12.15}$$

exponential increase of the scale factor The solution to (12.15) for $t > t_i$ is

$$R(t) = R(t_i)\,e^{H(t-t_i)} = R(t_i)\exp\left[\sqrt{\frac{8\pi G}{3}\varrho_v}\,(t - t_i)\right]$$

$$= R(t_i)\exp\left[\sqrt{\frac{\Lambda}{3}}\,(t - t_i)\right]. \tag{12.16}$$

That is, the scale factor increases exponentially in time.

More generally, the condition for a period of accelerating expansion can be seen by recalling the acceleration equation from Sect. 8.6,

acceleration equation

$$\frac{\ddot{R}}{R} = -\frac{4\pi G}{3}(\varrho + 3P) \,. \tag{12.17}$$

This shows that one will have an accelerating expansion, i.e., $\ddot{R} > 0$, as long as the energy density and pressure satisfy

$$\varrho + 3P < 0 \,. \tag{12.18}$$

w parameter, equation-of-state parameter That is, if the equation of state is expressed as $P = w\varrho$, one has an accelerating expansion for $w < -1/3$. The pressure is related to the derivative of the total energy U of a system with respect to volume V at constant entropy S (see Appendix B). If one takes $U/V = \varrho_v$ as constant, then the pressure is

$$P = -\left(\frac{\partial U}{\partial V}\right)_S = -\frac{U}{V} = -\varrho_v \,. \tag{12.19}$$

negative pressure Thus, a vacuum energy density leads to a negative pressure and an equation-of-state parameter $w = -1$.

One now needs to ask several further questions: "What could cause such a vacuum energy density?", "How and why did inflation stop?", "How does this solve any of the previously mentioned problems?", and "What are the observational consequences of inflation?".

12.5 Mechanisms for Inflation

> *"Using the forces we know now, you can't make the universe we know now."*
>
> George Smoot

To predict an accelerating scale factor, an equation of state relating pressure P and energy density ϱ of the form $P = w\varrho$ with $w < -1/3$ was needed. Vacuum energy with $w = -1$ satisfies this requirement, but what makes us believe that it should exist?

The idea of vacuum energy arises naturally in a quantum field theory. A good analogy for a quantum field is a lattice of atoms occupying all space. The system of atoms behaves like a set of coupled quantum-mechanical oscillators. A particular mode of vibration, for example, will describe a plane wave of a certain frequency and wavelength propagating in some direction through the lattice. Such a mode will carry a given energy and momentum, and in a quantum field theory this corresponds to a particle. This is in fact how *phonons* are described in a crystal lattice. They are quantized collective vibrations in a lattice that carry energy and momentum.

quantum field theories

phonons

The total energy of the lattice includes the energies of all of the atoms. But a quantum-mechanical oscillator contributes an energy $\hbar\omega/2$ even in its lowest energy state. This is the zero-point energy; it is the analogue of the vacuum energy in a quantum field theory.

zero-point energy

In a quantum field theory of elementary particles, one must dispense with the atoms in the analogy and regard the 'interatomic' spacing as going to zero. In the Standard Model of particle physics, for example, there is an electron field, photon field, etc., and all of the electrons in the universe are simply an enormously complicated excitation of the electron field.

particles as field excitations

The mathematical formalism of quantum field theory goes beyond the scope of this book, but in order to give some idea of how inflation can arise, simply some general results will be quoted. Suppose one has a scalar field $\phi(x, t)$. An

scalar field → spin-0 particle

excitation of this field corresponds to a spin-0 particle. The energy density ϱ and pressure P associated with ϕ are (in units with $\hbar = c = 1$) given by [38]

$$\varrho = \frac{1}{2}\dot{\phi}^2 + \frac{1}{2}(\nabla\phi)^2 + V(\phi) \, , \tag{12.20}$$

$$P = \frac{1}{2}\dot{\phi}^2 - \frac{1}{6}(\nabla\phi)^2 - V(\phi) \, . \tag{12.21}$$

The term V corresponds to a potential which could emerge from some particle physics theory. A particular form for V is, for example, predicted for a Higgs field (see below). For now $V(\phi)$ will be treated as a function which one can choose freely.

Now, for a field $\phi(\boldsymbol{x}, t)$, which is almost constant in time and space such that the $\dot{\phi}^2$ and $(\nabla\phi)^2$ terms can be neglected relative to $V(\phi)$, one then has

$$\varrho \approx -P \approx V(\phi) \, , \tag{12.22}$$

and therefore the equation-of-state parameter w is approximately -1. Since this fulfills the relation $w < -1/3$, it will give an accelerating universe, i.e., $\ddot{R} > 0$.

Now, in constructing a quantum field theory there is a fairly wide degree of freedom in writing down the potential term $V(\phi)$. For example, Fig. 12.1 shows the potential associated with the Higgs field. This is a scalar field in the Standard Model of particle physics needed to explain the masses of the known particles.

In 1981, Alan Guth [39] proposed that a scalar Higgs field associated with a Grand Unified Theory could be responsible for inflation. The potential of this field should have a dip around $\phi = 0$, as shown in Fig. 12.2. As a consequence of the local minimum at $\phi = 0$, there should be a classically stable configuration of the field at this position. In this state, the energy density of the field is given by the height of the potential at $\phi = 0$. Suppose the field goes into this state at a time t_i. The scale factor then follows an exponential expansion,

$$R(t) = R(t_i) \exp\left(\sqrt{\frac{8\pi G\varrho}{3}} \, (t - t_i) \right) \, , \tag{12.23}$$

with $\varrho = V(0)$.

In a classical field theory, if the field were to settle into the local minimum at $\phi = 0$, then it would stay there forever. In a quantum-mechanical theory, however, it can tunnel from

Higgs field

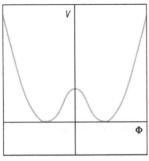

Fig. 12.1
Schematic illustration of the potential $V(\phi)$ associated with the Higgs field ϕ in the Standard Model of particle physics

Higgs potential

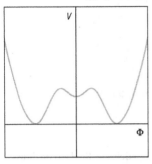

Fig. 12.2
Schematic illustration of the potential $V(\phi)$ first proposed to provide inflation

modified Higgs potential

this 'false vacuum' to the true vacuum with $V = 0$. When
this happens, the vacuum energy will no longer dominate
and the expansion will be driven by other contributions to ϱ
such as radiation.

In the quantum field theories such as the Standard Model
of particle physics, different fields interact. That is, energy
from one field can be transformed into excitations in another.
This is how reactions are described in which particles are
created and destroyed. So, when the *inflaton field* (see below   **inflaton field**
for more details) moves from the false to the true vacuum,
one expects its energy to be transformed into other 'normal'
particles such as photons, electrons, etc. Thus, the end of in-
flation simply matches onto the hot expanding universe of   **how inflation ends**
the existing Big Bang model. It may then appear that in-
flation has not made any predictions that one can verify by
observation, and as described so far no verification has been
given. Later in this chapter, however, it will be shown how
inflation provides explanations for the initial conditions of
the Big Bang, which otherwise would have to be imposed
by hand.

Shortly after the original inflaton potential was proposed, it was realized by Guth and others that the model
had important flaws. As the quantum-mechanical tunneling   **quantum-mechanical**
is a random process, inflation should end at slightly differ-   **tunneling problem**
ent times in different places. Therefore, some places will un-
dergo inflation for longer than others. And because these regions continue to inflate, their contribution to the total volume of the universe remains significant. Effectively, infla-
tion would never end. This has been called inflation's 'grace-   **graceful exit problem**
ful exit problem'.

It was pointed out by Linde [40] and Albrecht and Steinhardt [41] that by a suitable modification of the potential
$V(\phi)$, one could achieve a graceful ending to inflation. The
potential needed for this is shown schematically in Fig. 12.3.
The field it describes is called an *inflaton*, as its potential is   **inflaton**
not derived from particle physics considerations such as the
Higgs mechanism, but rather its sole motivation is to provide for inflation. The local minimum in V at $\phi = 0$ is replaced by an almost flat plateau. The field can settle down
into a metastable state near $\phi \approx 0$, and then effectively 'roll'
down the plateau to the true vacuum. One can show that in
this scenario, called *new inflation*, the exponential expansion   **new inflation**
ends gracefully everywhere.

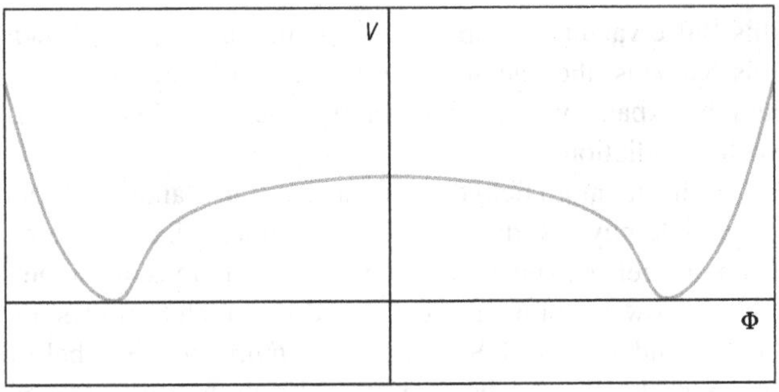

Fig. 12.3
Schematic illustration of the
potential $V(\phi)$ for 'new inflation'

As in Guth's original theory, however, new inflation does
not end everywhere at the same time. But now this feature
is turned to an advantage. It is used to explain the structure
or clumpiness of the universe currently visible on distance
scales less than around 100 Mpc. Because of quantum fluc-
quantum fluctuations   tuations, the value of the field ϕ at the start of the inflationary
phase will not be exactly the same at all places. As with the
randomness of quantum-mechanical tunneling, these quantum fluctuations lead to differences, depending on position,
in the time needed to move to the true vacuum state.

When a volume of space is undergoing inflation, the en-
ergy density $\varrho \approx V(0)$ is essentially constant. After inflation
end of inflation   ends, the energy is transferred to particles such as photons,
electrons, etc. Their energy density then decreases as the
universe continues to expand, e.g., $\varrho \sim R^{-4}$ for relativistic
particles. The onset of this decrease is therefore delayed in
density fluctuations   regions where inflation goes on longer. Thus, the variation in
the time of the end of inflation provides a natural mechanism
to explain spatial variations in energy density. These density
fluctuations are then amplified by gravity and finally result
in the structures that one sees today, e.g., galaxies, clusters,
and superclusters.

12.6 Solution to the Flatness Problem

> *"I have just invented an anti-gravity*
> *machine. It's called a chair."*
>
> Richard P. Feynman

It will now be shown that an early period of inflationary expansion can explain the flatness problem, i.e., why the en-
why is $\varrho = \varrho_c$?   ergy density today is so close to the critical density. Suppose
that inflation starts at some initial time t_i and continues until

a final time t_f, and suppose that during this time the energy density is dominated by vacuum energy ϱ_v, which could result from some inflaton field. The expansion rate H during this period is given by

$$H = \sqrt{\frac{8\pi G}{3}\varrho_v} \, . \tag{12.24}$$

So from t_i to t_f, the scale factor increases by a ratio

number of e foldings of the scale factor

$$\frac{R(t_f)}{R(t_i)} = e^{H(t_f - t_i)} \equiv e^N \, , \tag{12.25}$$

where $N = H(t_f - t_i)$ represents the number of e foldings of expansion during inflation.

Referring back to (12.3), it was shown that the Friedmann equation could be written as

$$\Omega - 1 = \frac{k}{\dot{R}^2} \, . \tag{12.26}$$

Now, during inflation, one has $R \sim e^{Ht}$, with $H = \dot{R}/R$ constant. Therefore, one finds

Ω during inflation

$$\Omega - 1 \sim e^{-2Ht} \tag{12.27}$$

during the inflationary phase. That is, during inflation, Ω is driven exponentially *towards* unity.

What one would like is to have this exponential decrease of $|\Omega - 1|$ to offset the divergence of Ω from unity during the radiation- and matter-dominated phases. To do this one can work out how far one expects Ω to differ from unity for a given number of e foldings of inflation. It has been shown that during periods of matter domination one has $|\Omega - 1| \sim t^{2/3}$, see (12.4), and during radiation domination $|\Omega - 1| \sim t$, see (12.6). Let us assume inflation starts at t_i, ends at t_f, then radiation dominates until $t_{mr} \approx 50\,000$ a, followed by matter domination to the present, $t_0 \approx 14 \times 10^9$ a. The current difference between Ω and unity is therefore

Ω is driven to unity during inflation

$$|\Omega(t_0) - 1| = |\Omega(t_i) - 1| e^{-2H(t_f - t_i)} \left(\frac{t_{mr}}{t_f}\right)\left(\frac{t_0}{t_{mr}}\right)^{2/3} \, . \tag{12.28}$$

In Chap. 11 it was seen that the WMAP data indicate $|\Omega(t_0) - 1| < 0.04$. If, for example, one assumes that the inflaton field is connected to the physics of a Grand Unified Theory, then one expects inflation to be taking place

inflationary period in time scales around 10^{-38} to 10^{-36} s. Suppose one takes $t_f = 10^{-36}$ s and uses the WMAP limit $|\Omega(t_0)-1| < 0.04$. If one assumes that before inflation $\Omega(t_i) - 1$ is of order unity, then this implies that the number of e foldings of inflation must be at least

$$N > \frac{1}{2} \ln \left[\frac{|\Omega(t_i) - 1|}{|\Omega(t_0) - 1|} \left(\frac{t_{mr}}{t_f} \right) \left(\frac{t_0}{t_{mr}} \right)^{2/3} \right] \approx 61 \ .$$

(12.29)

beginning of inflation Suppose that before inflation began, the universe was radiation dominated. This means that the expansion rate followed $H = 1/2t$. By the 'beginning of inflation' at time t_i, the time is meant when the vacuum energy began to dominate, so one should have roughly $H \approx 1/t_i$ during inflation. So, the number of e foldings is basically determined by the duration of the inflationary period,

$$N = H(t_f - t_i) \approx \frac{t_f - t_i}{t_i} \ .$$

(12.30)

$N \approx 100$? So, if $t_f \approx 10^{-36}$ s, then t_i could not have been much earlier than 10^{-38} s. Of course, inflation might have gone on longer. Equation (12.29) only gives the smallest number of e foldings needed to explain why Ω is within 0.04 of unity today.

12.7 Solution to the Horizon Problem

> *"There was no 'before' the beginning of our universe, because once upon a time there was no time."*
>
> John D. Barrow

It can also be shown that an early period of inflationary expansion can explain the horizon problem, i.e., why the entire sky appears at the same temperature. To be specific, suppose that inflation starts at $t_i = 10^{-38}$ s, ends at $t_f = 10^{-36}$ s, and has an expansion rate during inflation of $H = 1/t_i$. Assuming the universe was radiation dominated up to t_i, the

horizon problem particle-horizon distance d_H at this point was

$$d_H = 2ct_i \approx 2 \times 3 \times 10^8 \,\text{m/s} \times 10^{-38} \,\text{s} = 6 \times 10^{-30} \,\text{m} \ .$$

(12.31)

This is the largest region where one would expect to find the same temperature, since any region further away would not be in causal contact. Now, during inflation, a region of size d expands in proportion to the scale factor R by a factor e^N, where the number of e foldings is from (12.30) $N \approx 100$. After inflation ends, the region expands following $R \sim t^{1/2}$ up to the time of matter–radiation equality (50 000 a) and then in proportion to $t^{2/3}$ from then until the present, assuming matter domination. So, the current size of the region which would have been in causal contact before inflation is

region in causal contact

$$d(t_0) = d(t_i) \, e^N \left(\frac{t_{mr}}{t_f}\right)^{1/2} \left(\frac{t_0}{t_{mr}}\right)^{2/3} \approx 10^{38} \, \text{m} \ .$$

$$(12.32)$$

This distance can be compared to the size of the current Hubble distance, $c/H_0 \approx 10^{26}$ m. So, the currently visible universe, including the entire surface of last scattering, fits easily into the much larger region which one can expect to be at the same temperature. With inflation, it is not true that opposite directions of the sky were never in causal contact. So, the very high degree of isotropy of the CMB can be understood.

Hubble distance

12.8 Solution to the Monopole Problem

> "If you can't find them, dilute them."
>
> Anonymous

The solution to the monopole problem is equally straightforward. One simply has to arrange for the monopoles to be produced before or during the inflationary period. This arises naturally in models where inflation is related to the Higgs fields of a Grand Unified Theory and works, of course, equally well if inflation takes place after the GUT scale. The monopole density is then reduced by the inflationary expansion, leaving it with so few monopoles that one would not expect to see any of them.

To see this in numbers, let us suppose the monopoles are formed at a critical time $t_c = 10^{-39}$ s. Suppose, as in the previous example, the start and end times for inflation are $t_i = 10^{-38}$ s and $t_f = 10^{-36}$ s, and let us assume an expansion rate during inflation of $H = 1/t_i$. This gives $N \approx 100$ e foldings of exponential expansion. The volume containing a given number of monopoles increases in proportion to

monopole density

dilution during inflation

R^3, so that during inflation, the density is reduced by a factor $e^{3N} \approx 10^{130}$. Putting together the entire time evolution of the monopole density, one takes $R \sim t^{1/2}$ during radiation domination from t_c to t_i, then $R \sim e^{Ht}$ during inflation, followed again by a period of radiation domination up to the time of matter–radiation equality at $t_{mr} = 50\,000$ a, followed by matter domination with $R \sim t^{2/3}$ up to the present, $t_0 = 14 \times 10^9$ a. Inserting explicitly the necessary factors of c, one would therefore expect today a monopole number density of

dilution of magnetic monopoles

$$n_m(t_0) \approx \frac{1}{(2ct_c)^3} \left(\frac{t_i}{t_c}\right)^{-3/2} e^{-3(t_f - t_i)/t_i}$$

$$\times \left(\frac{t_{mr}}{t_f}\right)^{-3/2} \left(\frac{t_0}{t_{mr}}\right)^{-2} . \tag{12.33}$$

present-day monopole density

Using the relevant numbers one would get

$$n_m(t_0) \approx 10^{-114}\,\mathrm{m}^{-3} \approx 3 \times 10^{-38}\,\mathrm{Gpc}^{-3} . \tag{12.34}$$

So, the number of monopoles one would see today would be suppressed by such a huge factor that one would not even expect a single monopole in the observable part of the universe.

12.9 Inflation and the Growth of Structure

"The universe is not made, but is being made continuously. It is growing, perhaps indefinitely."

Henri Bergson

specific initial conditions

The primary success of inflationary models is that they provide a dynamical explanation for specific initial conditions, which otherwise would need to be imposed by hand. Furthermore, inflation provides a natural mechanism to explain the density fluctuations which grew into the structures such as galaxies and clusters as are seen today. But beyond this qualitative statement, one can ask whether the properties of the predicted structure indeed match what one observes.

The level of structure in the universe is usually quantified by considering the relative difference between the density at a given position, $\varrho(x)$, and the average density $\langle \varrho \rangle$,

$$\delta(x) = \frac{\varrho(x) - \langle \varrho \rangle}{\langle \varrho \rangle} . \tag{12.35}$$

The quantity $\delta(x)$ is called the *density contrast*. Suppose a cube of size L is considered, i.e., volume $V = L^3$, and the density contrast inside V is expanded into a Fourier series. Assuming periodic boundary conditions, this gives

density contrast

$$\delta(x) = \sum \delta(k)\, e^{i k \cdot x} , \qquad (12.36)$$

where the sum is taken over all values of $k = (k_x, k_y, k_z)$ which fit into the box, e.g., $k_x = 2\pi n_x / L$, with $n_x = 0, \pm 1, \pm 2, \ldots$, and similarly for k_y and k_z. Averaging over all directions provides the average magnitude of the Fourier coefficients as a function of $k = |k|$. Now the *power spectrum* is defined as

power spectrum

$$P(k) = \langle |\delta(k)|^2 \rangle . \qquad (12.37)$$

Its interpretation is simply a measure of the level of structure present at a wavelength $\lambda = 2\pi/k$. Different observations can provide information about the power spectrum at different distance scales. For example, at distances up to around 100 Mpc, surveys of galaxies such as the Sloan Digital Sky Survey (SDSS) can be used to measure directly the galaxy density. At larger distance scales, i.e., smaller values of k, the temperature variations in the CMB provide the most accurate information. Some recent measurements of $P(k)$ are shown in Fig. 12.4 [42].

Sloan Digital Sky Survey

In most cosmological theories of structure formation, one finds a power law valid for large distances (small k) of the form

$$P(k) \sim k^n . \qquad (12.38)$$

Using the value $n = 1$ for the *scalar spectral index* gives what is called the scale-invariant Harrison–Zel'dovich spectrum. Most inflationary models predict $n \approx 1$ to within around 10%. The exact value for a specific model is related to the form of the potential $V(\phi)$, i.e., how long and flat its plateau is (see, e.g., [43] and Fig. 12.3).

scalar spectral index

The curve shown in Fig. 12.4 is based on a model with $n = 1$ for large distances, and this agrees well with the data. Depending on the specific model assumptions, the best determined values of the spectral index are equal to unity to within around 10%. This is currently an area with a close interplay between theory and experiment, and future improvements in the measurement of the power spectrum should lead to increasingly tight constraints on models of inflation.

models of inflation

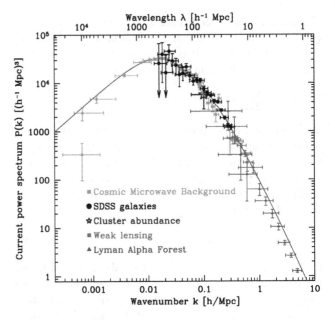

Fig. 12.4
Measurements of the power
spectrum $P(k)$ from several types
of observations. The curve shows
the prediction of a model with the
spectral index n equal to unity. The
parameter h is assumed to be 0.72
(from [42])

12.10 Outlook on Inflation

> *"The argument seemed sound enough,*
> *but when a theory collides with a fact,*
> *the result is tragedy."*
>
> *Louis Nizer*

tests of inflation

So far, inflation seems to have passed several important tests.
Its most generic predictions, namely an energy density equal
to the critical density, a uniform temperature for the observable universe, and a lack of relic particles created at any pre-
inflationary time, are well confirmed by observation. Furthermore, there are no serious alternative theories that come
close to this level of success. But inflation is not a single
theory but rather a class of models that include a period of
accelerating expansion. There are still many aspects of these
models that remain poorly constrained, such as the nature of
the energy density driving inflation and the time when it existed.

dark energy now

Concerning the nature of the energy density, it will be
seen in Chap. 13 that around 70% of the current energy density in the universe is in fact 'dark energy', i.e., something
with properties similar to vacuum energy. Data from type-
Ia supernovae and also information from CMB experiments
such as WMAP indicate that for the last several billion

years, the universe has been undergoing a renewed quasi-exponential expansion compatible with a value of $\Omega_{\Lambda,0} = \varrho_{v,0}/\varrho_{c,0} \approx 0.7$. (The subscript Λ refers to the relation between vacuum energy and a cosmological constant; the subscript 0 denotes as usual present-day values.) The critical density is $\varrho_{c,0} = 3H_0^2/8\pi G$, where $H_0 \approx 70\,\text{km}\,\text{s}^{-1}\,\text{Mpc}^{-1}$ is the Hubble constant. This gives a current vacuum energy density of around ($\hbar c \approx 0.2\,\text{GeV}\,\text{fm}$)

current vacuum energy density

$$\varrho_{v,0} \approx 10^{-46}\,\text{GeV}^4 \,. \tag{12.39}$$

In the example of inflation considered above, however, the magnitude of the vacuum energy density is related to the expansion rate during inflation by (12.15). Furthermore, it was argued that the expansion rate is approximately related to the start time of inflation by $H \approx 1/t_i$. So, if one believes that inflation occurred at the GUT scale with, say, $t_i \approx 10^{-38}$ s, then one needs a vacuum energy density of, see (9.40),

dark energy at inflation

$$\varrho_v = \frac{3m_{\text{Pl}}^2}{32\pi t_i^2} \approx 10^{64}\,\text{GeV}^4 \,. \tag{12.40}$$

The vacuum energies now and those which existed during an earlier inflationary period may well have a common explanation, but given their vast difference, it is by no means obvious what their relationship is.

One could try to argue that inflation took place at a later time, and therefore had a smaller expansion rate and correspondingly lower ϱ_v. The latest possible time for inflation would be just before the Big Bang Nucleosynthesis era, around $t \approx 1$ s. With inflation any later than this, the predictions of BBN for abundances of light nuclei would be altered and would no longer stand in such good agreement with observation. Even at an inflation time of $t_i = 1$ s, one would need a vacuum energy density of $\varrho_v \approx 10^{-12}\,\text{GeV}^4$, still 34 orders of magnitude greater than what is observed today.

inflation time

In addition to those observable features mentioned, inflation predicts the existence of gravitational waves. These would present a fossilized record of the first moments of time. In particular, the energy density of gravity waves is expected to be

gravitational waves from inflation

$$\varrho_{\text{gravity waves}} = \frac{h^2\omega^2}{32\pi G} \tag{12.41}$$

which gives

$$\Omega_{\text{gravity waves}} = \frac{\varrho_{\text{gravity waves}}}{\varrho_c} = \frac{h^2\omega^2}{12H^2} \,. \qquad (12.42)$$

gravity-wave detection

This would result in distortions of gravity-wave antennae on the order of $h = 10^{-27}$ for kHz gravity waves. Even though present gravitational-wave antennae do not reach this kind of sensitivity, a detection of the predicted gravitational-wave background at the level expected in inflationary models would result in a credibility comparable to that which was achieved for the Big Bang by the observation of the 2.7 K blackbody radiation.

new experimental possibilities?

Even though the detection of gravity waves has not been established experimentally, it might still be possible in the future to find means of testing these predictions of inflationary models with new techniques or new ideas. At the time of the discovery of the blackbody radiation in 1965 by Penzias and Wilson it seemed unrealistic to assume that one might be able to measure the spectrum at the level at which it is known now. With future satellites, like with the European Planck mission, even further improvements are envisaged.

12.11 Problems

1. A value of the cosmological constant Λ can be estimated from the Friedmann equation extended by the cosmological term. It is often said that this value disagrees by about 120 orders of magnitude with the result from the expected vacuum energy in a unified supergravity theory. How can this factor be illustrated?

2. Work out the time dependence of the size of the universe for a flat universe with a cosmological constant Λ.

3. Work out the time evolution of the universe if Λ were a dynamical constant ($\Lambda = \Lambda_0(1 + \alpha t)$) for a Λ-dominated flat universe.

4. Estimate the size of the universe at the end of the inflation period ($\approx 10^{-36}$ s). Consider that for a matter-dominated universe its size scales as $t^{\frac{2}{3}}$, while for a radiation-dominated universe one has $R \sim t^{\frac{1}{2}}$.

13 Dark Matter

"There is a theory, which states that if ever anyone discovers exactly what the universe is for and why it is here, it will instantly disappear and be replaced by something even more bizarre and inexplicable. There is another theory, which states that this has already happened."

Douglas Adams

Recent observations have shown that the universe is flat, i.e., $\Omega = 1$. The baryonic matter constitutes only a small fraction on the order of 4%. The main ingredients of the universe are presented by material and/or energy that one can only speculate about. 23% of the matter in the universe is dark and 73% is given by dark energy (see Fig. 13.1). There is no clear idea what kind of material dominates over the well-known baryonic matter. Even the visible matter constitutes only a small fraction of the total baryonic matter.

The following sections discuss some candidates for the unseen baryonic matter and offer some proposals for the missing non-baryonic matter and dark energy.

energy content of the universe

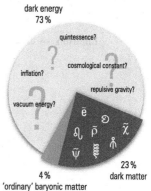

Fig. 13.1
Illustration of the relative fractions of dark matter, dark energy, and baryonic matter

13.1 Large-Scale Structure of the Universe

"There are grounds for cautious optimism that we may now be near the end of the search for the ultimate laws of nature."

Stephen W. Hawking

Originally it has been assumed that the universe is homogeneous and isotropic. However, all evidence speaks to the contrary. Nearly on all scales one observes inhomogeneities: stars form galaxies, galaxies form galactic clusters, and there are superclusters, filaments of galaxies, large voids, and great walls, just to name a few of them. The large-scale structure has been investigated up to distances of about 100 Mpc. On these large scales a surprising lumpiness was found. One has, however, to keep in mind that the spatial distribution of galaxies must not necessarily coincide with the distribution of matter in the universe.

inhomogeneities of the universe

early universe

The measurement of the COBE and WMAP satellites on the inhomogeneities of the 2.7 Kelvin blackbody radiation has shown that the early universe was much more homogeneous. On the other hand, small temperature variations of the blackbody radiation have served as seeds for structures, which one now observes in the universe.

development of large-scale structures

It is generally assumed that the large-scale structures have evolved from gravitational instabilities, which can be traced back to small primordial fluctuations in the energy density of the early universe. Small perturbations in the energy density are amplified due to the associated gravitational forces. In the course of time these gravitational aggregations collect more and more mass and thus lead to structure formation. The reason for the original microscopic small inhomogeneities is presumably to be found in quantum fluctuations. Cosmic inflation and the subsequent slow expansion have stretched these inhomogeneities to the presently observed size. One consequence of the idea of cosmic inflation is that the exponential growth has led to a smooth and flat universe which means that the density parameter Ω should be very close to unity. To understand the formation of the large-scale structure of the universe and its dynamics in detail, a sufficient amount of mass is required because otherwise the original fluctuations could never have been transformed into distinct mass aggregations. It is, however, true that the amount of visible matter only does not support the critical density of $\Omega = 1$ as required by inflation.

quantum fluctuations

inflation $\rightarrow \Omega = 1$

major rôle of non-baryonic dark matter

To get a flat universe seems to require a second Copernican revolution. Copernicus had noticed that the Earth was not the center of the universe. Cosmologists now conjecture that the kind of matter, of which man and Earth are made, only plays a minor rôle compared to the dark non-baryonic matter, which is – in addition to the dark energy – needed to understand the dynamics of the universe and to reach the critical mass density.

13.2 Motivation for Dark Matter

"If it's not DARK, it doesn't MATTER."
Anonymous

The idea that our universe contains dark matter is not entirely new. Already in the thirties of the last century Zwicky

[44] had argued that clusters of galaxies would not be gravitationally stable without additional invisible dark matter. Recent observations of high-z supernovae and detailed measurements of the cosmic microwave background radiation have now clearly demonstrated that large quantities of dark matter must exist, which fill up the universe. An argument for the existence of invisible dark matter can already be inferred from the Keplerian motion of stars in galaxies. Kepler had formulated his famous laws, based on the precision measurements of Tycho Brahe. The stability of orbits of planets in our solar system is obtained from the balance of centrifugal and the attractive gravitational force:

circumstantial evidence for dark matter

$$\frac{mv^2}{r} = G\frac{mM}{r^2} \qquad (13.1)$$

(m is the mass of the planet, M is the mass of the Sun, r is the radius of the planet's orbit assumed to be circular). The resulting orbital velocity is calculated to be

$$v = \sqrt{GM/r} \,. \qquad (13.2)$$

The radial dependence of the orbital velocity of $v \sim r^{-1/2}$ is perfectly verified in our solar system (Fig. 13.2).

The rotational curves of stars in galaxies, however, show a completely different pattern (Fig. 13.3). Since one assumes that the majority of the mass is concentrated at the center of a galaxy, one would at least expect for somewhat larger distances a Keplerian-like orbital velocity of $v \sim r^{-1/2}$. Instead, the rotational velocities of stars are almost constant even for large distances from the galactic center.

The flat rotational curves led to the conclusion that the galactic halo must contain nearly 90% of the mass of the galaxy. To obtain a constant orbital velocity, the mass of the galactic nucleus in (13.1) has to be replaced by the now dominant mass of invisible matter in the halo. This requirement leads to a radial dependence of the density of this mass of

$$\varrho \sim r^{-2} \,, \qquad (13.3)$$

because

$$\frac{mv^2}{r} = G\,\frac{m\int_0^r \varrho\,dV}{r^2} \sim G\,\frac{m\,\varrho\,V}{r^2} \sim G\,\frac{mr^{-2}r^3}{r^2}$$
$$\Rightarrow v^2 = \text{const}\,. \qquad (13.4)$$

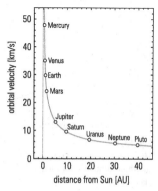

Fig. 13.2
Rotational curves of planets in our solar system, 1 Astronomical Unit (AU) = distance Earth to Sun

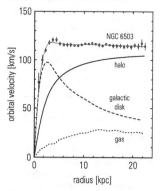

Fig. 13.3
Rotational curves of the spiral galaxy NGC 6503. The contributions of the galactic disk, the gas, and the halo are separately shown

parameterization
of the mass density
of a galaxy

Frequently the proportionality of (13.3) for larger r is parameterized in the following form:

$$\varrho(r) = \varrho_0 \frac{R_0^2 + a^2}{r^2 + a^2} \,, \tag{13.5}$$

where r is the galactocentric distance, $R_0 = 8.5\,\mathrm{kpc}$ (for the Milky Way) is the galactocentric radius of the Sun, and $a = 5\,\mathrm{kpc}$ is the radius of the halo nucleus. ϱ_0 is the local energy density in the solar system,

$$\varrho_0 = 0.3\,\mathrm{GeV/cm^3} \,. \tag{13.6}$$

If the mass density, like in (13.3) for flat rotational curves required, decreases like r^{-2}, the integral mass of a galaxy grows like $M(r) \sim r$, since the volume increases like r^3.

So far, three-dimensional rotationally symmetric mass distributions of galaxies have been considered. In another limiting case, one can study purely two-dimensional distributions, since the luminous matter is practically confined to a disk in a galactic plane with the exception of the bulk. If M denotes the mass and a the radius of the galaxy, the two-dimensional mass density

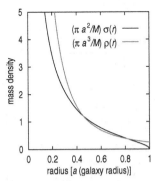

Fig. 13.4
Mass density of a galaxy for a two- and three-dimensional model of the mass density

$$\sigma(r) = \begin{cases} \frac{M}{2\pi a r}\arccos\left(\frac{r}{a}\right), & r < a \\ 0, & r \geq a \end{cases} \,, \tag{13.7}$$

see Fig. 13.4, also leads to a flat rotational curve in the galactic disk [45]. In contrast to the three-dimensional distribution the surface density has a $1/r$ singularity for $r \to 0$ and vanishes at the radius of the galaxy. The radial force on a mass m in the plane and the orbital velocity read

flat galaxies?

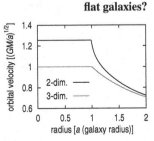

Fig. 13.5
Flat rotational curves of stars in a galactic disk for two- and three-dimensional rotationally symmetric mass distributions

$$F(r) = G\frac{mM}{ra}\begin{cases} \frac{\pi}{2}, & r < a \\ \arcsin\left(\frac{a}{r}\right), & r \geq a \end{cases} \,, \quad v(r) = \sqrt{\frac{rF(r)}{m}} \,, \tag{13.8}$$

the latter of which is shown in Fig. 13.5.

From elementary particle physics and nuclear physics it is well-known that the mass is essentially concentrated in atomic nuclei and thereby in baryons. In the framework of the primordial element synthesis the abundance of the light elements (D, ^{3}He, ^{4}He, ^{7}Li, ^{7}Be) formed by fusion can be determined. Based on these arguments and the findings on the cosmic microwave background radiation, the contribu-

contribution
of baryonic matter

tion of baryonic matter – expressed in terms of the critical density $\varrho_c = 3H^2/8\pi G$ and $\Omega_b = \varrho_b/\varrho_c$ – is obtained to be

$$\Omega_b \approx 0.04 . \tag{13.9}$$

This corresponds to just 0.3 baryons per m^3. The visible luminous matter only provides a density of

visible luminous matter

$$0.003 \leq \Omega_{lum} \leq 0.007 , \tag{13.10}$$

that is, $\Omega_{lum} < 1\%$. From the fact that the rotational curves of all galaxies are essentially flat, the contribution of galactic halos to the mass density of the universe can be estimated to be

contribution of galactic halos

$$0.2 \leq \Omega_{gal} \leq 0.4. \tag{13.11}$$

Since galaxies also have the tendency to form clusters, an estimation of the cosmic mass density can be obtained from the dynamics of galaxies in galactic clusters. The mass obtained from these considerations also surmounts the visible luminous matter by a large margin.

dynamics of galaxies in clusters

The existence of large amounts of invisible non-luminous dark matter seems to be established in a convincing fashion. Consequently, the question arises what this matter is and how it is distributed.

13.2.1 Dark Stars

The rotational curves of stars in galaxies require a considerable fraction of gravitating matter which is obviously invisible. The idea of primordial nucleosynthesis suggests that the amount of baryonic matter is larger than the visible matter, (13.9) and (13.10). Therefore, it is obvious to assume that part of the galactic halos consists of baryonic matter. Because of the experimental result of $\Omega_{lum} < 0.007$, this matter cannot exist in form of luminous stars. Also hot and therefore luminous gas clouds, galactic dust, and cold galactic gas clouds are presumably unlikely candidates, too, because they would reveal themselves by their absorption. These arguments only leave room for special stellar objects which are too small to shine sufficiently bright or celestial objects which just did not manage to initiate hydrogen burning. In addition, burnt-up stars in form of neutron stars, black holes, or dwarf stars could also be considered. Neutron stars and black holes are, however, unlikely candidates,

gas clouds
galactic dust

baryonic matter in dark stellar objects

neutron stars
black holes

because they would have emerged from supernova explosions. From the observed chemical composition of galaxies (dominance of hydrogen and helium) one can exclude that many supernova explosions have occurred, since these are a source of heavier elements.

Therefore, it is considered likely that galactic baryonic matter could be hidden in brown dwarves. Since the mass spectrum of stars increases to small masses, one would expect a significant fraction of small brown stars in our galaxy. The question is, whether one is able to find such massive, compact, non-luminous objects in galaxies (MACHO)[1]. As has already been shown in the introduction (Chap. 1), a dark star can reveal itself by its gravitational effect on light. A point-like invisible deflector between a bright star and the observer produces two star images (see Fig. 1.7). If the deflecting brown star is directly on the line of sight between the star and the observer, a ring will be produced, the Einstein ring (Fig. 13.6). The radius r_E of this ring depends on the mass M_d of the deflector like $r_E \sim \sqrt{M_d}$. Such phenomena have frequently been observed. If, however, the mass of the brown star is too small, the two star images or the Einstein ring cannot be resolved experimentally. Instead, one would observe an increase in brightness of the star, if the deflector passes through the line of sight between star and observer (*microlensing*). This brightness excursion is related to the fact that the light originally emitted into a larger solid angle (without the deflecting star) is focused by the deflector onto the observer. The brightness increase now depends on how close the dark star comes to the line of sight between observer and the bright star ('impact parameter' b). The expected apparent light curve is shown in Fig. 13.7 for various parameters. One assumes that b is the minimum distance of the dark star with respect to the line of sight between source and observer and that it passes this line of sight with the velocity v. A characteristic time for the brightness excursion is given by the time required to pass the Einstein ring ($t = r_E/v$).

To be able to find non-luminous halo objects by gravitational lensing, a large number of stars has to be observed over a long period of time to search for a brightness excursion of individual stars. Excellent candidates for such a search are stars in the Large Magellanic Cloud (LMC). The

brown dwarves

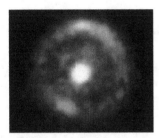

Fig. 13.6
Image of a distant background galaxy as Einstein ring, where the foreground galaxy in the center of the figure acts as gravitational lens {29}

microlensing

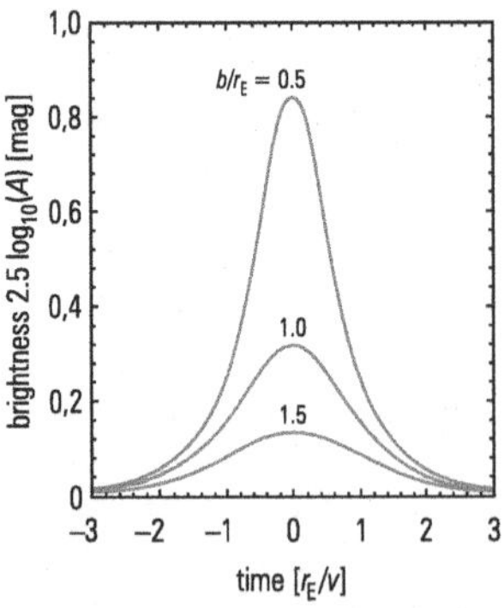

Fig. 13.7
Apparent light curve of a bright star produced by microlensing, when a brown dwarf star passes the line of sight between source and observer. The brightness excursion is given in terms of magnitudes generally used in astronomy {30}

[1] MACHO – MAssive Compact Halo Object, sometimes also called 'Massive Astrophysical Compact Halo Objects'

Large Magellanic Cloud is sufficiently distant so that the light of its stars has to pass through a large region of the galactic halo and therefore has a chance to interact gravitationally with many brown non-luminous star candidates. Furthermore, the Large Magellanic Cloud is just above the galactic disk so that the light of its stars really has to pass through the halo. From considerations of the mass spectrum of brown stars ('MACHOs') and the size of the Einstein ring one can conclude that a minimum of 10^6 stars has to be observed to have a fair chance to find some MACHOs.

The experiments MACHO, EROS[2], and OGLE[3] have found approximately a dozen MACHOs in the halo of our Milky Way. Figure 13.8 shows the light curve of the first candidate found by the MACHO experiment. The observed width of the brightness excursion allows to determine the mass of the brown objects.

If the deflector has a mass corresponding to one solar mass ($M_\odot$), one would expect an average brightness excursion of three months while for $10^{-6} M_\odot$ one would obtain only two hours. The measured brightness curves all lead to masses of approximately 0.5 $M_\odot$. The non-observation of short brightness signals already excludes a large mass range of MACHOs as candidates for dark halo matter. If the few seen MACHOs in a limited solid angle are extrapolated to the whole galactic halo, one arrives at the conclusion that possibly 20% of the halo mass, which takes an influence on the dynamics of the Milky Way, could be hidden in dark stars. Because of the low number of observed MACHOs, this result, however, has a considerable uncertainty ($(20^{+30}_{-12})\%$).

A remote possibility for additional non-luminous baryonic dark matter could be the existence of massive quark stars (several hundred solar masses). Because of their anticipated substantial mass the duration of brightness excursions would be so large that it would have escaped detection.

The Andromeda Galaxy with many target stars would be an ideal candidate for microlensing experiments. This galaxy is right above the galactic plane. Unfortunately, it is too distant that individual stars can be resolved. Still one could employ the 'pixel-lensing' technique by observing the apparent brightness excursions of individual pixels with a CCD camera. In such an experiment one pixel would cover

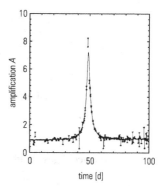

Fig. 13.8
Light curve of a distant star caused by gravitational lensing. Shown is the first brown object found by the MACHO experiment in the galactic halo {30}

expected durations
of brightness excursions

estimated mass fraction
of dark stars in the halo

massive quark stars

pixel-lensing technique

---

[2] EROS – Expérience pour la Recherche d'Objets Sombres

[3] OGLE – Optical Gravitational Lens Experiment

several non-resolved star images. If, however, one of these stars would increase in brightness due to microlensing, this would be noticed by the change in brightness of the whole pixel.

One generally assumes that MACHOs constitute a certain fraction of dark matter. However, it is not clear, which objects are concerned and where these gravitational lenses **NACHOs** reside. To understand the dynamics of galaxies there must be also some NACHOs (Not Astrophysical Compact Halo Objects).

A certain contribution to baryonic dark matter could also **ultracold gas clouds** be provided by ultracold gas clouds (temperatures < 10 K), which are very difficult to detect.

Recently a promising new technique of weak gravita- **weak gravitational lensing:** tional lensing has been developed for the determination of **image distortions** the density of dark matter in the universe. Weak gravitational lensing is based on the fact that images of distant galaxies will be distorted by dark matter between the observer and the galaxy. The particular pattern of the distortions mirrors the mass and its spatial distribution along the line of sight to the distant galaxy. First investigations on 145 000 galaxies in three different directions of observation have shown that $\Omega \leq 1$ and that the cosmological constant very likely plays a dominant rôle in our universe.

13.2.2 Neutrinos as Dark Matter

For a long time neutrinos were considered a good candidate for dark matter. A purely baryonic universe is in contradiction with the primordial nucleosynthesis of the Big Bang. Furthermore, baryonic matter is insufficient to explain the large-scale structure of the universe. The number of neutrinos approximately equals the number of blackbody photons. If, however, they had a small mass, they could provide a significant contribution for dark matter.

upper neutrino mass limits From direct mass determinations only limits for the neu-
from direct measurements trino masses can be obtained ($m_{\nu_e} < 3\,\text{eV}$, $m_{\nu_\mu} < 190\,\text{keV}$, $m_{\nu_\tau} < 18\,\text{MeV}$). The deficit of atmospheric muon neutrinos being interpreted as ν_μ–ν_τ oscillations, leads to a mass of the ν_τ neutrino of approximately 0.05 eV.

Under the assumption of $\Omega = 1$ the expected number density of primordial neutrinos allows to derive an upper limit for the total mass that could be hidden in the three neutrino flavours. One expects that there are approximately equal numbers of blackbody neutrinos and blackbody photons. With $N \approx 300$ neutrinos/cm^3 and $\Omega = 1$ (corresponding to the critical density of $\varrho_c \approx 1 \times 10^{-29}$ g/cm^3 at an age of the universe of approximately 1.4×10^{10} years) one obtains[4]

upper summative neutrino mass limit from flat universe

$$N \sum m_\nu \leq \varrho_c \,, \quad \sum m_\nu \leq 20 \,\text{eV} \,. \tag{13.12}$$

The sum extends over all sequential neutrinos including their antiparticles. For the three known neutrino generations one has $\sum m_\nu = 2(m_{\nu_e} + m_{\nu_\mu} + m_{\nu_\tau})$. The consequence of (13.12) is that for each individual neutrino flavour a mass limit can be derived:

upper individual neutrino mass limits from flat universe

$$m_\nu \leq 10 \,\text{eV} \,. \tag{13.13}$$

It is interesting to see that on the basis of these simple cosmological arguments the mass limit for the τ neutrino as obtained from accelerator experiments can be improved by about 6 orders of magnitude.

If the contribution of neutrino masses to dark matter is assumed to be $\Omega_\nu > 0.1$, a similar argument as before provides also a lower limit for neutrino masses. Under the assumption of $\Omega_\nu > 0.1$ (13.12) yields for the sum of masses of all neutrino flavours $\sum m_\nu > 2 \,\text{eV}$. If one assumes that in the neutrino sector the same mass hierarchy holds as with charged leptons ($m_e \ll m_\mu \ll m_\tau \to m_{\nu_e} \ll m_{\nu_\mu} \ll m_{\nu_\tau}$), the mass of the τ neutrino can be limited to the range

lower neutrino mass limit from minimum energy density

$$1 \,\text{eV} \leq m_{\nu_\tau} \leq 10 \,\text{eV} \,. \tag{13.14}$$

This conclusion, however, rests on the assumption of $\Omega_\nu > 0.1$. Recent estimates of Ω_ν indicate much smaller values ($\Omega_\nu < 1.5\%$) and hence the ν_τ mass can be substantially smaller. Neutrinos with low masses are relativistic and would constitute in the early universe in thermal equilibrium to the so-called 'hot dark matter'. With hot dark matter, however, it is extremely difficult to understand the structures in the universe on small scales (size of galaxies). Therefore, neutrinos are not considered a good candidate for dark matter.

hot dark matter

There is a possibility to further constrain the margin for neutrino masses. To contribute directly to the dark matter of a galaxy, neutrinos must be gravitationally bound to the

gravitational binding to a galaxy

---

[4] For simplification $c = 1$ has been generally used. If numbers, however, have to be worked out, the correct numerical value for the velocity of light, $c \approx 3 \times 10^8$ m/s, must be used.

galaxy, i.e., their velocity must be smaller than the escape velocity v_f. This allows to calculate a limit for the maximum momentum $p_{max} = m_\nu v_f$. If the neutrinos in a galaxy are treated as a free relativistic fermion gas in the lowest energy state (at $T = 0$), one can derive from the Fermi energy

neutrinos
as free relativistic fermion gas

$$E_F = \hbar c (3\pi^2 n_{max})^{1/3} = p_{max} c \qquad (13.15)$$

neutrino mass density

an estimation for the neutrino mass density (n_{max} – number density):

$$n_{max}\, m_\nu = \frac{m_\nu^4 v_f^3}{3\pi^2 \hbar^3} . \qquad (13.16)$$

Since $n_{max}\, m_\nu$ must be at least on the order of magnitude of the typical density of dark matter in a galaxy, if one wants to explain its dynamics with neutrino masses, these arguments lead to a *lower* limit for the neutrino mass. Using

lower neutrino mass limit
from galaxy dynamics

$v_f = \sqrt{2GM/r}$, where M and r are galactic mass and radius, one obtains under plausible assumptions about the neutrino mass density and the size and structure of the galaxy

$$m_\nu > 1\,\text{eV} . \qquad (13.17)$$

Again, this argument is based on the assumption that neutrino masses might contribute substantially to the matter density of the universe. These cosmological arguments leave only a relatively narrow window for neutrino masses.

These considerations are not necessarily in contradiction to the interpretation of results on neutrino oscillations, because in that case one does not directly measure neutrino masses but rather the difference of their masses squared. From the deficit of atmospheric muon neutrinos one obtains

$$\delta m^2 = m_1^2 - m_2^2 = 3 \times 10^{-3}\,\text{eV}^2 . \qquad (13.18)$$

If ν_μ–ν_τ oscillations are responsible for this effect, muon and tau neutrino masses could still be very close without getting into conflict with the cosmological mass limits. Only

neutrino masses
from oscillations

if the known mass hierarchy ($m_e \ll m_\mu \ll m_\tau$) from the sector of charged leptons is transferred to the neutrino sector and if one further assumes $m_{\nu_\mu} \ll m_{\nu_\tau}$, the result ($m_{\nu_\tau} \approx 0.05$ eV) would be in conflict with the cosmological limits, but then the cosmological limits have been derived under the assumption that neutrino masses actually play an important rôle for the matter density in the universe, which is now known not to be the case.

If light neutrinos would fill up the galactic halo, one would expect a narrow absorption line in the spectrum of high-energy neutrinos arriving at Earth. The observation of such an absorption line would be a direct proof of the existence of neutrino halos. Furthermore, one could directly infer the neutrino mass from the energetic position of this line. For a neutrino mass of 10 eV the position of the absorption line can be calculated from, see (3.16), $\nu + \bar{\nu} \to Z^0 \to$ to hadrons or leptons, **neutrino absorption line** **neutrino halos**

$$2m_\nu E_\nu = M_Z^2 \qquad (13.19)$$

to be

$$E_\nu = \frac{M_Z^2}{2m_\nu} = 4.2 \times 10^{20}\,\text{eV} . \qquad (13.20)$$

The verification of such an absorption line, which would result in a burst of hadrons or leptons from Z decay ('Z bursts') represents a substantial experimental challenge. The fact that recent fits to cosmological data indicate that the contribution of neutrino masses to the total matter density in the universe is rather small ($\Omega_\nu \leq 1.5\%$), makes the observation of such an absorption line rather unlikely. **Z bursts**

13.2.3 Weakly Interacting Massive Particles (WIMPs)

> *"In questions like this, truth is only to be had by laying together many variations of error."*
>
> *Virginia Wolf*

Baryonic matter and neutrino masses are insufficient to close the universe. A search for further candidates of dark matter must concentrate on particles, which are only subject to weak interactions apart from gravitation, otherwise one would have found them already. There are various scenarios which allow the existence of weakly interacting massive particles (WIMPs). In principle, one could consider a fourth generation of leptons with heavy neutrinos. However, the LEP[5] measurements of the Z width have shown that the number of light neutrinos is exactly three (see Chap. 2, Fig. 2.1, and Sect. 10.7) so that for a possible fourth generation the mass limit **fourth generation?** **heavy neutrinos**

$$m_{\nu_x} \geq m(Z)/2 \approx 46\,\text{GeV} \qquad (13.21)$$

---

[5] LEP – Large Electron–Positron Collider at CERN in Geneva

holds. Such a mass, however, is considered to be too large to expect that a sizable amount of so heavy particles could have been created in the Big Bang.

An alternative to heavy neutrinos is given by WIMPs, which would couple even weaker to the Z than sequential neutrinos. Candidates for such particles are provided in supersymmetric extensions of the Standard Model. Supersymmetry is a symmetry between fundamental fermions (leptons and quarks) and gauge bosons (γ, W^+, W^-, Z, gluons, Higgs bosons, gravitons). In supersymmetric models all particles are arranged in supermultiplets and each fermion is associated with a bosonic partner and each boson gets a fermionic partner.

supersymmetric extensions of the Standard Model

Bosonic quarks and leptons (called squarks and sleptons) are associated to the normal quarks and leptons. The counterparts of the usual gauge bosons are supersymmetric gauginos, where one has to distinguish between charginos (supersymmetric partners of charged gauge bosons: winos ($\widetilde{W}^+$, $\widetilde{W}^-$) and charged higgsinos ($\widetilde{H}^+$, $\widetilde{H}^-$)) and neutralinos (photino $\widetilde{\gamma}$, zino $\widetilde{Z}$, neutral higgsinos ($\widetilde{H}^0$, ...), gluinos $\widetilde{g}$, and gravitinos). The non-observation of supersymmetric particles at accelerators means that supersymmetry must be broken and the superpartners obviously are heavier than known particles and not in the reach of present-day accelerators. The theory of supersymmetry appears to be at least for the theoreticians so aesthetical and simple that they expect it to be true. A new quantum number, the R parity, distinguishes normal particles from their supersymmetric partners. If the R parity is a conserved quantity – and this is assumed to be the case in the most simple supersymmetric theories – the lightest supersymmetric particle (LSP) must be stable (it could only decay under violation of R parity). This lightest supersymmetric particle represents an ideal candidate for dark matter (Fig. 13.9). It is generally assumed that the Large Hadron Collider (LHC) under construction at CERN has a fair chance to produce and to be able to successfully reconstruct the creation of supersymmetric particles. The first collisions in LHC are expected for the year 2007.

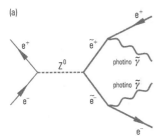

(a)

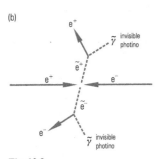

(b)

Fig. 13.9
Production and decay of supersymmetric particles as Feynman diagram **(a)** and in the detector **(b)**

WIMP, LSP detection possibilities

The low interaction strength of the LSP, however, is at the same time an obstacle for its detection. If, for example, supersymmetric particles would be created at an accelerator in proton–proton or electron–positron interactions, the final state would manifest itself by missing energy, because in decays of supersymmetric particles one lightest supersymmetric particle would always be created, which would leave the detector without significant interaction. Primordially produced supersymmetric particles would have decayed already a long time ago, apart from the lightest supersymmetric particles, which are expected to be stable if R parity is conserved. These would lose a certain amount of energy in collisions with normal matter, that could be used for their detection. Unfortunately, the recoil energy transferred to a target nucleus in a WIMP interaction (mass 10–100 GeV) is rather small, namely in the range of about 10 keV. Still one tries to measure the ionization or scintillation produced by the recoiling nucleus. Also a direct calorimetric measurement of the energy deposited in a bolometer is conceivable. Because of the low energy ΔQ to be measured and the related minute temperature rise

recoil energy

ionization and scintillation of recoiling nucleus

calorimetric measurement

bolometer

$$\Delta T = \Delta Q / c_{sp} m \qquad (13.22)$$

(c_{sp} is the specific heat and m the mass of the calorimeter), these measurements can only be performed in ultrapure crystals (e.g., in sapphire) at extremely low temperatures (milli-Kelvin range, $c_{sp} \sim T^3$). It has also been considered to use superconducting strips for the detection, which would change to a normal-conducting state upon energy absorption, thereby producing a usable signal.

Based on general assumptions on the number density of WIMPs one would expect a counting rate of at most one event per kilogram target per day. The main problem in these experiments is the background due to natural radioactivity and cosmic rays.

expected WIMP rate

Due to their high anticipated mass WIMPs could also be gravitationally trapped by the Sun or the Earth. They would occasionally interact with the material of the Sun or Earth, lose thereby energy, and eventually obtain a velocity below the escape velocity of the Sun or the Earth, respectively. Since WIMPs and in the same way their antiparticles would be trapped, they could annihilate with each other and provide proton–antiproton or neutrino pairs. One would expect to obtain an equilibrium between trapping and annihilation rate.

gravitational binding to the solar system

WIMP annihilation

The WIMP annihilation signal could be recorded in large existing neutrino detectors originally designed for neutrino astronomy. Also large neutrino detectors under construction (ANTARES[6] or IceCube) would have a chance to pick up a WIMP signal.

possible recording in neutrino detectors

[6] ANTARES – Astronomy with a Neutrino Telescope and Abyss environmental RESearch

**black hole
as WIMP candidate**

It is not totally inconceivable that particularly heavy WIMP particles could be represented by primordial black holes, which could have been formed in the Big Bang before the era of nucleosynthesis. They would provide an ideal candidate for cold dark matter. However, it is very difficult to imagine a mechanism by which such primordial black holes could have been produced in the Big Bang.

**gravitational binding
to the Milky Way:
possible seasonal-dependent
WIMP flux**

Recently the Italian–Chinese DAMA[7] collaboration has published a result, which could hint at the existence of WIMPs. Similarly to our Sun, the Milky Way as a whole could also trap WIMPs gravitationally. During the rotation of the solar system around the galactic center the velocity of the Earth relative to the hypothetical WIMP halo would change depending on the season. In this way the Earth while orbiting the Sun would encounter different WIMP fluxes depending on the season. In the month of June the Earth moves against the halo ($\rightarrow$ large WIMP collision rate) and in December it moves parallel to the halo ($\rightarrow$ low WIMP collision rate).

**results from DAMA
and CDMS collaborations**

The DAMA collaboration has measured a seasonal variation of the interaction rate in their 100 kg heavy sodium-iodide crystal. The modulation of the interaction rate with an amplitude of 3% is interpreted as evidence for WIMPs. The results obtained in the Gran Sasso laboratory would favour a WIMP mass of 60 GeV. This claim, however, is in contradiction to the results of an American collaboration (CDMS)[8], which only observes a seasonal-independent background due to neutrons in their highly sensitive low-temperature calorimeter.

13.2.4 Axions

> *"For every complex natural phenomenon there is a simple, elegant, compelling, but wrong explanation."*
>
> *Thomas Gold*

C, P, CP violation

Weak interactions not only violate parity P and charge conjugation C, but also the combined symmetry CP. The CP violation is impressively demonstrated by the decays of neutral kaons and B mesons. In the framework of quantum chromodynamics (QCD) describing strong interactions, CP-violating terms also originate in the theory. However, experimentally CP violation is not observed in strong interactions.

---

[7] DAMA – DArk MAtter search

[8] CDMS – Cryogenic Dark Matter Search

Based on theoretical considerations in the framework of QCD the electric dipole moment of the neutron should be on the same order of magnitude as its magnetic dipole moment. Experimentally one finds, however, that it is much smaller and even consistent with zero. This contradiction has been known as the so-called strong *CP* problem. The solution to this enigma presumably lies outside the Standard Model of elementary particles. A possible solution is offered by the introduction of additional fields and symmetries, which eventually require the existence of a pseudoscalar particle, called axion. The axion is supposed to have similar properties as the neutral pion. In the same way as the π^0 it would have a two-photon coupling and could be observed by its two-photon decay or via its conversion in an external electromagnetic field (Fig. 13.10).

Theoretical considerations appear to indicate that the axion mass should be somewhere between the µeV and meV range. To reach the critical density of the universe with axions only, the axion density – assuming a mass of 1 µeV – should be enormously high ($> 10^{10}$ cm^{-3}). Since the conjectured masses of axions are very small, they must possess non-relativistic velocities to be gravitationally bound to a galaxy, because otherwise they would simply escape from it. For this reason axions are considered as cold dark matter.

As a consequence of the low masses the photons generated in the axion decay are generally of low energy. For axions in the preferred µeV range the photons produced by axion interactions in a magnetic field would lie in the microwave range. A possible detection of cosmological axions would therefore involve the measurement of a signal in a microwave cavity, which would significantly stand out above the thermal noise. Even though axions appear to be necessary for the solution of the strong *CP* problem and therefore are considered a good candidate for dark matter, all experiments on their search up to date gave negative results.

strong *CP* problem

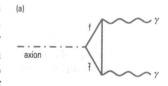

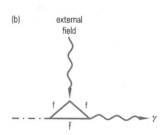

Fig. 13.10
Coupling of an axion to two photons via a fermion loop (**a**). Photons could also be provided by an electromagnetic field for axion conversion (**b**)

gravitational binding to the galaxy

cold dark matter

axion decay

13.2.5 The Rôle of the Vacuum Energy Density

To obtain a flat universe a non-zero cosmological constant is required, which drives the exponential expansion of the universe. The cosmological constant is a consequence of the finite vacuum energy. This energy could have been stored originally in a false vacuum, i.e., a vacuum, which is not at the lowest-energy state. The energy of the false vacuum

flat universe and cosmological constant

false vacuum

true vacuum could be liberated in a transition to the true vacuum (see Sect. 12.5).

Paradoxically, a non-zero cosmological constant was introduced by Einstein as a parameter in the field equations of **cosmological constant and static universe** the theory of general relativity to describe a static universe, which was popular at that time and to prevent a dynamic behaviour, which followed from his theory. Now it appears **cosmological constant as dominant energy** that the dominant energy, which determines the dynamics of the universe, is stored in the empty space itself. The question now arises, whether the cosmological constant is a practicable supplement to the required dark matter. This question was controversial just as the question of the existence of supersymmetric particles or axions over the last years. Recently, however, several experiments have found compelling evidence for a substantial amount of dark energy, which can be interpreted, e.g., in terms of a non-zero cosmological constant, see Chap. 12 on inflation.

There is a fundamental difference between classical dark matter in form of MACHOs, WIMPs, or axions and the effect of the cosmological constant Λ.

The potential energy on a test mass m created by matter and the vacuum energy density can easily be written after Newton to be

$$E_{\text{pot}} = -G\frac{mM_{\text{matter}}}{R} - G\frac{mM_{\text{vacuum energy}}}{R}$$

$$\sim -\frac{\varrho_{\text{matter}}R^3}{R} - \frac{\varrho_{\text{vacuum}}R^3}{R} \ . \tag{13.23}$$

matter density vs. vacuum energy density There is a fundamental difference between the matter density and the vacuum energy density during the expansion of the universe. For the vacuum-energy term in (13.23) the vacuum energy *density* remains constant, since this energy density is a property of the vacuum. In contrast to this the matter density does not remain constant during expansion, since only the *mass* is conserved leading to the dilution of the matter density. Therefore, the spatial dependence of the potential energy is given by

$$E_{\text{pot}} \sim -\frac{M_{\text{matter}}}{R} - \varrho_{\text{vacuum}}\, R^2 \ . \tag{13.24}$$

Since $\varrho_{\text{vacuum}} \sim \Lambda$, (13.24) shows that the radial dependence of the mass term is fundamentally different from the term containing the cosmological constant. Therefore, the question of the existence of dark matter (M_{matter}) and a finite vacuum energy density ($\Lambda \neq 0$) are not trivially coupled. Furthermore, Λ could be a dynamical constant, which is not only of importance for the development of the early universe. Since the Λ term in the field equations appears to dominate today, one would expect an accelerated expansion. The experimental determination of the acceleration parameter could provide evidence for the present effect of Λ. To do this one would have to compare the expansion rate of the universe in earlier cosmological times with that of the present.

cosmological constant and expanding universe

Such measurements have been performed with the Supernova Cosmology Project at Berkeley and with the High-z Supernova Search Team at Australia's Mount Stromlo Spring Observatory. The surprising results of these investigations was that the universe is actually expanding at a higher pace than expected (see also Sect. 8.8 and, in particular, Fig. 8.6). It is important to make sure that SN Ia upon which the conclusions depend explode the same way now and at much earlier times so that these supernovae can be considered as dependable standard candles. This will be investigated by looking at older and more recent supernovae of type Ia. Involved in these projects surveying distant galaxies are the Cerro Tololo Interamerican Observatory (CTIO) in the Chilean Andes, the Keck Telescope in Hawaii, and for the more distant supernovae the Hubble Space Telescope (HST).

Supernova Cosmology Project

High-z Supernova Search

13.2.6 Galaxy Formation

As already mentioned in the introduction to this chapter, the question of galaxy formation is closely related to the problem of dark matter. Already in the 18th century philosophers like Immanuel Kant and Thomas Wright have speculated about the nature and the origin of galaxies. Today it seems to be clear that galaxies originated from quantum fluctuations, which have been formed right after the Big Bang.

Kant, Wright

quantum fluctuations

With the Hubble telescope one can observe galaxies up to redshifts of $z = 3.5$ ($\lambda_{observed} = (1 + z)\lambda_{emitted}$); this corresponds to 85% of the total universe. The idea of cosmic inflation predicts that the universe is flat and expands forever, i.e., the Ω parameter is equal to unity.

Hubble telescope

Ω parameter

The dynamics of stars in galaxies and of galaxies in galactic clusters suggests that less than $\approx 5\%$ of matter is in form of baryons. Apart from the vacuum energy the main part of matter leading to $\Omega = 1$ has to be non-baryonic,

**properties
of non-baryonic matter**

which means it must consist of particles that do not occur in the Standard Model of elementary particle physics. The behaviour of this matter is completely different from normal matter. This dark matter interacts with other matter predominantly via gravitation. Therefore, collisions of dark-matter particles with known matter particles must be very rare. As a consequence of this, dark-matter particles lose only a small fraction of their energy when moving through the universe. This is an important fact when one discusses models of galaxy formation.

'cold' and 'hot' particles

Candidates for dark matter are subdivided into 'cold' and 'hot' particles. The prototype of hot dark matter is the neutrino ($m_\nu \neq 0$), which comes in at least in three different flavour states. Low-mass neutrinos are certainly insufficient to close the universe. Under cold dark matter one normally subsumes heavy weakly interacting massive particles (WIMPs) or axions.

models of galaxy formation

**quantum fluctuations
and gravitational instabilities**

The models of galaxy formation depend very sensitively on whether the universe is dominated by hot or cold dark matter. Since in all models one assumes that galaxies have originated from quantum fluctuations which have developed to larger gravitational instabilities, two different cases can be distinguished.

'protogalaxies'

'top–down' scenario

If the universe would have been dominated by the low-mass neutrinos, fluctuations below a certain critical mass would not have grown to galaxies because fast relativistic neutrinos could easily escape from these mass aggregations, thereby dispersing the 'protogalaxies'. For a neutrino mass of $20\,\text{eV}$ a critical mass of about 10^{16} solar masses is required, so that structure formation can really set in. With such large masses one lies in the range of the size of superclusters. Neutrinos as candidates for dark matter therefore would favour a scenario, in which first the large structures (superclusters), later clusters, and eventually galaxies would have been formed ('top–down' scenario). This would imply that galaxies have formed only for $z \leq 1$. However, from Hubble observations one already knows that even for $z \geq 3$ large populations of galaxies existed. This is also an argument to exclude a neutrino-dominated universe.

Massive, weakly interacting and mostly non-relativistic (i.e., slow) particles, however, will be bound gravitationally already to mass fluctuations of smaller size. If cold dark matter would dominate, initially small mass aggregations would collapse and grow by further mass attractions to form galaxies. These galaxies would then develop galactic clusters and later superclusters. Cold dark matter therefore favours a scenario in which smaller structures would be formed first and only later develop into larger structures ('bottom–up' scenario).

'bottom–up' scenario

In particular, the COBE and WMAP observations of the inhomogeneities of the 2.7 Kelvin radiation confirm the idea of structure formation by gravitational amplification of small primordial fluctuations. These observations therefore support a cosmogony driven by cold dark matter, in which smaller structures (galaxies) are formed first and later develop into galactic clusters.

inhomogeneities of blackbody radiation

The dominance of cold dark matter, however, does not exclude non-zero neutrino masses. The values favoured by the observed neutrino anomaly in the Super-Kamiokande and SNO experiments are compatible with a scenario of dominating cold dark matter. In this case one would have a cocktail – apart from baryonic matter – consisting predominantly of cold dark matter with a pinch of light neutrinos.

neutrino anomaly

13.2.7 Resumé on Dark Matter

It is undisputed that large quantities of dark matter must be hidden somewhere in the universe. The dynamics of galaxies and the dynamics of galactic clusters can only be understood, if the dominating part of gravitation is caused by non-luminous matter. In theories of structure formation in the universe in addition to baryonic matter (Ω_b) also other forms of matter or energy are required. These could be represented by hot dark matter (e.g., light relativistic neutrinos: Ω_{hot}) or cold dark matter (e.g., WIMPs: Ω_{cold}). At the present time cosmologists prefer a mixture of these three components. Recent measurements of distant supernovae and precise observations of inhomogeneities of the blackbody radiation lead to the conclusion that the dark energy, mostly interpreted in terms of the cosmological constant Λ, also has a very important impact on the structure of the universe. In general, the density parameter Ω can be presented as a sum of four contributions,

non-luminous matter
baryonic matter

hot dark matter
cold dark matter

cosmological constant

Ω parameter

$$\Omega = \Omega_b + \Omega_{hot} + \Omega_{cold} + \Omega_\Lambda. \tag{13.25}$$

The present state of the art in cosmology gives $\Omega_b \approx 0.04$, $\Omega_{hot} \leq 1\%$, $\Omega_{cold} \approx 0.23$, and $\Omega_\Lambda \approx 0.73$.

As demonstrated by observations of distant supernovae, the presently observed expansion is accelerated. Therefore,

accelerated expansion

it is clear that the cosmological constant plays a dominant rôle even today. Even if the repulsive force of the vacuum is small compared to the action of gravity, it will eventually win over gravity if there is enough empty space, see also (13.24). The present measurements (status 2005) indicate a value for Ω_A of ≈ 0.7. This means that the vacuum is filled with a weakly interacting hypothetical scalar field, which essentially manifests itself only via a repulsive gravitation and negative pressure, which consequently corresponds to a tension. In this model the field (e.g., 'quintessence', see glossary) produces a dynamical energy density of the vacuum in such a way that the repulsive gravity plays an important rôle also today. This idea is supported by the findings of the American Boomerang[9] and Maxima[10] experiments. These two balloon experiments measure the blackbody radiation over a limited region of the sky with a much higher precision than the COBE experiment. The observed temperature anisotropies are considered as a seed for galaxy formation.

quintessence

Boomerang

Maxima

These temperature variations throw light on the structure of the universe about 380 000 years after the Big Bang when the universe became transparent. The universe before that time was too hot for atoms to form and photons were trapped in a sea of electrically charged particles. Pressure waves pulsated in this sea of particles like sound waves in water. The response of this sea to the gravitational potential fluctuations allows to measure the properties of this fluid in an expanding universe consisting of ordinary and dark matter. When the universe cooled down, protons and electrons recombined to form electrically neutral hydrogen atoms, thereby freeing the photons. These hot photons have cooled since then to a temperature of around 2.7 K at present. Temperature variations in this radiation are fingerprints of the pattern of sound waves in the early universe, when it became transparent. The size of the spots in the thermal map is related to the curvature or the geometry of the early universe and gives informations about the energy density. Both experiments, Boomerang and Maxima, and the satellite experiment WMAP observed temperature clusters of a size of about one degree across. This information is experimentally obtained from the power spectrum of the primordial sound waves in the dense fluid of particles (see Chap. 11). The observed power spectrum corresponding to characteristic cluster sizes of the temperature

surface of last scattering

recombination

Boomerang, Maxima, WMAP
temperature clusters

---

[9] Boomerang – Balloon Observations of Millimetric Extragalactic Radiation and Geophysics

[10] Maxima – Millimeter Anisotropy Experiment Imaging Array

map of about 1° indicates – according to theory – that the universe is flat. If the results of the supernova project (see Sect. 8.3), the WMAP results, and the Boomerang and Maxima data are combined, a value for the cosmological constant corresponding to $\Omega_\Lambda \approx 0.7$ is favoured. Such a significant contribution would mean that the vacuum is filled with an incredibly weakly interaction substance, which reveals itself only via its repulsive gravitation (e.g., in terms of a quintessence model). The suggested large contribution of the vacuum energy to the Ω parameter would naturally raise the question, whether appreciable amounts of exotic dark matter (e.g., in the form of WIMPs) are required at all. However, it appears that for the understanding of the dynamics of galaxies and the interpretation of the fluctuations in the blackbody temperature a contribution of classical, nonbaryonic dark *matter* corresponding to $\Omega_{dm} \approx 0.23$ is indispensable.

flat universe

vacuum energy

In this scenario the long-term fate of the universe is characterized by eternal accelerated expansion. A Big Crunch event as anticipated for Ω larger than unity appears to be completely ruled out. Since the nature of the dark energy is essentially unknown, also its long-term properties are not understood. Therefore, one must be prepared for surprises. For example, it is possible that the dark energy becomes so powerful that the known forces are insufficient to preserve the universe. Galactic clusters, galaxies, planetary systems, and eventually atoms would be torn apart. In such theories one speculates about an apocalyptic Big Rip, which would shred the whole physical structure of the universe at the end of time.

Big Crunch

Big Rip

13.3 Problems

1. In the vicinity of the galactic center of the Milky Way celestial objects have been identified in the infrared and radio band which appear to rotate around an invisible center with high orbital velocities. One of these objects circles the galactic center with an orbital velocity of $v \approx 110\,\text{km/s}$ at a distance of approximately 2.5 pc. Work out the mass of the galactic center assuming Keplerian kinematics for a circular orbit!

2. What is the maximum energy that a WIMP ($m_W = 100$ GeV) of 1 GeV kinetic energy can transfer to an electron at rest?

3. Derive the Fermi energy of a classical and relativistic gas of massive neutrinos. What is the Fermi energy of the cosmological neutrino gas?

4. If axions exist, they are believed to act as cold dark matter. To accomplish the formation of galaxies they must be gravitationally bound; i.e., their velocities should not be too high.

   Work out the escape velocity of a $1\,\mu$eV-mass axion from a protogalaxis with a nucleus of 10^{10} solar masses of radius 3 kpc!

5. Consider a (not highly singular) spherically symmetric mass density $\varrho(r)$, the total mass of which is assumed to be finite.

   a) Determine the potential energy of a test mass m in the gravitational field originating from $\varrho(r)$. The force is given by Newton's formula in which only the mass $M(r)$ inside that sphere enters, whose radius is given by the position of m.

   b) With the previous result calculate the potential energy of the mass density, where the test mass is replaced by $dM = M'(r)\,dr$ and the r integration is performed to cover the whole space. Show with an integration by parts in the outer integral that the total potential energy can be written as

   $$E_{\text{pot}} = -G \int_0^\infty \frac{M^2(r)}{r^2}\, dr \ .$$

   c) Determine the potential energy of a massive spherical shell of radius R and mass M, i.e., $M(r) = 0$ for $r < R$ and $M(r) = M$ for $r > R$.

   d) Work out the potential energy of a sphere of radius R and mass M with homogeneous mass density.

   (This problem is a little difficult and tricky.)

6. Motivate the lower mass limit of neutrinos based on their maximum escape velocity from a galaxy (13.17), if it is assumed that they are responsible for the dynamics of the galaxy.

7. In the discussion on the search for MACHOs there is a statement that the radius of the Einstein ring varies as the square root of the mass of the deflector. Work out this dependence and determine the ring radius for
   - distance star–Earth = 55 kpc (LMC),
   - distance deflector–Earth = 10 kpc (halo),
   - Schwarzschild radius of the deflector = 3 km (corresponding to the Schwarzschild radius of the Sun).

14 Astrobiology

"In the beginning the universe was created.
This has made a lot of people very angry
and been widely regarded as a bad move."

Douglas Adams

Apart from considering the physical development of our universe, it is interesting to speculate about alternatives to our universe, i.e., whether the evolution in a different universe might have led to different physical laws, other manifestations of matter, or to other forms of life. Einstein had already pondered about such ideas and had raised the question whether the necessity of logical simplicity leaves any freedom for the construction of the universe at all.

The numerous free parameters in the Standard Model of elementary particles appear to have values such that stable nuclei and atoms, in particular, carbon, which is vital for the development of life, could be produced.[1] Also the coincidence of time scales for stellar evolution and evolution of life on planets is strange. In the early universe, hydrogen and helium were mainly created, whereas biology, as we know it, requires all elements of the periodic table. These other elements are provided in the course of stellar evolution and in supernova explosions. The long-lived, primordial, radioactive elements (^{238}U, ^{232}Th, ^{40}K) that have half-lives on the order of giga-years, appear to be particularly important for the development of life. Furthermore, a sufficient amount of iron must be created in supernovae explosions such that habitable planets can develop a liquid iron core. This is essential in the generation of magnetic fields which shield life on the planets against lethal radiation from solar and stellar winds and cosmic radiation.

free parameters

early universe

development of life

cosmic radiation

---

[1] α particles are formed in pp fusion processes. In $\alpha\alpha$ collisions ^{8}Be could be formed, but ^{8}Be is highly unstable and does not exist in nature. To reach ^{12}C, a triple collision of three α particles has to happen to get on with the production of elements. This appears at first sight very unlikely. Due to a curious mood of nature, however, this reaction has a resonance-like large cross section (see also Problem 2 in this chapter).

age of life on Earth

The oldest sedimentary rocks on Earth (3.9×10^9 years old) contain fossils of cells. It is known that bacteria even existed 3.5×10^9 years ago. If the biological evolution had required extremely long time scales, it is conceivable that the development of the higher forms of life would have been impossible, since the fuel of stars typically only lasts for 10^{10} years.

The origin of life is a question of much debate. In addition to the theory of a terrestrial origin also the notion of extra-terrestrial delivery of organic matter to the early Earth has gained much recognition. These problems are investigated in the framework of bioastronomy.

bioastronomy

The many free parameters in the Standard Model of electroweak and strong interactions can actually be reduced to a small number of parameters, whose values are of eminent importance for the astrobiological development of the universe. Among these important parameters are the masses of the u and d quarks and their mass difference. Experiments in elementary particle physics have established the fact that the mass of the u quark ($\approx 5\,\text{MeV}$) is smaller than that of the d quark ($\approx 10\,\text{MeV}$). The neutron, which has a quark content udd, is therefore heavier than the proton (uud), and can decay according to

masses of quarks

$$n \rightarrow p + e^- + \bar{\nu}_e \,. \tag{14.1}$$

unstable proton?

If, on the other hand, m_u were larger than m_d, the proton would be heavier than the neutron, and it would decay according to

$$p \rightarrow n + e^+ + \nu_e \,. \tag{14.2}$$

Stable elements would not exist, and there would be no chance for the development of life as we know it. On the other hand, if the d quark were much heavier than $10\,\text{MeV}$, deuterium ($d = {}^2\text{H}$) would be unstable and heavy elements could not be created. This is because the synthesis of elements in stars starts with the fusion of hydrogen to deuterium,

stability of deuterium

$$p + p \rightarrow d + e^+ + \nu_e \,, \tag{14.3}$$

and the formation of heavier elements requires stable deuterium. Life could also not develop under these circumstances. Also the value of the lifetime of the neutron takes an important influence on primordial chemistry and thereby

on the chances to create life (see also Sect. 10.5). There are
many other parameters which appear to be fine-tuned to the **fine-tuned parameters**
development of life in the form that we know.

One of the problems of the unification of forces in an
all-embracing Theory of Everything is the 'weakness' of the
gravitational force. If, however, gravitation would be much **gravitation**
stronger, we would live in a short-lived miniature universe.
No creatures could grow larger than a few centimeters, and
there would be hardly time for biological evolution.

The fact that we live in a flat universe with $\Omega = 1$ ap- **Ω parameter**
pears also to be important for us. If Ω were much larger than
unity, the universe would have recollapsed long ago, also
with the impossibility for the development of life. In this
context the exact value of the cosmological constant which **cosmological constant**
describes a repulsive gravity, plays an important rôle for the
development of the universe.

It is remarkable that the effect of Λ on microscopical
scales, just as that of attractive gravity, is negigible. The discrepancy between the measured small value of Λ in cosmology and the extremely large values of the vacuum energy in
quantum field theories tells us that important ingredients in
the understanding of the universe are still missing.

Another crucial parameter is the number of dimensions. **number of dimensions**
It is true that superstring theories can be formulated in
eleven dimensions, out of which seven are compactified, so
that we live in three spatial and one time dimension. But life
would be impossible, e.g., in two spatial dimensions. What
has singled out our case?

Also the efficiency of energy generation in stars is im- **efficiency**
portant for the production of different chemical elements. **of energy generation**
If this efficiency would be much larger than the value that
we know (0.7%), stars would exhaust their fuel in a much
shorter time compared to that needed for biological evolution.

It has become clear that the fine-tuning of parameters
in the Standard Model of elementary particle physics and
cosmology is very important for the development of stars,
galaxies, and life. If some of the parameters which describe **fine-tuning of parameters**
our world were not finely tuned, then our universe would
have grossly different properties. These different properties
would have made unlikely that life – in the form we know
it – would have developed. Consequently, physicists would
not be around to ask the question why the parameters have
exactly the values which they have. Our universe could have

plethora of universes been the result of a selection effect on the plethora of uni-
multiverse verses in a multiverse. It is also conceivable that there could
exist even a diversity of physical laws in other universes.
Only in universes where the conditons allow life to develop,
questions about the specialness of the parameters can be
anthropic principle posed. As a consequence of this *anthropic principle* there
is nothing mysterious about our universe being special. It
might just be that we are living in a most probable universe
which allows life to develop.

Nevertheless, it is the hope of particle theorists and cosmologists that a Theory of Everything can be found such
that all sensitive parameters are uniquely fixed to the values
that have been found experimentally. Such a theory might
eventually also require a deeper understanding of time. To
find such a theory and also experimental verifications of it,
cosmoparticle physics is the ultimate goal of cosmoparticle physics.

This is exactly what Einstein meant when he said: "What
I am really interested in, is whether God could have made
the world in a different way; that is, whether the necessity of
logical simplicity leaves any freedom at all."

14.1 Problems

1. Estimate the maximum stable size of a human in the
   Earth's gravitational field. What would happen to this
   result if the acceleration due to gravity would double?
   Hint: Consider that the weight of a human is proportional to the cube of a typical dimension, while its
   strength only varies as its cross section.
2. What are the conditions to synthesize carbon in stars?
   Hint: The detailed, quantitative solution of this problem
   is difficult. To achieve carbon fusion one has to have
   three alpha particles almost at the same time in the same
   place. Consequently high alpha-particle densities are required, and high temperatures to overcome the Coulomb
   barriers. Also the cross section must be large. Check
   with the web pages
   ```
   http://www.campusprogram.com/
       reference/en/wikipedia/t/tr/
       triple_alpha_process.html,
   http://en.wikipedia.org/wiki/
       Triple-alpha_process,
   ```
   and `nucl-th/0010052` on
   ```
   http://xxx.lanl.gov/.
   ```

3. What determines whether a planet has an atmosphere?
   Hint: Gas atoms or molecules will be lost from an atmosphere if their average velocity exceeds the escape
   velocity from the planet.
4. Rotationally invariant long-range radial forces: (non-
   relativistic) two-body problem in n dimensions.[2]
   a) How does such a force F between two bodies de-
      pend on the distance between two point-like bodies?
      Recall from three dimensions how the force depends
      on the surface or the solid angle of a sphere, re-
      spectively ('force flow'). The (hyper)surface $s_n(r)$
      of the n-dimensional (hyper)sphere can be written
      as $s_n(r) = s_n r^{g(n)}$, where $g(n)$ needs to be fixed.
   b) What is the corresponding potential energy and how
      does the effective potential (which includes the cen-
      trifugal barier) look like? Find conditions for stable
      orbits with attractive forces.
   c) Consider the introduction of a field-energy density
      $w \sim F^2$ (mediated by a field strength $\sim F$) and
      integrate w within the range of the two radii λ and
      Λ. The n-dimensional volume element is given by
      $dV_n = s_n(r) dr$, see also part (a). At what n does
      this expression diverge for $\lambda \to 0$ or $\Lambda \to \infty$?
      If quantum corrections are supposed to compensate
      the divergences, at what limit should they become
      relevant?

---

[2] This problem discusses unconventional dimensions. It is quite
tricky and mathematically demanding.

15 Outlook

"My goal is simple. It is complete understanding of the universe: why it is as it is
and why it exists at all."

Stephen Hawking

Astroparticle physics has developed from the field of cos- **cosmic rays**
mic rays. As far as the particle physics aspect is concerned,
accelerator experiments have taken over and played a leading rôle in this field since the sixties. However, physicists
have realized that the energies in reach of terrestrial accelerators represent only a tiny fraction of nature's wide window,
and it has become obvious that accelerator experiments will
never match the energies required for the solution of important astrophysical and cosmological problems. Actually,
recently cosmic ray physics has again taken the lead over
accelerator physics by discovering physics beyond the Standard Model, namely, by finding evidence for neutrino oscil- **physics**
lations in cosmic-ray experiments, which require non-zero **beyond the Standard Model**
neutrino masses.

The electromagnetic and weak interactions unify to a
common electroweak force at center-of-mass energies of
around $100\,\text{GeV}$, the scale of $W^{\pm}$ and Z masses. Already **electroweak scale**
the next unification of strong and electroweak interactions
at the GUT scale of $10^{16}\,\text{GeV}$ is beyond any possibility to **GUT scale**
reach in accelerator experiments even in the future. This is
even more true for the unification of all interactions which
is supposed to happen at the Planck scale, where quantum **Planck scale**
effects of gravitation become important ($\approx 10^{19}\,\text{GeV}$).

These energies will never be reached, not even in the
form of energies of cosmic-ray particles. The highest energy observed so far of cosmic-ray particles, which has been
measured in extensive-air-shower experiments ($3 \times 10^{20}\,\text{eV}$),
corresponds to a center-of-mass energy of about $800\,\text{TeV}$ in
proton–proton collisions, which is 60-fold the energy that
is going to be reached at the Large Hadron Collider LHC **LHC**
in the year 2007. It has, however, to be considered that the
rate of cosmic-ray particles with these high energies is extremely low. In the early universe ($< 10^{-35}\,\text{s}$), on the other
hand, conditions have prevailed corresponding to GUT- and

Theory of Everything

Planck-scale energies. The search for stable remnants of the GUT or Planck time probably allows to get information about models of the all-embracing theory (TOE – Theory of Everything). These relics could manifest themselves in exotic objects like heavy supersymmetric particles, magnetic monopoles, axions, or primordial black holes.

origin of cosmic rays

Apart from the unification of all interactions the problem of the origin of cosmic rays has still not been solved. There is a number of known cosmic accelerators (supernova explosions, pulsars, active galactic nuclei, M87(?), ...), but it is completely unclear how the highest energies ($> 10^{20}$ eV)

cosmic accelerators

are produced. Even the question of the identity of these particles (protons?, heavy nuclei?, photons?, neutrinos?, new particles?) has not been answered. It is conjectured that active galactic nuclei, in particular those of blazars, are able to accelerate particles to such high energies. If, however, protons or gamma rays are produced in these sources, our field of view into the cosmos is rather limited due to the short mean free path of these particles ($\lambda_{\gamma p} \approx 10$ Mpc, $\lambda_{\gamma\gamma} \approx 10$ kpc). Therefore, the community of astroparticle physicists is optimistic to be able to explore the universe

neutrino astronomy

with neutrino astronomy where cosmic neutrinos are produced in a similar way to γ rays in cosmic beam-dump experiments via pion decays. Neutrinos directly point back to the sources, they are not subject to deflections by magnetic fields and they are not attenuated or even absorbed by interactions.

The enigmatic neutrinos could also give a small contribution to the dark matter, which obviously dominates the universe. The investigations on the flavour composition of atmospheric neutrinos have shown that a deficit of muon neutrinos exists, which obviously can only be explained by oscillations. Such neutrino oscillations require a non-zero neutrino mass, which carries elementary particle physics already

beyond the Standard Model

beyond the well-established Standard Model. It was clear already for a long time that the Standard Model of elementary particles with its 26 free parameters cannot be the final answer, however, first hints for a possible extension of the Standard Model do not originate from accelerator experiments but rather from investigations of cosmic rays.

Neutrinos alone are by far unable to solve the problem of dark matter. To which extent exotic particles (WIMPs, SUSY particles, axions, quark nuggets, ...) contribute to the invisible mass remains to be seen. In addition, there is

also the cosmological constant, not really beloved by Ein- **cosmological constant**
stein, which provides an important contribution to the struc-
ture and to the expansion of the universe via its associated
vacuum energy density. In the standard scenario of the classical Big Bang model the presently observed expansion of
the universe is expected to be decelerated due to the pull of
gravity. Quite on the contrary, the most recent observations
of distant supernovae explosions indicate that the expansion
has changed gear. It has turned out that the cosmological
constant – even at present times – plays a dominant rôle. A
precise time-dependent measurement of the acceleration pa-
rameter – via the investigation of the expansion velocity in **acceleration parameter**
different cosmological epochs, i.e., distances – has shown
that the universe expands presently at a faster rate than at
earlier times.

Another input to cosmology which does not come from
particle physics experiments is due to precise observations
of the cosmic microwave background radiation. The results
of these experiments in conjunction with the findings of the
distant supernova studies have shown that the universe is
flat, i.e., the Ω parameter is equal to unity. Given the low **Ω parameter**
amount of baryonic mass (about 4%) and that dark matter **baryonic mass**
constitutes only about 23%, there is large room for dark
energy (73%) to speculate about. These results from non- **dark energy**
accelerator experiments have led to a major breakthrough in
cosmology.

Eventually the problem of the existence of cosmic anti-
matter remains to be solved. From accelerator experiments **cosmic antimatter**
and investigations in cosmic rays it is well-known that the
baryon number is a sacred conserved quantity. If a baryon
is produced, an antibaryon is always created at the same
time. In the same way with each lepton an antilepton is produced. The few antiparticles ($\bar{p}$, e^{+}) measured in astroparticle physics are presumably of secondary origin. Since one
has to assume that equal amounts of quarks and antiquarks
had been produced in the Big Bang, a mechanism is urgently
required, which acts in an asymmetric fashion on particles
and antiparticles. Since it is known that weak interactions **CP violation**
violate not only parity P and charge conjugation C but also
CP, a CP-violating effect could produce an asymmetry in
the decay of the parent particles of protons and antiprotons
so that the numbers of protons and antiprotons developed in
a slightly different way. In subsequent $p\bar{p}$ annihilations very **$p\bar{p}$ annihilation**
few particles, which we now call protons, would be left over.

matter dominance

Whether a small *CP* violation is sufficient to accomplish this remains to be seen. Since in $p\bar{p}$ annihilations a substantial amount of energy is transformed into photons, the observed γ/p ratio ($\hat{=} n_\gamma/n_p$) of about 10^9 indicates that even a minute difference in the decay properties of the parent particles of protons and antiprotons is sufficient to explain the matter dominance in the universe. Such an asymmetry could have its origin in Grand Unified Theories of electroweak and strong interactions. The matter–antimatter asymmetry could then be explained by different decay properties of heavy X and $\bar{X}$ particles which are supposed to have existed in the early universe. Equal numbers of X and $\bar{X}$ particles decaying into quarks, antiquarks, and leptons could lead to different numbers of quarks and antiquarks if baryon-number and lepton-number conservation were violated, thereby leading to the observed matter dominance of the universe.

primordial antimatter

The non-observation of primordial antimatter is still not a conclusive proof that the universe is matter dominated. If galaxies made of matter and 'antigalaxies' made of antimatter would exist, one would expect that they would attract each other occasionally by gravitation. This should lead to a spectacular annihilation event with strong radiation. A clear signal for such a catastrophe would be the emission of the 511 keV line due to e^+e^- annihilation. It is true that such a 511 keV γ-ray line has been observed – also in our galaxy – but its intensity is insufficient to be able to understand it as an annihilation of large amounts of matter and antimatter. On the other hand, it is conceivable that the annihilation

annihilation radiation

radiation produced in interactions of tails of galaxies with tails of antigalaxies would establish such a radiation pressure that the galaxies would be driven apart and the really giant spectacular radiation outburst would never happen.

Questions about a possible dominance of matter are unlikely to be answered in accelerator experiments alone. On the other hand, the early universe provides a laboratory, which might contain – at least in principle – the answers.

However, the step to bridge the gap from the Planck era to present-day theories is still hidden in the mist of space and time.

The investigations in the framework of astroparticle physics represent an important tool for a deeper understanding of the universe, from its creation to the present and future state.

15.1 Problems

1. It is generally believed that the cosmological constant represents the energy of the vacuum. On the other hand, quantum field theories lead to a vacuum energy which is larger by about 120 orders of magnitude. This dicrepancy can be solved by an appropriate quantization of Einstein's theory of general relativity. Start from the Standard Model of particle physics (see Chap. 2) and use the Friedmann equation (see Sect. 8.4) extended by a term for the vacuum energy ϱ_v to solve the discrepancy between the value of the cosmological constant obtained from astroparticle physics measurements and the vacuum energy as obtained from quantum field theories!

16 Glossary

> "A good notation has a subtlety and suggestiveness, which at times makes it almost seem like a live teacher."
>
> *Bertrand Russell*

A

The total luminosity emitted from an astrophysical object. **absolute brightness**

Lowest possible temperature, 0 Kelvin, at which all motion comes to rest except for quantum effects (0 Kelvin = $-273.15°$ Celsius). **absolute zero**

Percentage of an element occurring on Earth, in the solar system, or in primary cosmic rays in a stable isotopic form. **abundance**

It describes the acceleration of the universe in its dependence on the energy density and negative pressure (i.e., due to the cosmological constant). **acceleration equation**

A measurement for the change of the expansion rate with time. For an increased expansion rate the acceleration parameter is positive. It was generally expected that the present expansion rate is reduced by the attractive gravitation (negative acceleration parameter = deceleration). The most recent measurements of distant supernovae, however, contradict this expectation by finding an accelerated expansion. **acceleration parameter**

A machine used to accelerate charged particles to high speeds. There are, of course, also cosmic accelerators. **accelerator**

Accumulation of dust and gas into a disk, normally rotating around a compact object. **accretion disk**

After decoupling of matter from radiation the pattern of acoustic oscillations of the primordial baryon fluid became frozen as structures in the power spectrum of the CMB. **acoustic peaks**

Galaxy with a bright central region, an active galactic nucleus. Seyfert galaxies, quasars, and blazars are active galaxies, which are presumably powered by a black hole. **active galaxy**

In the Big Bang model the inverse of the Hubble constant is identified with the age of the universe. The present estimate of the age of the universe is about 14 billion years. **age of the universe**

AGN	Active Galactic Nucleus. If one assumes that black holes powered by infalling matter reside at centers of AGNs and if these AGNs emit polar jets, then the different AGN types (Seyfert galaxies, BL-Lacertae objects, radioquasars, radiogalaxies, quasars) can be understood as a consequence of the random direction of view from Earth.
air-shower Cherenkov technique	Measurement of extensive air showers via their produced Cherenkov light in the atmosphere.
alpha decay	Nuclear decay consisting of the emission of an α particle (helium-4 nucleus).
AMANDA	Antarctic Muon And Neutrino Detector Array for the measurement of high-energy cosmic neutrinos.
Andromeda Nebula	M31, one of the main galaxies of the local group; distance from the Milky Way: 700 kpc.
Ångström	The unit of length equal to 10^{-10} m.
annihilation	A process, in which particles and antiparticles desintegrate, e.g., $e^+ e^- \to \gamma\gamma$.
anthropic principle	One might ask why the universe is as it is. If some of the physics constants which describe our universe were not finely tuned, then it would have grossly different properties. It is considered highly likely that life forms as we know them would not develop under an even slightly different set of basic parameters. Therefore, if the universe were different, nobody would be around to ask why the universe is as it is.
antigravitation	Repulsive gravitation caused by a negative pressure as a consequence of a finite non-zero cosmological constant.
antimatter	Matter consisting only of antiparticles, like positrons, antiprotons, or antineutrinos. When antimatter particles and ordinary matter particles meet, they annihilate mostly into γ rays (e.g., $e^+ e^- \to \gamma\gamma$).
antiparticles	For each particle there exists a different particle type of exactly the same mass but opposite values for all other quantum numbers. This state is called antiparticle. For example, the antiparticle of an electron is a particle with positive electric unit charge, which is called positron. Also bosons have antiparticles apart from those for which all charge quantum numbers vanish. An example for this is the photon or a composite boson consisting of a quark and its corresponding antiquark. In this particular case there is no distinction between particle and antiparticle, they are one and the same object.

The antiparticle of a quark. **antiquark**

The point of greatest separation of two stars as in a binary star orbit. **apastron**

The point in the orbit of a planet, which is farthest from the Sun. **aphelion**

The point in the orbit of an Earth satellite, which is farthest from Earth. **apogee**

The brightness of a star as it appears to the observer. It is measured in units of magnitude. See magnitude. **apparent magnitude**

Considerations on the formation and evolution of life in universes with different fundamental parameters (different physical laws, quark masses, etc.) **astrobiology**

The average distance from Earth to Sun. 1 AU $\approx$ 149 597 870 km. **astronomical unit (AU)**

At large momenta quarks will be eventually deconfined and become asymptotically free. **asymptotic freedom**

The mass of a neutral atom or a nuclide. The atomic weight of an atom is the weight of the atom based on the scale where the mass of the carbon-12 nucleus is equal to 12. For natural elements with more than one isotope, the atomic weight is the weighted average for the mixture of isotopes. **atomic mass**

The number of protons in the nucleus. **atomic number**

The northern (aurora borealis) or southern (aurora australis) lights are bright emissions of atoms and molecules in the polar upper atmosphere around the south and north geomagnetic poles. The aurora is associated with luminous processes caused by energetic solar cosmic-ray particles incident on the upper atmosphere of the Earth. **aurora**

Advanced X-ray Astrophysics Facility; now named Chandra. **AXAF**

Hypothetical pseudoscalar particle, which was introduced as quantum of a field to explain the non-observation of *CP* violation in strong interactions. **axion**

B

See cosmic background radiation. **background radiation**

Formation of baryons out of the primordial quark soup in the very early universe. **baryogenesis**

baryon–antibaryon asymmetry   An asymmetry created by some, so far unknown baryon-number-violating process in the early universe (see Sakharov conditions).

baryon number   The baryon number B is unity for all strongly interacting particles consisting of three quarks. Quarks themselves carry baryon number $1/3$. For all other particles $B = 0$.

baryons   Elementary particles consisting of three quarks like, e.g., protons and neutrons.

BATSE   Burst And Transient Source Experiment on board of the CGRO satellite.

beam-dump experiment   If a high-energy particle beam is stopped in a sufficiently thick target, all strongly and electromagnetically interacting particles are absorbed. Only neutral weakly interacting particles like neutrinos can escape.

beta decay   In nuclear β decay a neutron of the nucleus is transformed into a proton under the emission of an electron and an electron antineutrino. The transition of the proton in the nucleus into a neutron under emission of a positron and an electron neutrino is called β^+ decay, contrary to the aforementioned β^- decay. The electron capture which is the reaction of proton and electron into neutron and an electron neutrino ($p + e^- \rightarrow n + \nu_e$) is also considered as beta decay.

Bethe–Bloch formula   Describes the energy loss of charged particles by ionization and excitation when passing through matter.

Bethe–Weizsäcker cycle   Carbon–nitrogen–oxygen cycle: nuclear fusion process in massive stars, in which hydrogen is burnt to helium with carbon as catalyzer (CNO cycle).

Bethe–Weizsäcker formula   Describes the nuclear binding energy in the framework of the liquid-drop model.

Big Bang   Beginning of the universe when all matter and radiation emerged from a singularity.

Big Bang theory   The theory of an expanding universe that begins from an infinitely dense and hot primordial soup. The initial instant is called the Big Bang. The theory says nothing at all about time zero itself.

Big Crunch   If the matter density in the universe is larger than the critical density, the presently observed expansion phase will turn over into a contraction, which eventually will end in a singularity.

The Big Rip is a cosmological hypothesis about the ultimate fate **Big Rip**
of the universe. The universe has a large amount of dark energy
corresponding to a repulsive gravity, and in the long run it will win
over the classical attractive gravity anyhow. Therefore, all matter
will be finally pulled apart. Galaxies in galactic clusters will be
separated, the gravity in galaxies will become too weak to hold the
stars together, and also the solar system will become gravitationally
unbound. Eventually also the Earth and the atoms themselves will
be destroyed with the consequence that the Big Rip will shred the
whole physical structure of the universe.

Binary stars are two stars that orbit around their common center of **binary stars**
mass. An X-ray binary is the special case where one of the stars
is a collapsed object such as a neutron star or black hole. Matter
is stripped from the normal star and falls onto the collapsed star
producing X rays or also gamma rays.

The energy that has to be invested to desintegrate a nucleus into its **binding energy**
single constituents (protons and neutrons, nuclear binding energy).

Scientific branch of astronomy that deals with the question of the **bioastronomy**
origin of life, e.g., whether organic material has been delivered to ·
Earth from exraterrestrial sources. The techniques are to look for
biomolecules with spectroscopic methods used in astronomy.

A hypothetical body that absorbs all the radiation falling on it. The **blackbody**
emissivity of a blackbody depends only on its temperature.

The radiation produced by a blackbody. The blackbody is a perfect **blackbody radiation**
radiator and absorber of heat or radiation, see also Planck distribution.

A massive star that has used up all its hydrogen can collapse under **black hole**
its own gravity to a mathematical singularity. The size of a black
hole is characterized by the event horizon. The event horizon is the
radius of that region, in which gravity is so strong that even light
cannot escape (see also Schwarzschild radius).

Short for variable active galactic nuclei, which are similar to BL- **blazars**
Lacertae objects and quasars except that the optical spectrum is
almost featureless.

Reduction of the wavelength of electromagnetic radiation by the **blueshift**
Doppler effect as observed, e.g., during the contraction of the universe.

Variable extragalactic objects in the nuclei of some galaxies, which **BL-Lacertae objects**
outshine the whole galaxy. BL stands for radio loud emission in the
B band ($\approx 100\,\mathrm{MHz}$); lacerta = lizard.

bolometer Sensitive resistance thermometer for the measurement of small energy depositions.

Boltzmann constant A constant of nature which describes the relation between the temperature and kinetic energy for atoms or molecules in an ideal gas.

Bose–Einstein distribution The energy distribution of bosons in its dependence on the temperature.

boson A particle with integer spin. The spin is measured in units of $\hbar$ (spin $s = 0, 1, 2, \ldots$). All particles are either fermions or bosons. Gauge bosons are associated with fundamental interactions or forces, which act between the fermions. Composite particles with an even number of fermion constituents (e.g., consisting of a quark and antiquark) are also bosons.

bottom quark b; the fifth quark flavour (if the quarks are ordered with increasing mass). The b quark has the electric charge $-\frac{1}{3}e$.

Brahe Tycho Brahe, a Danish astronomer, whose accurate astronomical observations form the basis for Johannes Kepler's laws of planetary motion. The supernova remnant SNR 1572 (Tycho) is named after Brahe.

brane Branes are higher-dimensional generalisations of strings. A p brane is a space-time object with p spatial dimensions.

bremsstrahlung Emission of electromagnetic radiation when a charged particle is decelerated in the Coulomb field of a nucleus. Bremsstrahlung can also be emitted during the deceleration of a charged particle in the Coulomb field of an electron or any other charge.

brown dwarves Low-mass stars (< 0.08 solar masses), in which thermonuclear reactions do not ignite, which, however, still shine because gravitational energy is transformed into electromagnetic radiation during a slow shrinking process of the objects.

C

Calabi–Yau space Complex space of higher dimension, which has become popular for the compactification of extraspatial dimensions in the framework of string theories.

carbon cycle See Bethe–Weizsäcker cycle.

cascade See shower.

Casimir effect A reduction in the number of virtual particles between two flat parallel metal plates in vacuum compared to the surrounding leads to a measurable attractive force between the plates.

Stars which have rapid and unpredictable changes in brightness, mostly associated with a binary star system.	**cataclysmic variables**
See cosmic background radiation.	**CBR**
Charge-Coupled Device, a solid-state camera.	**CCD**
Strong galactic radio source in the constellation Centaurus.	**Centaurus A**
Type of pulsating variable stars where the period of oscillation is proportional to the average absolute magnitude; often used as 'standard' candles.	**Cepheid variables**
Centre Européen pour la Recherche Nucléaire. The major European center for particle physics located near Geneva in Switzerland.	**CERN**
Compton Gamma-Ray Observatory. Satellite with four experiments on board for the measurement of galactic and extragalactic gamma rays.	**CGRO**
X-ray satellite of the NASA (original name: AXAF) started in July 1999; named after Subrahmanyan Chandrasekhar.	**Chandra**
Limiting mass for white dwarves (1.4 solar masses). If a star exceeds this mass, gravity will eventually defeat the pressure of the degenerate electron gas leading to a compact neutron star.	**Chandrasekhar mass**
Quantum number carried by a particle. The charge quantum number determines whether a particle can participate in a special interaction process. Particles with electric charge participate in electromagnetic interactions, such with strong charge undergo strong interactions, and those with weak charge are subject to weak interactions.	**charge**
The principle of charge invariance claims that all processes again constitute a physical reality if particles are exchanged by their antiparticles. This principle is violated in weak interactions.	**charge conjugation**
The observation that the charge of a system of particles remains unchanged in an interaction or transformation. In this context charge stands for electric charge, strong charge, or also weak charge.	**charge conservation**
Interaction process mediated by the exchange of a virtual charged gauge boson.	**charged current**
c; the fourth quark flavour (if quarks are ordered with increasing mass). The c quark has the electric charge $+\frac{2}{3}e$.	**charm quark**
For a given component in a particle mixture the chemical potential describes the change in free energy with respect to a change in the amount of the component at fixed pressure and temperature.	**chemical potential**

Cherenkov effect   Cherenkov radiation occurs, if the velocity of a charged particle in a medium with index of refraction n exceeds the velocity of light c/n in that medium.

closed universe   A Friedmann–Lemaître model of the universe with positive curvature of space. Such a universe will eventually contract leading to a Big Crunch.

cluster of galaxies   A set of galaxies containing from a few tens to several thousand member galaxies, which are all gravitationally bound together. The Virgo cluster includes 2500 galaxies.

CMB   Cosmic microwave background, see cosmic background radiation.

COBE   Cosmic Background Explorer. Satellite with which the minute temperature inhomogeneities ($\frac{\Delta T}{T} \approx 10^{-5}$) of the cosmic background radiation were first discovered.

cold dark matter   Type of dark matter that was moving at much less than the velocity of light some time after the Big Bang. It could consist of Weakly Interacting Massive Particles (WIMPs, such as supersymmetric particles) or axions.

collider   An accelerator in which two counter-rotating beams are steered together to provide a high-energy collision between the particles from one beam with those of the other.

colour   The strong 'charge' of quarks and gluons is called colour.

colour charge   The quantum number that determines the participation of hadrons in strong interactions. Quarks and gluons carry non-zero colour charges.

compactification   The universe may have extra dimensions. In string theories there are 10 or more dimensions. Since our world appears to have only three plus one dimension, it is assumed that the extra dimensions are curled up into sizes so small that one can hardly detect them directly, which means they are compactified.

COMPTEL   Compton telescope on board the CGRO satellite.

Compton effect   Compton effect is the scattering of a photon off a free electron. The scattering off atomic electrons can be considered as pure Compton effect, if the binding energy of the electrons is small compared to the energy of the incident photon.

confinement   The property of strong interactions, which says that quarks and gluons can never be observed as single free objects, but only inside colour-neutral objects.

A quantity is conserved, if it does not change in the course of a reaction between particles. Conserved quantities are, for example, electric charge, energy, and momentum. **conserved quantity**

A group of stars that produce in projection a shape often named after mythological characters or animals. **constellation**

An alternative process to X-ray emission during the de-excitation of an excited atom or an excited nucleus. If the excitation energy of the atomic shell is transferred to an atomic electron and if this electron is liberated from the atom, this electron is called an Auger electron. If the excitation energy of a nucleus is transferred to an atomic electron of the K, L, or M shell, this process is called internal conversion. Auger electron emission or emission of characteristic X rays is often a consequence of internal conversion. **conversion electron**

A very hot outer layer of the Sun's atmosphere consisting of highly diffused ionized gas and extending into interplanetary space. The hot gas in the solar corona forms the solar wind. **corona**

European gamma-ray satellite started in 1975. **COS-B**

Standard composition of elements in the universe as determined from terrestrial, solar, and extrasolar matter. **cosmic abundance**

CBR; nearly isotropic blackbody radiation originating from the Big Bang ('echo of the Big Bang'). The cosmic background radiation has now a temperature of 2.7 K. **cosmic background radiation**

Nuclear and subatomic particles moving through space at velocities close to the speed of light. Cosmic rays originate from stars and, in particular, from supernova explosions. The origin of the highest-energy cosmic rays is an unsolved problem. **cosmic rays**

One-dimensional defects, which might have been created in the early universe. **cosmic strings**

Topological defect which involves a kind of twisting of the fabric of space-time. **cosmic textures**

The branch of particle physics and astronomy that tries to dig out results on elementary particles and cosmology from experimental evidence obtained from data about the early universe. **cosmoarcheology**

The science of the origins of galaxies, stars, planets, and satellites, and, in particular, of the universe as a whole. **cosmogony**

Two-dimensinonal representation of the universe in the light of the 2.7 K blackbody radiation showing temperature variations as seed for galaxy formation. **cosmographic map**

cosmological constant (Λ) The cosmological constant was introduced by hand into the Einstein field equations with the intention to allow an aesthetic cosmological solution, i.e., a steady-state universe which was believed to represent the correct picture of the universe at the time. The cosmological constant is time independent and spatially homogeneous. It is physically equivalent to a non-zero vacuum energy. In an expanding universe the repulsive gravity due to Λ will eventually win over the classical attractive gravity.

cosmological principle Hypothesis that the universe at large scales is homogeneous and isotropic.

cosmological redshift Light from a distant galaxy appears redshifted by the expansion of the universe.

cosmology Science of the structure and development of the universe.

cosmoparticle physics The study of elementary particle physics using informations from the early universe.

CP invariance Nearly all interactions are invariant under simultaneous interchange of particles with antiparticles and space inversion. The *CP* invariance, however, is violated in weak interactions (e.g., in the decay of the neutral kaon and *B* meson).

CPT invariance All interactions lead again to a physically real process, if particles are replaced by antiparticles and if space and time are reversed. It is generally assumed that *CPT* invariance is an absolutely conserved symmetry.

Crab Nebula Crab; supernova explosion of a star in our Milky Way (observed by Chinese astronomers in 1054). The masses ejected from the star form the Crab Nebula in whose center the supernova remnant resides.

critical density Cosmic matter density ϱ_c leading to a flat universe. In a flat universe without cosmological constant ($\Lambda = 0$) the presently observed expansion rate will asymptotically tend to zero. For $\varrho > \varrho_c$ the expansion will turn into a contraction (see Big Crunch); for $\varrho < \varrho_c$ one would expect eternal expansion.

cross section The cross section is that area of an atomic nucleus or particle, which has to be hit to induce a certain reaction.

curvature The curvature of space-time determines the evolution of the universe. A universe with zero curvature is Euclidean. Positive curvature is the characteristic of a closed universe, while an open universe has negative curvature.

cyclotron mechanism Acceleration of charged particles on circular orbits in a transverse magnetic field.

X-ray binary consisting of a blue supergiant and a compact object, which is considered to be a black hole. — **Cygnus X1**

X-ray and gamma-ray binary system consisting of a pulsar and a companion. The pulsar appears to be able to emit occasionally gamma rays with energies up to 10^{16} eV. — **Cygnus X3**

D

A special form of energy that creates a negative pressure and is gravitationally repulsive. It appears to be a property of empty space. Dark energy appears to contribute about 70% to the energy density of the universe. The vacuum energy, the cosmological constant, and light scalar fields (like quintessence) are particular forms of dark-energy candidates. — **dark energy**

Unobserved non-luminous matter, whose existence has been inferred from the dynamics of the universe. The nature of the dark matter is an unsolved problem. — **dark matter**

Historically first experiment for the measurement of solar neutrinos. — **Davis experiment**

Quantum-mechanical wavelength λ of a particle: $\lambda = h/p$ (p – momentum, h – Planck's constant). — **de Broglie wavelength**

A process in which a particle disappears and in its place different particles appear. The sum of the masses of the produced particles is always less than the mass of the original decaying particle. — **decay**

A measure of how far an object is above or below the celestial equator (in degrees); similar to latitude on Earth. — **declination**

The pressure in a degenerate gas of fermions caused by the Pauli exclusion principle and the Heisenberg uncertainty principle. — **degeneracy pressure**

Fermi gas (electrons, neutrons), whose stability is guaranteed by the Pauli principle. In a gas of degenerate particles quantum effects become important. — **degenerate matter**

The number of values of a system that are free to vary. — **degrees of freedom**

Process during supernova explosions where electrons and protons merge to form neutrons and neutrinos ($e^- + p \to n + \nu_e$). — **deleptonization**

Local increase or decrease of the mass or radiation density in the early universe leading to galaxy formation. — **density fluctuations**

Isotope of hydrogen with one additional neutron in the nucleus. — **deuterium**

differential gravitation	Different forces of gravity acting on two different points of a body in a strong gravitational field lead to a stretching of the body.
dipole anisotropy	The apparent change in the cosmic background radiation temperature caused by the motion of the Earth through the CBR.
disk	The visible surface of any heavenly body projected against the sky.
distance ladder	A number of techniques (e.g., redshift, standard candles, ...) used by astronomers to obtain the distances to progressively more distant astronomical objects.
domain wall	Topological defect that might have been created in the early universe. Domain walls are two-dimensional defects.
Doppler effect	Change of the wavelength of light caused by a relative motion between source and observer.
double pulsar	Predictions of the general theory of relativity about the perihelion rotation and the energy loss by emission of gravitational waves have been confirmed for the binary pulsar system PSR1913+16 with extreme precision.
down quark	d; the second quark flavour, (if quarks are ordered according to increasing mass). The d quark has the electric charge $-\frac{1}{3}e$.

E

east–west effect	Geomagnetically caused asymmetry in the arrival direction of primary cosmic rays, which is related to the fact that most primary cosmic-ray particles carry positive charge.
EGRET	Energy Gamma Ray Telescope Experiment on board the CGRO satellite.
electromagnetic interactions	The interactions of particles due to their electric charge. This type includes also magnetic interactions.
electron	e; the lightest electrically charged particle. As a consequence of this it is absolutely stable, because there are no lighter electrically charged particles in which it could decay. The electron is the most frequent lepton with the electric charge -1 in units of the elementary charge.
electron capture	Nuclear decay by capture of an atomic electron. If the decay energy is larger than twice the electron mass, positron emission can also occur in competition with electron capture.

The electron and its associated electron neutrino are assigned the
electronic lepton number +1, its associated antiparticles the electronic lepton number −1. All other elementary particles have the
electron number 0. The electron number is generally a conserved
quantity. Only in neutrino oscillations it is violated.

electron number

eV; unit of energy and also of mass of particles. One eV is the energy that an electron (or, more generally, a singly charged particle)
gets, if it is accelerated in a potential difference of one volt.

electron volt

Standard Model of elementary particles in which the electromagnetic and weak interactions are unified.

electroweak theory

Galaxy of smooth elliptical structure without spiral arms.

elliptical galaxy

Radiation density in Joule/cm^3 or eV/cm^3.

energy density

Expérience pour la Recherche d'Objets Sombres. Experiment
searching for dark objects; see also MACHO.

EROS

The minimum velocity of a body to escape a gravitational field
caused by a mass M from a distance R from its center is $v = \sqrt{\frac{2GM}{R}}$. For the Earth the escape velocity is $v_E = 11.2\,\mathrm{km/s}$.

escape velocity

Surface of a black hole. Particles or light originating from inside
the event horizon cannot escape from the black hole.

event horizon

Theory in which all galaxies fly apart from each other. The expansion behaviour looks the same from every galaxy.

expansion model

Our universe has three spatial and one time dimension. In string
theories and supersymmetric theories there are in general p extended spatial dimensions, some of which may be compactified.

extended dimension

Large particle cascade in the atmosphere initiated by a primary
cosmic-ray particle of high energy.

extensive air shower (EAS)

Radiation originating from outside our galaxy.

extragalactic radiation

F

A metastable state describing a quantum field which is zero, even
though its corresponding vacuum energy density does not vanish.
A decaying false vacuum can in principle liberate the stored energy.

false vacuum

Organization of matter particles into three groups with each group
being known as a family. There is an electron, a muon, and a tau
family on the lepton side and equivalently there are three families
of quarks. LEP has shown that there are only three families (generations) of leptons with light neutrinos.

families

Fermi–Dirac distribution   The energy distribution of fermions in its dependence on the temperature.

Fermi energy   Energy of the highest occupied electron level at absolute zero for a free Fermi gas (e.g., electron, neutron gas).

Fermi mechanism   Acceleration mechanism for charged particles by shock waves (the 1.-order process) or extended magnetic clouds (the 2.-order process).

fermion   A particle with half integer spin ($\frac{1}{2}$, $\frac{3}{2}$, etc.) if the spin is measured in units of $\hbar$. An important consequence of this particular angular momentum is that fermions are subject to the Pauli principle. The Pauli principle says that no two fermions can exist in the same quantum-mechanical state at the same time. Many properties of ordinary matter originate from this principle. Electrons, protons, and neutrons – all of them are fermions – just as the fundamental constituents of matter, i.e., quarks and leptons.

Feynman diagrams   Feynman diagrams are pictorial shorthands to describe the interaction processes in space and time. With the necessary theoretical tools Feynman diagrams can be translated into cross sections. In this book the agreement is such that time is plotted on the horizontal and spatial coordinates on the vertical axis. Particles move forward in space-time, while antiparticles move backwards.

fission   The splitting of heavier atomic nuclei into lighter ones. Fission is the way how nuclear power plants produce energy.

fixed-target experiment   An experiment in which the beam of particles from an accelerator is aimed at a stationary target.

flare   Short duration outburst from stars in various spectral ranges.

flatness problem   In classical cosmology any value of Ω in the early universe even slightly different from one would be amplified with time and driven away from unity. In contrast, in inflationary cosmology Ω is approaching exponentially a value of one, thus explaining naturally the presently observed value of $\Omega = 1.02 \pm 0.02$. In classical cosmology such a value would have required an unbelievably careful fine-tuning.

flavour   Characterizes the assignment of a fermion (lepton or quark) to a particle family or a particle generation.

fluid equation   It describes the evolution of the energy density in its dependence on the expansion rate, pressure, and density. It can be derived using classical thermodynamics.

Particle detector for the measurement of large extensive air showers based on the measurement of the emitted scintillation light in the atmosphere.	**Fly's Eye**
Vector with four components comprising energy and the three momentum components.	**four-momentum**
Vector with four components mostly comprising time in the first compoment and three additional spatial components.	**four-vector**
Break-up of a heavy nucleus into a number of lighter nuclei in the course of a collision.	**fragmentation**
A differential equation that expresses the evolution of the universe depending on its energy density. This equation can be derived from Einstein's field equations of general relativity. Surprisingly, a classical treatment leads to the same result.	**Friedmann equation**
Standard Big Bang models with negative ($\Omega < 1$), positive ($\Omega > 1$), or flat ($\Omega = 1$) space curvature.	**Friedmann–Lemaître universes**
A particle with no observable inner structure. In the Standard Model of elementary particles quarks, leptons, photons, gluons, W^+, W^- bosons, and Z bosons are fundamental. All other objects are composites of fundamental particles.	**fundamental particle**
In fusion lighter elements are combined into heavier ones. Fusion is the way how stars produce energy.	**fusion**

G

Aggregation of galaxies in a spatially limited region.	**galactic cluster**
A spherical region mainly consisting of old stars surrounding the center of a galaxy.	**galactic halo**
Magnetic field in our Milky Way with an average strength of $3\,\mu G = 3 \times 10^{-10}$ Tesla.	**galactic magnetic field**
Coordinate transformation for frames of reference in which force-free bodies move in straight lines with constant speed without consideration of the fact that there exists a limiting velocity (velocity of light). This theorem of the addition of velocities leads to conflicts with experience, in particular, for velocities close to the velocity of light.	**Galilei transformation**
Gallium experiment in the Gran Sasso laboratory for the detection of solar neutrinos from the pp cycle.	**GALLEX experiment**

gamma astronomy   Astronomy in the gamma energy range ($> 0.1\,\text{MeV}$).

gamma burster   Extragalactic gamma-ray sources flaring up only once. Spectacular supernova explosions (hypernova) or colliding neutron stars are possible candidates for producing gamma-ray bursts.

gamma rays   Short-wavelength electromagnetic radiation corresponding to energies $\geq 0.1\,\text{MeV}$.

general principle of relativity   The principle of relativity argues that accelerated motion and being exposed to acceleration by a gravitational field is equivalent.

general relativity   Einstein's general formulation of gravity which shows that space and time communicate the gravitational force through their curvature.

general theory of relativity   Generalization of Newton's theory of gravity to systems with relative acceleration with respect to each other. In this theory gravitational mass (caused by gravitation) and inertial mass (mass resisting acceleration) are equivalent.

generation   A set of two quarks and two leptons which form a family. The order parameter for families is the mass of the family members. The first generation (family) contains the up and down quark, the electron and the electron neutrino. The second generation contains the charm and strange quark as well as the muon and its associated neutrino. The third generation comprises the top and bottom quark and the tau with its neutrino.

GLAST   Gamma-ray Large Area Space Telescope. GLAST will measure γ rays in the energy range $10\,\text{keV}$–$300\,\text{GeV}$. The space experiment is expected to be launched in 2005.

gluino   Supersymmetric partner of the gluon.

gluon g; the gluon is the carrier of strong interactions. There are altogether eight different gluons, which differ from each other by their colour quantum numbers.

grand unification   A theory which combines the strong, electromagnetic, and weak interactions into one unified theory (GUT).

gravitational collapse   If the gas and radiation pressure of a star can no longer resist the inwardly directed gravitational pressure, the star will collapse under its own gravity.

gravitational instability   Process by which density fluctuations of a certain size grow by self-gravitation.

The interaction of particles due to their mass or energy. Also particles without rest mass are subject to gravitational interactions if they have energy, since energy corresponds to mass according to $m = E/c^2$.

gravitational interaction

A large mass causes a strong curvature of space thereby deflecting the light of a distant radiation source. Depending on the relative position of source, deflector, and observer, multiple images, arcs, or rings are produced as images of the source.

gravitational lens

Increase of the wavelength of electromagnetic radiation in the course of emission against a gravitational field.

gravitational redshift

Supersymmetric partner of the graviton.

gravitino

Massless boson with spin $2\hbar$ mediating the interaction between masses.

graviton

Attractive force between two bodies caused by their mass. The gravitation between two elementary particles is negligibly small due to their low masses.

gravity

In the same way as an accelerated electrical charge emits electromagnetic radiation (photons), accelerated masses emit gravitational waves. The quantum of the gravity wave is called the graviton.

gravity waves

A 100 Mpc structure seen in the distribution of galaxies.

Great Wall

Threshold energy of energetic protons for pion production off photons of the cosmic background radiation via the Δ resonance.

Greisen–Zatsepin–Kuzmin cutoff (GZK)

Unified theory of strong, electromagnetic, and weak interactions.

GUT (Grand Unified Theory)

The very early universe when the strong, weak, and electromagnetic forces were unified and of equal importance.

GUT epoch

Energy scale at which strong and electroweak interactions merge into one common force, $E \approx 10^{16}$ GeV.

GUT scale

A charged particle moving in a magnetic field will orbit around the magnetic field lines. The radius of this orbit is called the gyroradius or sometimes also the Lamor radius.

gyroradius

H

A particle consisting of strongly interacting constituents (quarks or gluons). Mesons and baryons are hadrons. All these particles participate in strong interactions.

hadron

Hawking radiation	The gravitational field energy of a black hole allows to create particle–antiparticle pairs of which, for example, one can be absorbed by the black hole while the other can escape if the process occurs outside the event horizon. By this quantum process black holes can evaporate because the escaping particle reduces the energy of the system. The time scales for the evaporation of large black holes exceeds, however, the age of the universe considerably.
Hawking temperature	The temperature of a black hole manifested by the emission of Hawking radiation.
HEAO	High Energy Astronomy Observatory, X-ray satellite.
Heisenberg's uncertainty principle	Position and momentum of a particle cannot be determined simultaneously with arbitrary precision, $\Delta x \, \Delta p \geq \hbar/2$. This uncertainty relation refers to all complementary quantities, e.g., also to the energy and time uncertainty.
heliocentric	Centered on the Sun.
heliosphere	The large region starting at the Sun's surface and extending to the limits of the solar system.
Hertzsprung–Russell diagram	Representation of stars in a colour–luminosity diagram. Stars with large luminosity are characterized by a high colour temperature, i.e., shine strongly in the blue spectral range.
Higgs boson	Hypothetical particle as a quantum of the Higgs field predicted in the framework of the Standard Model to give masses to fermions via the mechanism of spontaneous symmetry breaking. Named after the Scottish physicist Peter Higgs.
Higgs field	A hypothetical scalar field which is assumed to be responsible for the generation of masses in a process called spontaneous symmetry breaking.
horizon	The observable range of our universe. The radius of the observable universe corresponds to the distance that light has travelled since the Big Bang.
horizon problem	In classical cosmology regions in the sky separated by more than $\approx 2°$ are causally disconnected. Experimentally it is found that they still have the same temperature. This can naturally be explained by an exponential expansion in the early universe which has smoothed out existing temperature fluctuations.
hot dark matter	Relativistic dark matter. Neutrinos are hot-dark-matter candidates.

High Resolution Imager, focal detector of the X-ray satellite ROSAT. **HRI**

Constant of proportionality between the receding velocity v of galaxies and their distance r, $v = H\,r$. According to recent measurements the value of the Hubble constant H is about $70\,(\text{km/s})/\text{Mpc}$. The inverse Hubble constant corresponds to the age of the universe. **Hubble constant H**

The receding velocity of galaxies is proportional to their distance. **Hubble law**

Space telescope with a mirror diameter of 2.2 m (HST – Hubble Space Telescope). **Hubble telescope**

Hydrogen burning is the fusion of four hydrogen nuclei into a single helium nucleus. It starts with the fusion of two protons into a deuterium, positron, and electron neutrino ($p + p \rightarrow d + e^+ + \nu_e$). **hydrogen burning**

I

Irvine–Michigan–Brookhaven collaboration for the search of nucleon decay and for neutrino astronomy. **IMB**

A violent inward-bound collapse. **implosion**

Property of matter that requires a force to act on it to change the way it is moving. **inertia**

Hypothetical exponential phase of expansion in the early universe starting at 10^{-38} s after the Big Bang and extending up to 10^{-36} s. The presently observed isotropy and uniformity of the cosmic background radiation can be understood by cosmic inflation. **inflation**

At low momenta quarks are confined in hadrons. They cannot escape their hadronic prison. **infrared slavery**

Astronomy in the light of infrared radiation emitted from celestial objects. **infrared astronomy**

A process in which a particle decays or responses to a force due to the presence of another particle as in a collision. **interaction**

Characteristic collision length for strongly interacting particles in matter. **interaction length λ**

The gas and dust which exist in the space between the stars. **interstellar medium**

Energy transfer by an energetic electron to a low-energy photon. **inverse Compton scattering**

ionization	Liberation of atomic electrons by photons or charged particles, namely photo ionization and ionization by collision.
isomers	Long-lived excited states of a nucleus.
isospin	Hadrons of equal mass (apart from electromagnetic effects) are grouped into isospin multipletts, in analogy to the spin. Individual members of the multipletts are considered to be presented by the third component of the isospin. Nucleons form an isospin doublet ($I = \frac{1}{2}$, $I_3(p) = +\frac{1}{2}$, $I_3(n) = -\frac{1}{2}$) and pions an isospin triplet ($I = 1$, $I_3(\pi^+) = +1$, $I_3(\pi^-) = -1$, $I_3(\pi^0) = 0$).
isotope	Nuclei of fixed charge representing the same chemical element albeit with different masses. The chemical element is characterized by the atomic number. Therefore, isotopes are nuclei of fixed proton number, but variable neutron number.

J

Jeans mass	An inhomogeneity in a matter distribution will grow due to gravitational attraction and contribute to the formation of structure in the universe (e.g., galaxy formation), if it exceeds a certain critical size (Jeans mass).
jets	Long narrow streams of matter emerging from radiogalaxies and quasars, or bundles of particles in particle–antiparticle annihilations.

K

Kamiokande	Nucleon Decay Experiment in the Japanese Kamioka mine for the measurement of cosmic and terrestrial neutrinos.
kaon	K; a meson consisting of a strange quark and an anti-up or anti-down quark or, correspondingly, an anti-strange quark and an up or down quark.
kpc	Kilo parsec (see parsec).

L

large attractor	Hypothetical supercluster complex with a mass of $> 10^4$ galaxies at a distance of several 100 million light-years which impresses an oriented proper motion on the local group of galaxies.
latitude effect	Increase of the cosmic-ray intensity to the geomagnetic poles caused by the interaction of charged primary cosmic rays with the Earth's magnetic field.

Short for Large Electron–Positron Collider, the e^+e^- storage ring at CERN with a circumference of 27 km. **LEP**

The production of leptons in the early universe. **leptogenesis**

A fundamental fermion which does not participate in strong interactions. The electrically charged leptons are the electron (e), the muon (μ), and the tau (τ), and their antiparticles. Electrically neutral leptons are the associated neutrinos (ν_e, ν_μ, ν_τ). **lepton**

Quantum number describing the lepton flavour. The three generations of charged leptons (e^-, μ^-, τ^-) have different individually conserved lepton numbers, while their neutrinos are allowed to oscillate, thereby violating lepton-number conservation. **lepton number**

Large Hadron Collider at CERN in Geneva. In the LHC protons will collide with a center-of-mass energy of approximately 14 TeV. LHC is expected to start operation in 2007. LHC will be the accelerator with the highest center-of-mass energy in the world. It is generally hoped that the physics of the Large Hadron Collider will help to clarify some open questions of particle physics, for example, the question whether the Higgs boson responsible for the generation of masses of fundamental fermions exists and which mass it has. **LHC**

The distance traveled by light in vacuum during one year; $1\,\mathrm{ly} = 9.46 \times 10^{15}\,\mathrm{m} = 0.307\,\mathrm{pc}$. **light-year**

The model that explains pulsars as flashes of light from a rotating pulsar. **lighthouse model**

Irregular galaxy in the southern sky at a distance of 170 000 light-years ($\hat{=} 52$ kpc). **LMC – Large Magellanic Cloud**

A system of galaxies comprising among others our Milky Way, the Andromeda Nebula, and the Magellanic Clouds. The diameter of the local group amounts to approximately 5 million light-years. **local group**

Virgo supercluster; 'Milky Way' of galaxies to which also the local group and the Virgo cluster belong. Its extension is approximately 30 Mpc. **local supercluster**

Feature emerging from special relativity in which a moving object appears shortened along its direction of motion. **Lorentz contraction**

Transformation of kinematical variables for frames of reference which move at linear uniform velocity with respect to each other under the consideration that the speed of light in vacuum is a maximum velocity. **Lorentz transformation**

luminosity	Total light emission of a star or a galaxy in all ranges of wavelengths. The luminosity therefore corresponds to the absolute brightness.
luminosity distance	The distance to an astronomical object derived by comparing its apparent brightness to its total known or presumed luminosity by assuming isotropic emission. If the luminosity is known, the source can be used as standard candle.

M

M87	Galaxy in the Virgo cluster at a distance of 11 Mpc. M87 is considered an excellent candidate for a source of high-energy cosmic rays.
MACHO	Massive Compact Halo Object. Experiment for the search for dark compact objects in the halo of our Milky Way, see also EROS.
Magellanic Cloud	Nearest galaxy neighbour to our Milky Way consisting of the Small (SMC) and the Large Magellanic Cloud (LMC); distance 52 kpc.
magic numbers	Nuclei whose atomic number Z or neutron number N is one of the magic numbers 2, 8, 20, 28, 50, or 126. In the framework of the shell model of the nucleus these magic nucleon numbers form closed shells. Nuclei whose proton and also neutron number are magic are named doubly magic.
magnetar	Special class of neutron stars having a superstrong magnetic field (up to 10^{11} Tesla). Magnetars emit sporadic γ-ray bursts.
magnetic monopole	Hypothetical particle which constitutes separate sources and charges of the magnetic field. Magnetic monopoles are predicted by Grand Unified Theories.
magnitude	A measure of the relative brightness of a star or some other celestial object. The brightness differences between two stars with intensities I_1 and I_2 are fixed through the magnitude $m_1 - m_2 = -2.5 \log_{10} (I_1/I_2)$. The zero of this scale is defined in such a way that the apparent brightness of the polestar comes out to be $2\overset{m}{.}12$. An apparent brightness difference of $\Delta m = 2.5$ corresponds to an intensity ratio of 10 : 1. On this scale the planet Venus has the magnitude $m = -4.4$. With the naked eye one can see stars down to the magnitude $m = 6$. The strongest telescopes are able to observe stars down to the 28th magnitude.
main-sequence star	Star of average age lying on the main sequence in the Hertzsprung–Russell diagram.

Distant galaxies with a bright active galactic nucleus emitting in the blue and UV range. The Markarian galaxies are also sources of $\geq$ TeV gamma rays.	**Markarian galaxies**
Rest mass; the rest mass m_0 of a particle is that mass, which one obtains, if the energy of an isolated free particle in the state of rest is divided by the square of the velocity of light. When particle physicists use the name mass, they always refer to the rest mass of a particle.	**mass**
The sum (A) of the number of neutrons (N) and protons (Z) in a nucleus.	**mass number**
An asymmetry created by some, so far unknown baryon- and lepton-number-violating process in the early universe (see Sakharov conditions).	**matter–antimatter asymmetry**
Flavour oscillations which can be amplified by matter effects in a resonance-like fashion; e.g., suppression of solar neutrinos in $\nu_e e^-$ interactions in the Sun.	**matter oscillation**
Moment when the energy density of radiation dropped below that of matter at around 50 000 years after the Big Bang.	**matter–radiation equality**
A hadron consisting of a quark and an antiquark.	**meson**
A small dark object in the line of sight to a bright background star can give rise to a brightness excursion of the background star due to a bending of the light rays by the dark body.	**microlensing**
Extremely small ($\approx$ μg) black holes could have been formed in the early universe. These primordial black holes are not final states of collapsing or evaporating stars.	**mini black holes**
Mass or energy in cosmology which must be present because of its gravitational force it exerts on other matter. The missing mass does not emit detectable electromagnetic radiation (see dark matter). Missing mass is also encountered in relativistic kinematics, where in many experiments, in which total energy is fully constrained by the center-of-mass energy, the mass, energy, and momentum of particles escaping from the detector without interaction (like neutrinos or SUSY particles), can be reconstructed due to the detection of all other particles.	**missing mass**
See Friedmann–Lemaître universes.	**models of the universe**
All grand unified theories predict large numbers of massive magnetic monopoles in contrast to observation. Inflation during the era of monopole production would have diluted the monopole density to a negligible level, thereby solving the problem.	**monopole problem**

Mpc	Megaparsec (see parsec).
MSW effect	Matter oscillations first proposed by Mikheyev, Smirnov, and Wolfenstein; see matter oscillation.
M theory	Supersymmetric string theory ('superstring theory') unifying all interactions. In particular, the M theory also contains a quantum theory of gravitation. The smallest objects of the M theory are p-dimensional membranes. It appears now that all string theories can be embedded into an 11-dimensional theory, called M theory. Out of the 10 spatial dimensions 7 are compactified into a Calabi–Yau space.
multiplicity	Number of secondary particles produced in an interaction.
multiverse	Hypothetical enlargement of the universe in which our universe is only one of an enormous number of separate and distinct universes.
muon	μ; the muon belongs to the second family of charged leptons. It has the electric charge -1.
muon number	The muon μ^- and its associated muon neutrino have the muonic lepton number $+1$, their antiparticles the muonic lepton number -1. All other elementary particles have the muon number 0. The muon number – except for neutrino oscillations – is a conserved quantity.

N

negative pressure	In the same way as a positive pressure increases gravitation through its field, negative pressure (like, for example, in a spring) leads to a repulsive gravity. Dark energy represents a form of energy that is gravitationally repulsive, due to its negative effective pressure.
neutral current	Interaction mechanism mediated by the exchange of a virtual neutral gauge boson.
neutralino	Candidate for a non-baryonic dark-matter particle. In the framework of supersymmetry fermions and bosons come in supermultiplets. The neutralino is the bosonic partner of the neutrino. The lightest supersymmetric partner is expected to be stable and would therefore be a good dark-matter candidate.
neutrino	A lepton without electric charge. Neutrinos only participate in weak and gravitational interactions and therefore are very difficult to detect. There are three known types of neutrinos, which are all very light. These leptons are ν_e, ν_μ, and ν_τ for the electron, muon, and tau family. At the LEP experiments it could be shown that apart from the three already known neutrino generations there is no further generation with light neutrinos ($m_\nu < 45\,\mathrm{GeV}/c^2$).

Transmutation of a neutrino flavour into another by vacuum or matter oscillations.   **neutrino oscillations**

Neutrino flavour oscillation in vacuum which can occur if neutrinos have mass and if the weak eigenstates are superpositions of different mass eigenstates.   **neutrino vacuum oscillation**

n; a baryon with electric charge zero. A neutron is a fermion with an internal structure consisting of two down quarks and one up quark, which are bound together by gluons. Neutrons constitute the neutral component of atomic nuclei. Different isotopes of the same chemical element differ from each other only by a different number of neutrons in the nucleus.   **neutron**

Neutron decay process into a proton, an electron, and an electron antineutrino.   **neutron decay**

Nuclear decay by emission of a neutron.   **neutron evaporation**

Star of extremely high density consisting predominantly of neutrons. Neutron stars are remnants of supernova explosions where the gravitational pressure in the remnant star is so large that the electrons and protons are merged to neutrons and neutrinos (e^- + $p \rightarrow n + \nu_e$). Neutron stars have a diameter of typically 20 km. If the gravitational pressure surmounts the degeneracy pressure of neutrons, the neutron star will collapse to a black hole.   **neutron star**

For an external observer the black hole has only three properties: mass, electric charge, and angular momentum. If two black holes agree in these properties they are indistinguishable.   **'no hair' theorem**

Star with a sudden increase in luminosity ($\approx 10^6$ fold). The star explosion is not as violent as in supernova explosions. A nova can also occur several times on the same star.   **nova**

The energy, which is required to desintegrate an atomic nucleus into its constituents; ≈ 8 MeV/nucleon.   **nuclear binding energy**

Fission of a nucleus in two fragments of approximately equal size. In most cases fission leads to asymmetric fragments. It can occur spontaneously or be induced by nuclear reactions.   **nuclear fission**

Fusion of light elements to heavier elements. In a fusion reactor protons are combined to produce helium via deuterium. Our Sun is a fusion reactor.   **nuclear fusion**

O

OGLE   Optical Gravitational Lens Experiment looking for dark stars in the galactic halo of the Milky Way.

Olbert's paradox   If the universe were infinite, uniform, and unchanging, the sky at night would be bright since in whatever direction one looked, one would eventually see a star. The number of stars would increase in proportion to the square of the distance from Earth while its intensity is inversely proportional to the square of the distance. Consequently, the whole sky should be about as bright as the Sun. The paradox is resolved by the fact that the universe is not infinite, not uniform, and not unchanging. Also light from distant stars and galaxies displays an extreme redshift and sometimes ceases to be visible.

open universe   A Friedmann–Lemaître model of an infinite, permanently expanding universe.

orbit   The path of an object (e.g., satellite) that is moving around a second object or point.

OSSE   Oriented Scintillation Spectroscopy Experiment on board the CGRO satellite.

P

pair production   Creation of fermion–antifermion pairs by photons in the Coulomb field of atomic nuclei. Pair production – in most cases electron–positron pair production – can also occur in the Coulomb field of an electron or any other charged particle.

parallax   An apparent displacement of a distant object with respect to a more distant background when viewed from two different positions.

parity   The property of a wave function that determines its behaviour, when all its spatial coordinates are reversed. If a wave function ψ satisfies the equation $\psi(r) = +\psi(-r)$, it is said to have even parity. If, however, $\psi(r) = -\psi(-r)$, the parity of the wave function is odd. Experimentally only the square of the absolute value of the wave function is observable. If parity were conserved, there would be no fundamental way of distinguishing between left and right. Parity is conserved in electromagnetic and strong interactions, but it is violated in weak interactions.

parsec   Parallax second; the unit of length used to express astronomical distances. It is the distance at which the mean radius of the Earth's orbit around the Sun appears under an angle of 1 arc second. 1 pc $=$ 3.086×10^{16} m $= 3.26$ light-years.

particle horizon   Largest region that can be in causal contact.

See fermion.	**Pauli principle**
The point in an orbit of a star in a double-star system in which the body describing the orbit is nearest to the star.	**periastron**
The point in the orbit of the Moon or an artificial Earth satellite, at which it is closest to Earth.	**perigee**
The point in the orbit of a planet, comet, or artificial satellite in a solar orbit, at which it is nearest to the Sun.	**perihelion**
Positron Electron Tandem Ring Accelerator, an electron–positron storage ring at DESY in Hamburg.	**PETRA**
Intensity maximum of secondary cosmic rays at an altitude of approximately 15 km produced by interactions of primary cosmic rays in the atmosphere.	**Pfotzer maximum**
Supersymmetric partner of the photon.	**photino**
Liberation of atomic electrons by photons.	**photoelectric effect**
The gauge boson of electromagnetic interactions.	**photon**
π; the lightest meson; pions constitute an isospin triplet the members of which have electric charges of $+1$, -1, or 0.	**pion**
Intensity distribution of blackbody radiation of a blackbody of temperature T as a function of the wavelength following Planck's radiation law.	**Planck distribution**
Scale, at which the quantum nature of gravitation becomes visible, $L_{Pl} = \sqrt{G\hbar/c^3} \approx 1.62 \times 10^{-35}$ m.	**Planck length**
Energy scale, at which all forces including gravity can be described by a unified theory; $m_{Pl} = \sqrt{\hbar c/G} \approx 1.22 \times 10^{19}$ GeV/c^2.	**Planck mass**
A law giving the distribution of energy radiated by a blackbody. The frequency-dependent radiation density depends on the temperature of the blackbody. See also Planck distribution.	**Planck's radiation law**
The tension of a typical string in string theories.	**Planck tension**
The time taken for a photon to pass the distance equal to the Planck length: $t_{Pl} = \sqrt{G\hbar/c^5} \approx 5.39 \times 10^{-44}$ s.	**Planck time**
A shell of gas ejected from and expanding about a certain kind of extremely hot star.	**planetary nebula**

positron	e^+; antiparticle of the electron.
positron annihilation	Positron decay in matter by annihilation with an electron ($e^+e^- \rightarrow \gamma\gamma$).
power spectrum	Describes a measure of the level of structure related to the density difference with respect to the average density in the early universe which became frozen during the growth of structure.
primary cosmic rays	Radiation of particles originating from our galaxy and beyond. It consists mainly of protons and α particles, but also elements up to uranium are present in primary cosmic rays.
primordial black holes	See Mini Black Holes.
primordial nucleosynthesis	Production of atomic nuclei occurring during the first three minutes after the Big Bang.
primordial particles	Particles from the sources.
proton	p; the most commonly known hadron, a baryon of electric charge $+1$. Protons are made up of two up quarks and one down quark bound together by gluons. The nucleus of a hydrogen atom is a proton. A nucleus with electric charge Z contains Z protons. Therefore, the number of protons determines the chemical properties of the elements.
proton–proton chain	Nuclear reaction, in which hydrogen is eventually fused to helium. The pp cycle is the main energy source of our Sun.
protostar	Very dense regions or aggregations of gas clouds, from which stars are formed.
Proxima Centauri	Nearest neighbour of our Sun at a distance of 4.27 light-years.
pseudoscalar	Scalar quantity, which changes sign under spatial inversion.
PSPC	Position Sensitive Proportional Chamber on board of ROSAT.
pulsar	Rotating neutron star with characteristic, pulsed emission in different spectral ranges (radio, optical, X-ray, gamma-ray emission; 'pulsating radiostar').

Q

quantum	A quantum is the minimum discrete amount by which certain properties such as energy or angular momentum of a system can change. Planck's constant is the smallest quantity of a physical action. The elementary charge is the smallest charge of freely observable particles.

Quantum anomalies can arise if a classical symmetry is broken in the process of quantization and renormalization.

quantum anomalies

QCD; theory of strong interactions of quarks and gluons.

quantum chromodynamics

Quantum-mechanical theory applied to systems that have an infinite number of degrees of freedom. Quantum field theory also describes processes, in which particles can be created or annihilated.

quantum field theory

Frothy character of the space-time fabrique on ultramicroscopic scales.

quantum foam

Quantum theory of gravitation aiming at the unification with the other types of interactions.

quantum gravitation

Quantum mechanics describes the laws of physics which hold at very small distances. The main feature of quantum mechanics is that, for example, the electric charge, the energies, and the angular momenta come in discrete amounts, which are called quanta.

quantum mechanics

q; a fundamental fermion subject to strong interactions. Quarks carry the electric charge of either $+\frac{2}{3}$ (up, charm, top) or $-\frac{1}{3}$ (down, strange, bottom) in units, in which the electric charge of a proton is $+1$.

quark

Quasistellar radio sources; galaxies at large redshifts with an active nucleus which outshines the whole galaxy and therefore makes the quasar appear like a bright star.

quasars

A scalar field model for the dark energy. In contrast to the cosmological constant the energy density in the quintessence field represents a time-varying inhomogeneous component with a negative pressure which satisfies $-1 < w < 0$, where $w \equiv P/\varrho$, and P and ϱ are the pressure and energy density of the quintessence field. The decay of the quintessence field could liberate energy into space-time which could drive the expansion of the universe.

quintessence

R

Solar-wind particles can be trapped in the Earth's magnetic field. They form the Van Allen belts in which charged particles spiral back and forth. There are separate radiation belts for electrons and protons. See also Van Allen belt.

radiation belt

Era up to 380 000 years after the Big Bang when radiation dominated the universe.

radiation era

Characteristic attentuation length for high-energy electrons and gamma rays.

radiation length X_0

Force exerted by photons if they are absorbed or scattered on small dust or matter particles or absorbed by atoms.

radiation pressure

radio astronomy   Astronomy in the radioband; it led to the discovery of the cosmic blackbody radiation.

radio galaxy   A galaxy with high luminosity in the radioband.

recombination   Capture of an electron by a positive ion, frequently in connection with radiation emission.

recombination temperature   The number of neutral and ionized atoms become equal at the recombination temperature (3500 K).

red giant   If a main-sequence star has used up its hydrogen supply, its nucleus will contract leading to a temperature increase so that helium burning can set in. This causes the star to expand associated with an increase in luminosity. The diameter of a red giant is large compared to the size of the original star.

redshift   Increase of the wavelength of electromagnetic radiation by the Doppler effect, the expansion of the universe, or by strong gravitational fields:

$$z = \frac{\Delta\lambda}{\lambda_0} = \frac{\lambda - \lambda_0}{\lambda_0}$$

(λ_0 – emitted wavelength, λ – observed wavelength),

$$z = \begin{cases} v/c & \text{, classical} \\ \sqrt{\frac{c+v}{c-v}} - 1 & \text{, relativistic} \end{cases}.$$

relativity principle   Equivalence of observation of the same event from different frames of reference having different velocities and acceleration. There is no frame of reference that is better or qualitatively different from any other.

residual interaction   Interaction between objects that do not carry a charge but do contain constituents that have that charge. The residual strong interaction between protons and neutrons due to the strong charges of their quark constituents is responsible for the binding of the nucleus. In the thirties it was believed that the binding of protons and neutrons in a nucleus was mediated by the exchange of pions.

rest mass   The rest mass of a particle is the mass defined by the energy of an isolated free particle at rest divided by the speed of light squared.

right ascension   A coordinate which along with declination may be used to locate any object in the sky. Right ascension is the angular distance measured eastwards along the celestial equator from the vernal equinox to the intersection of the hour circle passing through the body. It is the celestial equivalent to longitude.

The metric describing an isotropic and homogeneous space-time of the universe. **Robertson–Walker Metric**

German–British–American Roentgen satellite (launched 1990). **ROSAT**

Quantum number which distinguishes supersymmetric particles from normal particles. **R parity**

S

Soviet–American Gallium Experiment for the measurement of solar neutrinos from the pp cycle. **SAGE experiment**

Necessary conditions first formulated by Sakharov which are required to create a matter-dominated universe: a baryon-number-violating process, violation of C or CP symmetry, and departure from thermal equilibrium. **Sakharov conditions**

Small Astronomy Satellite; gamma-ray satellites launched by NASA, 1972 (SAS-2) resp. 1975 (SAS-3). **SAS-2, SAS-3**

The scale factor denotes an arbitrary distance which can be used to describe the expansion of the universe. The ratio of scale factors between two different epochs indicates by how much the size of the universe has grown. **scale factor**

Event horizon of a spherical black hole, $R = 2GM/c^2$. **Schwarzschild radius**

Excitation of atoms and molecules by the energy loss of charged particles with subsequent light emission. **scintillation**

Secondary cosmic rays are a complex mixture of elementary particles, which are produced in interactions of primary cosmic rays with the atomic nuclei of the atmosphere. **secondary cosmic rays**

Supersymmetric partner of the electron. **selectron**

Member of a small class of galaxies with brilliant nuclei and inconspicuous spiral arms. Seyfert galaxies are strong emitters in the infrared and are also detectable as radio and X-ray sources. The nuclei of Seyfert galaxies possess many features of quasars. **Seyfert galaxy**

Soft Gamma-Ray Repeaters are objects with repeated emission of γ bursts. Presumably soft gamma-ray repeaters are neutron stars with extraordinary strong magnetic fields (see magnetar). **SGR objects**

A sudden pressure, density, and temperature gradient. **shock front**

shock wave   A very narrow region of high pressure and temperature. Particles passing through a shock front can be effectively accelerated, if the velocity of the shock front and that of the particle have opposite direction.

shower   Or cascade. High-energy elementary particles can generate numerous new particles in interactions, which in turn can produce particles in further interactions. The particle cascade generated in this way can be absorbed in matter. The energy of the primary initiating particle can be derived from the number of particles observed in the particle cascade. One distinguishes electromagnetic (initiated by electrons and photons) and hadronic showers (initiated by strongly interacting particles, e.g., p, α, Fe, $\pi^{\pm}$, ...).

singularity   A space-time region with infinitely large curvature – a space-time point.

slepton   Supersymmetric partner of a lepton.

SMC   Small Magellanic Cloud. Galaxy in the immediate vicinity of the Large Magellanic Cloud (LMC).

SN 1987A   Supernova explosion in the Large Magellanic Cloud in 1987. The progenitor star was Sanduleak.

SNAP   SuperNova/Acceleration Probe. A space-based experiment to measure the properties of the accelerating universe, which depend on the amounts of dark energy and dark matter. The proposed SNAP satellite is expected to be launched in ≈ 2014.

SNO   Sudbury Neutrino Observatory.

SNR   Remnant after a supernova explosion; mostly a rotating neutron star (pulsar).

solar flare   Violent eruption of gas from the Sun's surface.

solar wind   Flux of solar particles (electrons and protons) streaming away from the Sun.

spaghettification   A body falling into a black hole will be stretched because of the differential gravitation.

spallation   Nuclear transmutation by high-energy particles in which – contrary to fission – a large number of nuclear fragments, α particles, and neutrons are produced.

This theory refers to inertial non-accelerated frames of reference. It assumes that physical laws are identical in all frames of reference and that the speed of light in vacuum c is constant throughout the universe and independent of the speed of the observer. The theorem of the addition of velocities is modified to account for the deviation from the Galilei transformation.	**special theory of relativity**
Intrinsic angular momentum of a particle quantized in units of $\hbar$, where $\hbar = \frac{h}{2\pi}$ and h is Planck's constant.	**spin**
Emergence of different properties of a system at low energies (e.g., weak and electromagnetic interaction) which do not exist at high energies where these interactions are described by the Unified Theory.	**spontaneous symmetry breaking**
Supersymmetric partner of a quark.	**squark**
An astronomical object with well-known absolute luminosity which can be used to determine distances.	**standard candle**
Theory of fundamental particles and their interactions. Originally the Standard Model described the unification of weak and electromagnetic interactions, the electroweak interaction. In a more general sense it is used for the common description of weak, electromagnetic, and strong interactions.	**Standard Model**
Galaxies with a high star-formation rate.	**starburst galaxies**
A bunch of stars which are bound to each other by their mutual gravitational attraction.	**star cluster**
A quake in the crust of a neutron star.	**star quake**
Older model of the universe in which matter is continuously produced to fill up the empty space created by expansion, thereby maintaining a constant density ('steady-state universe').	**steady-state universe**
This law states that the amount of energy radiated by a blackbody is proportional to the fourth power of its temperature.	**Stefan–Boltzmann law**
The ejection of gas off the surface of a star.	**stellar wind**
Synchrotron in which counter-rotating particles and antiparticles are stored in a vacuum pipe. The particles usually stored in bunches collide in interaction points, where the center of mass energy is equal to twice the beam energy for a head-on collision. Storage rings are used in particle physics and for experiments with synchrotron radiation.	**storage ring**

strangeness	The strangeness of an s quark is -1. Strangeness is conserved in strong and electromagnetic interactions, however, violated in weak interactions. In weak decays or weak interactions the strangeness changes by one unit.
strange quark	s; the third quark flavour (if quarks are ordered with increasing mass). The strange quark has the electric charge $-\frac{1}{3}e$.
string	In the framework of string theories the known elementary particles are different excitations of elementary strings. The length of strings is given by the Planck scale.
string theory	Theory, which unifies the general theory of relativity with quantum mechanics by introducing a microscopic theory of gravitation.
strong interaction	Interaction responsible for the binding between quarks, antiquarks, and gluons, the constituents of hadrons. The residual interaction of strong interactions is responsible for nuclear binding.
sunspot cycle	The recurring 11-year rise and fall in the number of sunspots.
sunspots	A disturbance of the solar surface which appears as a relatively dark center surrounded by less dark area. Sunspots appear dark because part of the thermal energy is transformed into magnetic field energy.
supercluster	See galactic cluster.
supergalactic plane	Concentration of many galaxies around the Virgo cluster into a plane of diameter of about 30 Mpc.
Super-Kamiokande detector	Successor experiment of the Kamiokande detector (see Kamiokande).
superluminal speed	Phenomenon of apparent superluminal speed caused by a geometrical effect related to the finite propagation time of light.
supermassive black hole	Black hole at the center of a galaxy containing about 10^9 solar masses.
supernova	Star explosion initiated by a gravitational collapse, if a star has exhausted its hydrogen and helium supply and collapses under its own gravity. In a supernova explosion the luminosity of a star is increased by a factor of 10^9 for a short period of time. The remnant star of a supernove explosion is a neutron star or pulsar.
supernova remnant	Remnant of a supernova explosion; mostly a neutron star or pulsar.

Particles whose spin differs by $1/2$ unit from normal particles. Superpartners are paired by supersymmetry. **superpartners**

In supersymmetric theories each fermion is associated with a bosonic partner and each boson has a fermionic partner. In this way the number of elementary particles is doubled. The bosonic partners of leptons and quarks are sleptons and squarks. The fermionic partners, for example, of the photon, gluon, Z, and W are photino, gluino, zino, and wino. Up to now no supersymmetric particles have been found. **supersymmetry (SUSY)**

A reduction in the amount of symmetry of a system, usually associated with a phase transition. **symmetry breaking**

Circular accelerator in which charged particles travel in bunches synchronized with external and magnetic fields at a fixed radius. The orbit is stabilized by synchronizing an external magnetic guiding field with the increasing momentum of the accelerated particles. **synchrotron**

Electromagnetic radiation emitted by an accelerated charged particle in a magnetic field. **synchrotron radiation**

T

Particle that moves faster than the speed of light. Its mass squared is negative. Its presence in a theory generally yields inconsistencies. **tachyon**

τ; the third flavour of charged leptons (if the leptons are arranged according to increasing mass). The tau carries the electric charge -1. **tau**

τ^- and its associated tau neutrino have the tau-lepton number $+1$, the antiparticles τ^+ and $\bar{\nu}_\tau$ the tau-lepton number -1. All other elementary particles have the tau-lepton number 0. The tau-lepton number is a conserved quantity except for neutrino oscillations. **tau-lepton number**

The ultimate theory in which the different phenomena of electroweak, strong, and gravitational interactions are unified. **Theory of Everything (TOE)**

Nuclear fusion of light elements to heavier elements at high temperatures (e.g., pp fusion at $T \approx 10^7$ K). **thermonuclear reaction**

Stretching of time explained by special relativity; also called time dilation. **time dilatation**

time-reversal invariance	T invariance or simply time reversal. The operation of replacing time t by time $-t$. As with CP violation, T violation occurs in weak interactions of kaon decays. The CPT operation which is a succession of charge conjugation, parity transformation, and time reversal is regarded as a conserved quantity.
topological defect	Topological defects like magnetic monopoles, cosmic strings, domain walls, or cosmic textures might have been created in the early universe.
top quark	t; the sixth quark flavour (if quarks are arranged according to increasing mass). The top quark carries the electric charge $+\frac{2}{3}e$. The mass of the top quarks is comparable to the mass of a gold nucleus ($\approx 175\,\mathrm{GeV}/c^2$).
triple-alpha process	Reaction in which three α particles are fused to carbon. Such a process requires high α-particle densities and a large resonance-like cross section.
tritium	Hydrogen isotope with two additional neutrons in the nucleus.

U

uncertainty relation	The quantum principle first formulated by Heisenberg which states that it is not possible to know exactly both the position x and the momentum p of an object at the same time. The same is true for the complementary quantities energy and time.
Unified Theory	Any theory that describes all four forces and all of matter within a single all-encompassing framework (see also Grand Unified Theory, GUT).
up quark	u; the lightest quark flavour with electric charge $+\frac{2}{3}e$.

V

vacuum energy density	Quantum fields in the lowest energy state describing the vacuum must not necessarily have the energy zero.
Van Allen belts	Low-energy particles of the solar wind are trapped in certain regions of the Earth's magnetic field and stored.
Vela pulsar	Vela X1, supernova remnant in the constellation Vela at a distance of about 1500 light-years; the Vela supernova explosion was observed by the Sumerians 6000 years ago.

(*c*) The value of the velocity of light forms the basis for the definition of the length unit meter. It takes $1/299\,792\,458$ seconds for light in vacuum to pass 1 meter. The velocity of light in vacuum is the same in all frames of reference.

velocity of light

Concentration of galaxies in the direction of the constellation Virgin.

Virgo cluster

A particle that exists only for an extremely brief instant of time in an intermediate process. For virtual particles the Lorentz-invariant mass does not coincide with the rest mass. Virtual particles with negative mass squared are called space-like, those with a positive mass squared time-like. Virtual particles can exist for times allowed by the Heisenberg's uncertainty principle.

virtual particle

W

Charged gauge quanta of weak interactions. These quanta participate in so-called charged-current processes. These are interactions in which the electric charge of the participating particles changes.

W^+, W^- **boson**

Probability waves upon which quantum mechanics is founded.

wave function

The weak interaction acts on all fermions, e.g., in the decay of hadrons. It is responsible for the beta decay of particles and nuclei. In charged-current interactions the quark flavour is changed, while in neutral-current interactions the quark flavour remains the same.

weak interaction

A star that has exhausted most or all of its nuclear fuel and has collapsed to a very small size. The stability of a white dwarf is not maintained by the radiation or gas pressure as with normal stars but rather by the pressure of the degenerate electron gas. If the white dwarf has a mass larger than 1.4 times the solar mass, also this pressure is overcome and the star will collapse to a neutron star.

white dwarf

Weakly Interacting Massive Particles are candidates for dark matter.

WIMPs

Wilkinson Microwave Anisotropy Probe launched in 2001 to measure the fine structure of the cosmic background radiation.

WMAP

A proposed channel of space-time connecting distant regions of the universe. It is not totally inconceivable that wormholes might provide the possibility of time travel.

wormhole

The w parameter is defined as the ratio of pressure over energy density ($w \equiv P/\varrho$). In models with a cosmological constant the pressure of the vacuum density P equals exactly the negative of the energy density ϱ; i.e., $w = -1$, in contrast to quintessence models, where $-1 < w < 0$.

w **parameter**

X

X boson   See *Y* boson.

XMM-Newton   European X-ray space observatory mission, with its X-ray Multi-Mirror design using three telescopes. Named in honour of Sir Isaac Newton.

X-ray astronomy   Astronomy in the X-ray range (0.1 keV–100 keV).

X-ray burster   X-ray source with irregular sudden outbursts of X rays.

Y

Y boson   The observed baryon–antibaryon asymmetry in the universe requires, among others, a baryon-number-violating process. Hypothetical heavy *X* and *Y* bosons, existing at around the GUT scale with masses comparable to the GUT energy scale ($\approx 10^{16}$ GeV), could have produced this matter–antimatter asymmetry in their decay to quarks and antiquarks, resp. baryons and antibaryons. The couplings of the *X* and *Y* bosons to fermion species is presently the simplest idea for which baryon-number violation appears possible.

ylem   Name for the state of matter before the Big Bang. Gamow proposed that the matter of the universe originally existed in a promordial state, which he coined 'ylem' (from the Greek ΰλη which stands for 'matter', 'wood' via the medieval latin *hylem* meaning *the primordial elements of life*). According to his idea all elements were formed shortly after the Big Bang from this primary substance.

Yukawa particle   Yukawa predicted a particle which should mediate the binding between protons and neutrons. After its discovery the muon was initially mistaken to be this particle. The situation was resolved with the discovery of the pion as Yukawa particle.

Z

Z boson   Neutral boson of weak interactions. It mediates weak interactions in all those processes, where electric charge and flavour do not change.

zero-point energy   The Heisenberg uncertainty relation does not allow a finite quantum-mechanical system to have a definite position and definite momentum at the same time. Thus any system even in the lowest energy state must have a non-zero energy. Zero energy would lead to zero momentum and thereby to an infinite position uncertainty.

17 Solutions

"The precise statement of any problem is the most important step in its solution."

Edwin Bliss

17.1 Chapter 1

1. a) The centrifugal force is balanced by the gravitational force between the Earth and the satellite:

$$\frac{mv^2}{R_\oplus} = G\frac{mM_\oplus}{R_\oplus^2} \approx mg \;,$$

where it has been taken into account that the altitude is low and $R \approx R_\oplus$. As a result,

$$v \approx \sqrt{g\,R_\oplus} \approx 7.9\,\text{km/s}\;.$$

b) The escape velocity is found from the condition that at infinity the total energy equals zero. Since

$$E_{\text{tot}} = \frac{1}{2}mv^2 + E_{\text{pot}}$$

and the potential energy is given by

$$E_{\text{pot}} = -G\frac{mM_\oplus}{R_\oplus}\;,$$

one obtains

$$\frac{1}{2}mv^2 = G\frac{mM_\oplus}{R_\oplus} \quad\Rightarrow\quad v = \sqrt{2M_\oplus G/R_\oplus} \approx 11.2\,\text{km/s}\;.$$

c) The centrifugal force is balanced by the gravitational force between Earth and the satellite. From the equality of the magnitudes of these two forces one gets:

$$\frac{v^2}{r} = G\frac{M_\oplus}{r^2}\;.$$

From the relation between the velocity v and revolution frequency ω one obtains:

$$v^2 = r^2\omega^2 = G\frac{M_\oplus}{r} \quad \Rightarrow \quad r^3 = \frac{GM_\oplus}{\omega^2} \;.$$

Taking into account that for the geostationary satellite the revolution period $T_\oplus = 1\,\text{day} = 86\,400\,\text{s}$ and $\omega_\oplus = 2\pi/T_\oplus$, one obtains

$$r = \sqrt[3]{\frac{GM_\oplus\, T_\oplus^2}{4\pi^2}} \approx 42\,241\,\text{km} \;.$$

The altitude above ground level therefore is $H = r - R_\oplus = 35\,871$ km.

Geostationary satellites can only be positioned above the equator because only there the direction of the centrifugal force can be balanced by the direction of the gravitational force. In other words, the centrifugal force for a geostationary object points outward perpendicular to the Earth's axis and only in the equator plane the gravitational force is collinear to the former.

2. The centrifugal force is balanced by the Lorentz force, so

$$\frac{mv^2}{\varrho} = evB \quad \Rightarrow \quad p = eB\varrho \quad \Rightarrow \quad \varrho = \frac{p}{eB} \;.$$

Since the kinetic energy of the proton (1 MeV) is small compared to its mass (≈ 938 MeV), a classical treatment for the energy–momentum relation, $E = p^2/2m_0$, is appropriate from which one obtains $p = \sqrt{2m_0 E_{kin}}$.

Then the bending radius is given by

$$\varrho = \frac{p}{eB} = \frac{\sqrt{2m_0 E_{kin}}}{eB} \quad \text{or} \quad \varrho \approx 2888\,\text{m} \;.$$

Dimensional analysis:

$$p = eB\varrho\,, \quad p\left\{\frac{\text{kg m}}{\text{s}}\right\} c\left\{\frac{\text{m}}{\text{s}}\right\} = (pc)\,\{\text{J}\} = e\,\{\text{A s}\}\,B\,\{\text{T}\}\,\varrho\,\{\text{m}\}\,c\,\{\text{m s}^{-1}\}\,,$$

$$(pc)\,[\text{J}] \times 1.6 \times 10^{-19}\,\text{J/eV} = 1.6 \times 10^{-19} \times B\,[\text{T}] \times \varrho\,[\text{m}] \times 3 \times 10^8\,,$$

$$(pc)[\text{eV}] = 3 \times 10^8\,B[\text{T}]\,\varrho[\text{m}] = 300\,B\,[\text{Gauss}]\,\varrho\,[\text{cm}]$$

$$\varrho = \frac{(pc)\,[\text{eV}]}{300\,B\,[\text{Gauss}]}\,\text{cm} = 2.888 \times 10^5\,\text{cm} \;.$$

3. From Fig. 4.2 one can read the average energy loss of muons in air-like materials to be $\approx 2\,\text{MeV}/(\text{g/cm}^2)$. This number can also be obtained from (4.6). The column density of the atmosphere from the production altitude can be read from Fig. 7.3. It is approximately $940\,\text{g/cm}^2$. Finally, the average energy loss in the atmosphere is

$$-\frac{dE}{dx}\,\Delta x \approx 2\,\text{MeV}/(\text{g/cm}^2) \times 940\,\text{g/cm}^2 = 1.88\,\text{GeV} \;.$$

4. By definition (see the Glossary) the following relation holds between the ratio of intensities I_1, I_2 and the difference of two star magnitudes m_1, m_2:

$$m_1 - m_2 = -2.5 \log_{10}(I_1/I_2) \,,$$

so that

$$I_1/I_2 = 10^{-0.4(m_1-m_2)} \,,$$

and from $\Delta m = 1$ one gets $I_1/I_2 \approx 0.398$ or $I_2 \approx 2.512\, I_1$.

5. Let N be the number of atoms making up the celestial body, μ the mass of the nucleon, and A the average atomic number of the elements constituting the celestial object. Gravitational binding dominates if

$$\frac{GM^2}{R} > N\varepsilon \,,$$

where ε is a typical binding energy for solid material (≈ 1 eV per atom). Here numerical factors of order unity are neglected; for a uniform mass distribution the numerical factor would be 6/5, see Problem 13.5. The mass of the object is $M = N\mu A$, so that the condition above can be written as

$$\frac{GM^2}{R} = \frac{GM}{R} N\mu A > N\varepsilon \quad \text{or} \quad \frac{M}{R} > \frac{\varepsilon}{\mu A G} \,.$$

M/R can be rewritten as

$$\frac{M}{R} = \frac{4}{3}\pi R^2 \varrho = \left(\frac{4}{3}\pi\right)^{1/3} \underbrace{\left(\frac{4}{3}\pi\right)^{2/3} R^2 \varrho^{2/3}}_{M^{2/3}} \varrho^{1/3} = \left(\frac{4}{3}\pi\right)^{1/3} M^{2/3} \varrho^{1/3} \,,$$

which leads to

$$\left(\frac{4}{3}\pi\right)^{1/3} M^{2/3} \varrho^{1/3} > \frac{\varepsilon}{\mu A G} \quad \text{or} \quad M > \left(\frac{\varepsilon}{\mu A G}\right)^{3/2} \frac{1}{\sqrt{\frac{4}{3}\pi\varrho}} \,.$$

The average density can be estimated to be (again neglecting numerical factors of order unity; for spherical molecules the optimal arrangement of spheres without overlapping leads to a packing fraction of 74%)

$$\varrho = \frac{\mu}{\frac{4}{3}\pi r_{\mathrm B}^3} \,,$$

where $r_{\mathrm B} = r_e/\alpha^2$ is the Bohr radius ($r_{\mathrm B} = 0.529 \times 10^{-10}$ m), r_e the classical electron radius, and α the fine-structure constant. This leads to

$$M > \left(\frac{\varepsilon}{\mu A G}\right)^{3/2} \frac{1}{\sqrt{\frac{4}{3}\pi\mu/(\frac{4}{3}\pi r_{\mathrm B}^3)}} = \left(\frac{\varepsilon}{\mu A G}\right)^{3/2} \frac{r_{\mathrm B}^{3/2}}{\sqrt{\mu}} = \frac{1}{\mu^2}\left(\frac{\varepsilon\, r_{\mathrm B}}{A G}\right)^{3/2} \,.$$

With

$$\mu = 1.67 \times 10^{-27} \, \text{kg} \,, \quad \varepsilon = 1 \, \text{eV} = 1.6 \times 10^{-19} \, \text{J} \,,$$
$$G = 6.67 \times 10^{-11} \, \text{m}^3 \, \text{kg}^{-1} \, \text{s}^{-2} \,, \quad A = 50$$

one gets the condition

$$M > 4.58 \times 10^{22} \, \text{kg} \,.$$

This means that our moon is gravitationally bound, while the moons of Mars (Phobos and Deimos) being much smaller are bound by solid-state effects, i.e., essentially by electromagnetic forces.

6. In special relativity the redshift is given by

$$z = \frac{\lambda - \lambda_0}{\lambda_0} = \sqrt{\frac{1 + \beta}{1 - \beta}} - 1 \,.$$

Let us assume that this relation can be applied even for the young universe, where a treatment on the basis of general relativity would be more appropriate.
Solve $z(\beta)$ for β or v:

$$\beta = \frac{(z + 1)^2 - 1}{(z + 1)^2 + 1} \quad \Rightarrow \quad v = \frac{(z + 1)^2 - 1}{(z + 1)^2 + 1} \, c = H \, d \,,$$

$$t = \frac{d}{c} = \frac{1}{H} \frac{(z + 1)^2 - 1}{(z + 1)^2 + 1} \approx \frac{0.967}{H} \,.$$

Since $\frac{1}{H}$ is the age of the universe, we see the distant quasar when it was $\approx 3.3\%$ of its present age.

17.2 Chapter 2

1. a) lepton-number conservation is violated: not allowed,

   b) possible,

   c) possible, a so-called Dalitz decay of the π^0,

   d) both charge and lepton-number conservation violated, not allowed,

   e) kinematically not allowed ($m_{K^-} + m_p > m_\Lambda$),

   f) possible,

   g) possible,

   h) possible.

2. For any unstable elementary particle like, e.g., a muon, the quantity 'lifetime' should be considered in the average sense only. In other words, it does not mean that a particle with lifetime τ will decay exactly the time τ after it was produced. Its actual lifetime t is a random number distributed with a probability density function,

$$f(t; \tau) \, dt = \frac{1}{\tau} e^{-\frac{t}{\tau}} \, dt \ ,$$

giving a probability that the lifetime t lies between t and $t + dt$. It can easily be checked that the mean value of t equals τ. For an unstable relativistic particle, a mean range before it decays is given by the product of its velocity βc and lifetime $\gamma \tau$ to $\beta \gamma c \tau$. For muons $c\tau = 658.653$ m. Therefore, to survive to sea level from an altitude of 20 km, the average range should equal $l = 20$ km or $\beta \gamma c \tau_\mu = l = 20$ km and $\beta \gamma = l / c \tau_\mu$. From $\beta \gamma = \sqrt{\gamma^2 - 1}$ one gets $\gamma^2 = (l/c\tau_\mu)^2 + 1$.
Since $l/c\tau \gg 1$, one finally obtains

$$\gamma \approx l/c\tau_\mu = \frac{20 \times 10^3 \text{ m}}{658.653 \text{ m}} \approx 30.4 \ .$$

Then the total energy is

$$E_\mu = \gamma m_\mu c^2 \approx 3.2 \,\text{GeV}$$

and the kinetic energy is

$$E_\mu^{\text{kin}} = E_\mu - m_\mu c^2 \approx 3.1 \,\text{GeV} \ .$$

3. The Coulomb force is

$$F_{\text{Coulomb}} = \frac{1}{4\pi\varepsilon_0} \frac{q_1 q_2}{r^2} \approx \frac{1}{4\pi \times 8.854 \times 10^{-12} \,\text{F} \text{m}^{-1}} \frac{(1.602 \times 10^{-19} \,\text{A s})^2}{(10^{-15} \,\text{m})^2}$$
$$\approx 230.7 \,\text{N}$$

and the gravitational force is

$$F_{\text{gravitation}} = G \frac{m_1 m_2}{r^2} \approx 6.674 \times 10^{-11} \frac{\text{m}^3}{\text{kg s}^2} \frac{(2.176 \times 10^{-8} \,\text{kg})^2}{(10^{-15} \,\text{m})^2} \approx 31\,600 \,\text{N} \ .$$

4. Energy–momentum conservation requires

$$q_{e^+} + q_{e^-} = q_f \ ,$$

where q_{e^+}, q_{e^-}, and q_f are the four-momenta of the positron, electron, and final state, respectively. To produce a Z, the invariant mass of the initial state squared should be not less than the invariant mass of the required final state squared or

$$(q_{e^+} + q_{e^-})^2 \geq m_Z^2 \ .$$

Since $q_{e^+} = (E_{e^+}, \boldsymbol{p}_{e^+})$ and $q_{e^-} = (m_e, 0)$, one obtains

$$2m_e(E_{e^+} + m_e) \geq m_Z^2$$

or

$$E_{e^+} \geq \frac{m_Z^2}{2m_e} - m_e \approx 8.1 \times 10^{15}\,\text{eV} = 8.1\,\text{PeV} .$$

17.3 Chapter 3

1. Similarly to Problem 2.4, for the reaction $\gamma + \gamma \rightarrow \mu^+ + \mu^-$ the threshold condition is

$$(q_{\gamma_1} + q_{\gamma_2})^2 \geq (2m_\mu)^2 .$$

For photons $q_{\gamma_1}^2 = q_{\gamma_2}^2 = 0$. Then

$$(q_{\gamma_1} + q_{\gamma_2})^2 = 2(E_{\gamma_1} E_{\gamma_2} - \boldsymbol{p}_{\gamma_1} \cdot \boldsymbol{p}_{\gamma_2}) = 4E_{\gamma_1} E_{\gamma_2}$$

for the angle π between the photon 3-momenta. Finally,

$$E_{\gamma_1} \geq \frac{(m_\mu c^2)^2}{E_{\gamma_2}} \approx 1.1 \times 10^{19}\,\text{eV} .$$

2. The number of collisions is obtained by subtracting the number of unaffected particles from the initial number of particles:

$$\Delta N = N_0 - N = N_0(1 - e^{-x/\lambda}) .$$

For thin targets $x/\lambda \ll 1$ and the expansion in the Taylor series gives

$$\Delta N = N_0(1 - (1 - x/\lambda + \cdots)) = \frac{N_0 x}{\lambda} = N_0 N_A \sigma_N x .$$

For the numerical example one gets

$$\Delta N = 10^8 \times 6.022 \times 10^{23}\,\text{g}^{-1} \times 10^{-24}\,\text{cm}^2 \times 0.1\,\text{g/cm}^2 \approx 6 \times 10^6 .$$

3. The threshold condition is

$$(q_{\bar{\nu}_e} + q_p)^2 \geq (m_n + m_e)^2 .$$

Taking into account that $q_{\bar{\nu}_e}^2 = 0$ and $q_p = (m_p, 0)$, one obtains

$$m_p^2 + 2E_{\bar{\nu}_e} m_p \geq (m_n + m_e)^2$$

and

$$E_{\bar{\nu}_e} \geq \frac{(m_n + m_e)^2 - m_p^2}{2m_p} \approx 1.8\,\text{MeV} ,$$

where the following values of the particles involved were used: $m_n = 939.565\,\text{MeV}$, $m_p = 938.272\,\text{MeV}$, $m_e = 0.511\,\text{MeV}$.

4.

Force $F = \dfrac{zeZe}{r^2} \dfrac{r}{r}$,

$$p_b = \int_{-\infty}^{+\infty} |F_b| \, dt = \int_{-\infty}^{+\infty} \frac{zZe^2}{r^2} \frac{b}{r} \frac{dx}{\beta c} \qquad \begin{array}{l} \text{momentum transfer} \\ \text{perpendicular to } p \end{array}$$

$$= \frac{zZe^2}{\beta c} \int_{-\infty}^{+\infty} \frac{b \, dx}{(\sqrt{x^2 + b^2})^3} = \frac{zZe^2}{\beta cb} \underbrace{\int_{-\infty}^{+\infty} \frac{d(x/b)}{\left(\sqrt{1 + \left(\frac{x}{b}\right)^2}\right)^3}}_{= 2}$$

$$= \frac{2zZe^2}{\beta cb} = \frac{2r_e m_e c}{b\beta} zZ \ ,$$

where the classical electron radius is $r_e = \dfrac{e^2}{m_e c^2}$.

5. Previously the transverse momentum transfer was obtained to be ($z = 1$)

$$p_b = 2 \, Z \frac{r_e m_e c}{b\beta} = 2p \frac{r_e Z}{b\beta^2} \ ,$$

where $p = m_e \, v$ was assumed (classical treatment). The transverse momentum transfer is given by $p_b = p \sin \vartheta$. Since

$$\sin 2\gamma = 2 \sin \gamma \cos \gamma = 2 \tan \gamma \cos^2 \gamma = \frac{2 \tan \gamma}{1 + \tan^2 \gamma} \ ,$$

one gets, using Rutherford's scattering formula:

$$p_b = 2p \frac{r_e Z/b\beta^2}{1 + (r_e Z/b\beta^2)^2} \ .$$

17.4 Chapter 4

1. $\phi = \dfrac{N_A}{A} \sigma_A = N_A \, \sigma_N \ , \quad \left. \begin{array}{l} [N_A] = \text{mol}^{-1} \\ [A] \ \ = \text{g mol}^{-1} \\ [\sigma_A] = \text{cm}^2 \end{array} \right\} \quad \Rightarrow \quad [\phi] = (\text{g/cm}^2)^{-1} \ .$

2. The relative energy resolution is determined by the fluctuations of the number N of produced particles. If W is the energy required for the production of a particle (pair), the relative energy resolution is

$$\frac{\Delta E}{E} = \frac{\Delta N}{N} = \frac{\sqrt{N}}{N} = \frac{1}{\sqrt{N}} = \sqrt{\frac{W}{E}} \, ,$$

since $N = E/W$. Here E and ΔE are the energy of the particle and its uncertainty,

$$\frac{\Delta E}{E} = \begin{cases} \text{a) } 10\% \\ \text{b) } 5.5\% \\ \text{c) } 1.9\% \\ \text{d) } 3.2 \times 10^{-4} \end{cases} .$$

3. $R = \displaystyle\int_E^0 \frac{\mathrm{d}E}{\mathrm{d}E/\mathrm{d}x} = \int_0^E \frac{\mathrm{d}E}{a + bE} = \frac{1}{b} \ln(1 + \frac{b}{a}E) \, .$

The numerical calculation gives

$$R \approx \frac{1}{4.4 \times 10^{-6}} \ln\left(1 + \frac{4.4 \times 10^{-6}}{2} \, 10^5\right) \frac{\mathrm{g}}{\mathrm{cm}^2} \approx 45\,193 \, \frac{\mathrm{g}}{\mathrm{cm}^2}$$

or $R = 181\,\mathrm{m}$ of rock taking into account that $\varrho_{\mathrm{rock}} = 2.5\,\mathrm{g/cm^3}$.

4. The Cherenkov angle θ_C is given by the relation

$$\cos\theta_C = \frac{1}{n\beta} \, .$$

From the relation between the momentum p and β:

$$p = \gamma m_0 \beta c \quad \Rightarrow \quad \beta = \frac{p}{\gamma m_0 c} \quad \text{and} \quad \cos\theta_C = \frac{\gamma m_0 c}{np} \, .$$

Then

$$\frac{np\cos\theta_C}{m_0 c} = \frac{E}{m_0 c^2} = \frac{c\sqrt{p^2 + m_0^2 c^2}}{m_0 c^2}$$

and

$$(np\cos\theta_C)^2 = p^2 + m_0^2 c^2 \quad \Rightarrow \quad p^2(n^2\cos^2\theta_C - 1) = m_0^2 c^2 \, .$$

Finally,

$$m_0 = \frac{p}{c}\sqrt{n^2\cos^2\theta_C - 1} \, .$$

5. From the expression for the change of the thermal energy, $\Delta Q = c_{sp} m \Delta T$, one can find the temperature rise:

$$\Delta T = \frac{\Delta Q}{c_{sp} m} = \frac{10^4 \, \text{eV} \times 1.6 \times 10^{-19} \, \text{J/eV}}{8 \times 10^{-5} \, \text{J/(g K)} \times 1 \, \text{g}} = 2 \times 10^{-11} \, \text{K} \, .$$

6. $q_\gamma + q_e = q'_\gamma + q'_e \quad \Rightarrow \quad q_\gamma - q'_\gamma = q'_e - q_e \quad \Rightarrow$

$q_\gamma^2 + q_{\gamma'}^2 - 2 q_\gamma q'_\gamma = -2(E_\gamma E'_\gamma - \boldsymbol{p}_\gamma \boldsymbol{p}'_\gamma) = q_e'^2 + q_e^2 - 2 q'_e q_e \quad \Rightarrow$

$-2 E_\gamma E'_\gamma (1 - \cos\theta) = 2 m_e^2 - 2 E'_e m_e \quad ; \quad (p_e = 0)$

$$= -2 m_e (E'_e - m_e) = -2 m_e E_e^{\text{kin}} = -2 m_e (E_\gamma - E'_\gamma) \quad \Rightarrow$$

$$\frac{E_\gamma - E'_\gamma}{E'_\gamma} = \frac{E_\gamma}{E'_\gamma} - 1 = \frac{E_\gamma}{m_e}(1 - \cos\theta) \quad \Rightarrow$$

$$\frac{E'_\gamma}{E_\gamma} = \frac{1}{1 + \frac{E_\gamma}{m_e}(1 - \cos\theta)} = \frac{1}{1 + \varepsilon(1 - \cos\theta)} \, .$$

7. Start from (4.13). The maximum energy is transferred for backscattering, $\theta = \pi$;

$$\frac{E'_\gamma}{E_\gamma} = \frac{1}{1 + 2\varepsilon} \, ,$$

$$E_e^{\text{max}} = E_\gamma - E'_\gamma = E_\gamma - \frac{E_\gamma}{1 + 2\varepsilon} = E_\gamma \frac{2\varepsilon}{1 + 2\varepsilon} = \frac{2\varepsilon^2}{1 + 2\varepsilon} m_e c^2 \, ; \qquad (*)$$

with numbers: $E_e^{\text{max}} = 478 \, \text{keV}$ ('Compton edge').
For $\varepsilon \to \infty$ (*) yields $E_e^{\text{max}} = E_\gamma$.
For $\theta_\gamma = \pi$ one has

$$E'_\gamma = E_\gamma \frac{1}{1 + 2\varepsilon}$$

and consequently

$$E'_\gamma = \frac{m_e c^2 \, \varepsilon}{1 + 2\varepsilon} = \frac{m_e c^2}{2 + 1/\varepsilon} \, .$$

For $\varepsilon \gg 1$ this fraction approaches $m_e c^2 / 2$.

8. Momentum: $p = m \, v$; $m = \gamma \, m_0$, where m_0 is the rest mass and $\gamma = \dfrac{1}{\sqrt{1 - \beta^2}}$:

$p = \gamma \, m_0 \, \beta \, c \quad \Rightarrow \quad \gamma \beta = p / m_0 c$.

9. In the X-ray region the index of refraction is $n = 1$, therefore there is no dispersion and consequently no Cherenkov radiation.

17.5 Chapter 5

1. a) In the classical non-relativistic case ($v \ll c$) the kinetic energy is just

$$E^{\text{kin}} = \frac{1}{2}m_0 v^2 = \frac{1}{2}m_0 \left(\frac{eRB}{m_0}\right)^2 = \frac{1}{2}\frac{e^2 R^2 B^2}{m_0} \,,$$

where the velocity v was found from the usual requirement that the centrifugal force be balanced by the Lorentz force:

$$\frac{m_0 v^2}{R} = evB \quad \Rightarrow \quad v = \frac{eBR}{m_0} \,.$$

b) In the relativistic case

$$E^{\text{kin}} = \gamma m_0 c^2 - m_0 c^2 = m_0 c^2 \left(\frac{1}{\sqrt{1 - \frac{v^2}{c^2}}} - 1\right) = c\sqrt{p^2 + m_0^2 c^2} - m_0 c^2$$

$$= c\sqrt{e^2 R^2 B^2 + m_0^2 c^2} - m_0 c^2 = ecRB\sqrt{1 + \frac{m_0^2 c^2}{e^2 R^2 B^2}} - m_0 c^2$$

$$\approx 5.95 \times 10^7 \text{ eV} \,.$$

Alternatively, with an early relativistic approximation

$$E^{\text{kin}} = c\sqrt{p^2 + m_0^2 c^2} - m_0 c^2 \approx cp = ecRB \approx 6 \times 10^7 \text{ eV} \,.$$

2. Let us first find the number of electrons N_e in the star:

$$N_e = M_{\text{star}}/m_p = 10\, M_\odot/m_p \approx \frac{2 \times 10^{34}\,\text{g}}{1.67 \times 10^{-24}\,\text{g}} \approx 1.2 \times 10^{58} \,.$$

Here it is assumed that the star consists mainly of hydrogen ($m_p \approx 1.67 \times 10^{-24}$ g) and is electrically neutral, so that $N_e = N_p$.
The pulsar volume is

$$V = \frac{4}{3}\pi r^3 \approx 4.19 \times 10^{18}\,\text{cm}^3$$

and the electron density is $n = 0.5\, N_e/V \approx 1.43 \times 10^{39}/\text{cm}^3$. Then the Fermi energy is

$$E_\text{F} = \hbar c(3\pi^2 n)^{1/3}$$

$$\approx 6.582 \times 10^{-22}\,\text{MeV s} \times 3 \times 10^{10}\,\frac{\text{cm}}{\text{s}}\,(3 \times 3.1416^2 \times 1.43 \times 10^{39})^{1/3}\,\text{cm}^{-1}$$

$$\approx 688\,\text{MeV} \,.$$

Consequences:

The electrons are pressed into the protons,

$$e^- + p \rightarrow n + \nu_e .$$

The neutrons cannot decay, since in free neutron decay the maximum energy transfer to the electron is only $\approx 780\,\text{keV}$, and all energy levels in the Fermi gas are occupied, even if only 1% of the electrons are left: ($n^* = 0.01\,n \Rightarrow E_F \approx 148\,\text{MeV}$).

3. Event horizon of a black hole with mass $M = 10^6\,M_\odot$:

$$R_S = \frac{2GM}{c^2} \approx 2.96 \times 10^9\,\text{m} ,$$

$$\Delta E = -\int_\infty^{R_S} G\frac{m_p M}{r^2}\,dr = G\frac{m_p M}{R_S} = \frac{1}{2}m_p c^2 \approx 469\,\text{MeV} \approx 7.5 \times 10^{-8}\,\text{J} .$$

The result is independent of the mass of the black hole. This classical calculation, however, is not suitable for this problem and just yields an idea of the energy gain. In addition, the value of the energy depends on the frame of reference. In the vicinity of black holes only formulae should be used which hold under general relativity.

4. a) Conservation of angular momentum

$$\boldsymbol{L} = \boldsymbol{r} \times \boldsymbol{p} \Rightarrow L \approx mrv = mr^2\omega$$

requires $r^2\omega$ to be constant (no mass loss):

$$R_\odot^2\omega_\odot = R_{NS}^2\omega_{NS} \Rightarrow \omega_{NS} = \left(\frac{R_\odot}{R_{NS}}\right)^2\omega_\odot \approx 588\,\text{Hz} .$$

Then the rotational energy is

$$E_{rot} = \frac{1}{2}\frac{2}{5}M_{NS}R_{NS}^2\omega_{NS}^2 \approx 0.4 \times 10^{30}\,\text{kg} \times (5 \times 10^4)^2\,\text{m}^2 \times 588^2\,\text{s}^{-2}$$
$$\approx 0.35 \times 10^{45}\,\text{J} .$$

b) Assume that the Sun consists of protons only (plus electrons, of course). Four protons each are fused to He with an energy release of $26\,\text{MeV}$ corresponding to a mass–energy conversion efficiency of $\approx 0.7\%$:

$$E = M_\odot c^2 \times 7 \times 10^{-3} = 2 \times 10^{30}\,\text{kg} \times (3 \times 10^8)^2\,\text{m}^2/\text{s}^2 \times 7 \times 10^{-3}$$
$$= 1.26 \times 10^{45}\,\text{J} ,$$

which is comparable to the rotational energy of the neutron star.

5. A dipole field is needed to compensate the centrifugal force

$$\frac{mv^2}{R} = evB_{guide} . \tag{*}$$

Equation (5.38) yields $p = mv = \frac{1}{2}eRB$. Comparison with (*) gives

$$B_{guide} = \frac{1}{2}B \quad \text{(Wideroe condition)} .$$

17.6 Chapter 6

Sect. 6.1

1. C, O, and Ne are even–even nuclei, oxygen is even doubly magic, while F as even–odd and N, Na as odd–odd configurations are less tightly bound. The pairing term δ in the Bethe–Weizsäcker formula gives the difference in the nuclear binding energies for even–even, even–odd, and odd–odd nuclei:

$$m(Z, A) = Zm_p + (A - Z)m_n - a_v A + a_s A^{2/3} + a_C \frac{Z^2}{A^{1/3}} + a_a \frac{(Z - A/2)^2}{A} + \delta$$

$a_v A$ – volume term,

$a_s A^{2/3}$ – surface term,

$a_C \frac{Z^2}{A^{1/3}}$ – Coulomb repulsion,

$a_a \frac{(Z - A/2)^2}{A}$ – asymmetry term,

$$\delta = \begin{cases} -a_p A^{-3/4} & \text{for even–even} \\ 0 & \text{for even–odd/odd–even} \\ +a_p A^{-3/4} & \text{for odd–odd} \end{cases} \quad , \quad \delta - \text{pairing term} .$$

2. From Fig. 7.9 the rate of primary particles can be estimated as

$$\Phi \approx 0.2 \, (\text{cm}^2 \, \text{s} \, \text{sr})^{-1} .$$

The surface of the Earth is

$$S = 4\pi R_\oplus^2 \approx 4\pi (6370 \times 10^5)^2 \, \text{cm}^2 \approx 5.10 \times 10^{18} \, \text{cm}^2 .$$

The age of the Earth is $T = 4.5 \times 10^9$ years or 1.42×10^{17} s. The total charge accumulated in the solid angle of 2π during the time T is

$$\int \Phi(x, t) \, dx \, dt \approx 1.45 \times 10^{35} \times 2\pi$$

$$\approx 9.1 \times 10^{35} \text{ equivalent protons} \cong 1.5 \times 10^{17} \text{ Coulomb} .$$

Still, there is no charge-up. Primary particles mentioned in Sect. 6.1 are mainly those of high energy (typically > 1 GeV). In this high-energy domain positively charged particles actually dominate. If all energies are considered, there are equal amounts of positive and negative particles. Our Sun is also a source of large numbers of protons, electrons, and α particles. In total, there is no positive charge excess if particles of all energies are considered. This is not in contrast to the observation of a positive charge excess of sea-level muons because they are the result of cascade processes in the atmosphere initialized by energetic (i.e., mainly positively charged) primaries.

3. a) The average column density traversed by primary cosmic rays is $\approx 6\,g/cm^2$ (see Sect. 6.1). The interaction rate Φ is related to the cross section by, see (3.57),

$$\Phi = \sigma\, N_A \;.$$

Because of $\Phi[(6\,g/cm^2)^{-1}] \approx 0.1$ and assuming that the collision partners are nucleons, one gets

$$\sigma \approx 0.1 \times (6\,g/cm^2)^{-1}/N_A \approx 1.66 \times 10^{-25}\,cm^2 = 166\,mbarn \;.$$

b) If the iron–proton fragmentation cross section is $\approx 170\,mb$, the cross section for iron–air collisions can be estimated to be $\sigma_{frag}(iron–air) \approx 170\,mb \times A^{\alpha}$, where $A \approx 0.8\,A_N + 0.2\,A_O = 11.2 + 3.2 = 14.4$. With $\alpha \approx 0.75$ one gets $\sigma_{frag}(iron–air) \approx 1.26\,b$. The probability to survive to sea level is $P = \exp(-\sigma\,N_A d/A) = 1.3 \times 10^{-23}$, where it has been used that the thickness of the atmosphere is $d \approx 1000\,g/cm^2$.

Sect. 6.2

1. The neutrino flux ϕ_ν is given by the number of fusion processes $4p \to {}^4He + 2e^+ + 2\nu_e$ times 2 neutrinos per reaction chain:

$$\phi_\nu = \frac{\text{solar constant}}{\text{energy gain per reaction chain}} \times 2$$

$$\approx \frac{1400\,W/m^2}{28.3\,MeV \times 1.6 \times 10^{-13}\,J/MeV} \times 2 \approx 6.2 \times 10^{10}\,cm^{-2}\,s^{-1} \;.$$

2. $(q_{\nu_\alpha} + q_{e^-})^2 = (m_\alpha + m_{\nu_e})^2 \;, \quad \alpha = \mu, \tau \;;$

assuming m_{ν_α} to be small ($\ll m_e, m_\mu, m_\tau$) one gets

$$2E_{\nu_\alpha} m_e + m_e^2 = m_\alpha^2 \quad \Rightarrow \quad E_{\nu_\alpha} = \frac{m_\alpha^2 - m_e^2}{2 m_e} \quad \Rightarrow$$

$$\alpha = \mu : E_{\nu_\mu} = 10.92\,GeV \;, \quad \alpha = \tau : E_{\nu_\tau} = 3.09\,TeV \;;$$

since solar neutrinos cannot convert into such high-energy neutrinos, the proposed reactions cannot be induced.

3. a) The interaction rate is

$$R = \sigma_N N_A\, d\, A\, \phi_\nu$$

where σ_N is the cross section per nucleon, $N_A = 6.022 \times 10^{23}\,g^{-1}$ is the Avogadro constant, d the area density of the target, A the target area, and ϕ_ν the solar neutrino flux. With $d \approx 15\,g\,cm^{-2}$, $A = 180 \times 30\,cm^2$, $\phi_\nu \approx 7 \times 10^{10}\,cm^{-2}\,s^{-1}$, and $\sigma_N = 10^{-45}\,cm^2$ one gets $R = 3.41 \times 10^{-6}\,s^{-1} = 107\,a^{-1}$.

b) A typical energy of solar neutrinos is $100\,keV$, i.e., $50\,keV$ are transferred to the electron. Consequently, the total annual energy transfer to the electrons is

$$\Delta E = 107 \times 50\,keV = 5.35\,MeV = 0.86 \times 10^{-12}\,J \;.$$

c) With the numbers used so far the mass of the human is 81 kg. Therefore, the equivalent annual dose comes out to be

$$H_v = \frac{\Delta E}{m} w_R = 1.06 \times 10^{-14}\, \text{Sv} ,$$

actually independent of the assumed human mass. The contribution of solar neutrinos to the normal natural dose rate is negligible, since

$$H = \frac{H_v}{H_0} = 5.3 \times 10^{-12} .$$

4. a) The time evolution of the electron neutrino is

$$|v_e; t\rangle = \cos\theta\, e^{-iE_{v_1}t}|v_1\rangle + \sin\theta\, e^{-iE_{v_2}t}|v_2\rangle .$$

This leads to

$$\langle v_\mu|v_e; t\rangle = \sin\theta \cos\theta \left(e^{-iE_{v_2}t} - e^{-iE_{v_1}t} \right) ,$$

because

$$(-\sin\theta|v_1\rangle + \cos\theta|v_2\rangle)\, (\cos\theta\, e^{-iE_{v_1}t}|v_1\rangle + \sin\theta\, e^{-iE_{v_2}t}|v_2\rangle)$$

$$= -\sin\theta \cos\theta\, e^{-iE_{v_1}t} + \sin\theta \cos\theta\, e^{-iE_{v_2}t}$$

since $|v_1\rangle$; $|v_2\rangle$ are orthogonal states.

Squaring the time-dependent part of this relation, i.e., multiplying the expression by its complex conjugate, yields as an intermediate step

$$\left(e^{-iE_{v_2}t} - e^{-iE_{v_1}t} \right)\left(e^{+iE_{v_2}t} - e^{+iE_{v_1}t} \right) = 1 - e^{i(E_{v_2}-E_{v_1})t} - e^{i(E_{v_1}-E_{v_2})t} + 1$$

$$= 2 - (e^{i(E_{v_2}-E_{v_1})t} + e^{-i(E_{v_2}-E_{v_1})t}) = 2 - 2\cos[(E_{v_2} - E_{v_1})t]$$

$$= 2\,(1 - \cos[(E_{v_2} - E_{v_1})t]) = 4\sin^2[(E_{v_2} - E_{v_1})t/2] .$$

Using $2\sin\theta \cos\theta = \sin 2\theta$ one finally gets

$$P_{v_e \to v_\mu}(t) = |\langle v_\mu|v_e; t\rangle|^2 = \sin^2 2\theta \sin^2[(E_{v_1} - E_{v_2})t/2] .$$

Since the states $|v_e\rangle$ and $|v_\mu\rangle$ are orthogonal one has $P_{v_e \to v_e}(t) = 1 - P_{v_e \to v_\mu}(t)$. In the approximation of small masses, $E_{v_i} = p + m_i^2/2p + O(m_i^4)$, one gets $E_{v_1} - E_{v_2} \approx (m_1^2 - m_2^2)/2p$.
Since $t = E_{v_i}x/p$ and because of $m_i \ll E_{v_i}$, the momentum p can be identified with the neutrino energy E_v. If, finally, also the correct powers of $\hbar$ and c are introduced, one obtains

$$P_{v_e \to v_e}(x) = 1 - \sin^2 2\theta \sin^2\left(\frac{1}{4\hbar c}\delta m^2 \frac{x}{E_v} \right)$$

with $1/4\hbar c = 1.27 \times 10^9\, \text{eV}^{-1}\,\text{km}^{-1}$, which is the desired relation.

b) On both sides of the Earth a number of N muon neutrinos are created. The ones from above, being produced at an altitude of 20 km almost have no chance to oscillate, while the ones crossing the whole diameter of the Earth have passed a distance of more than 12 000 km, so that from below only $N P_{\nu_\mu \to \nu_\mu}(2R_E)$ muon neutrinos will arrive, leading to a ratio of

$$S = 0.54 = \frac{N P_{\nu_\mu \to \nu_\mu}(2R_E)}{N} = 1 - \sin^2(1.27 \times 12\,700 \times \delta m^2) .$$

Solving for δm^2 one obtains $\delta m^2 = 4.6 \times 10^{-5}\,\mathrm{eV^2}$ and $m_{\nu_\tau} \approx 6.8 \times 10^{-3}\,\mathrm{eV}$.

c) No. The masses of the lepton flavour eigenstates are quantum-mechanical expectation values of the mass operator $M = \sqrt{H^2 - p^2}$. For an assumed $(\nu_\mu \leftrightarrow \nu_\tau)$ mixing with a similar definition of the mixing angle as for the $(\nu_e \leftrightarrow \nu_\mu)$ mixing, one gets

$$m_{\nu_\mu} = \langle \nu_\mu | M | \nu_\mu \rangle = m_1 \cos^2 \theta + m_2 \sin^2 \theta ,$$

$$m_{\nu_\tau} = \langle \nu_\tau | M | \nu_\tau \rangle = m_1 \sin^2 \theta + m_2 \cos^2 \theta .$$

For maximum mixing ($\theta = 45°$, $\cos^2 \theta = \sin^2 \theta = 1/2$) one obtains

$$m_{\nu_\mu} = m_{\nu_\tau} = (m_1 + m_2)/2 .$$

Sect. 6.3

1. $I = I_0\,e^{-\mu x}$ photons survive, $I_0 - I = I_0(1 - e^{-\mu x})$ get absorbed; from Fig. 6.37 one reads $\mu = 0.2\,\mathrm{cm^{-1}}$;

   detection efficiency: $\quad \eta = \dfrac{I_0(1 - e^{-\mu x})}{I_0} = 1 - e^{-\mu x} \approx 0.45 = 45\%$.

2. The duration of the brightness excursion cannot be shorter than the time span for the light to cross the cosmological object. Figure 6.48 shows $\Delta t = 1\,\mathrm{s} \Rightarrow$ size $\approx c\,\Delta t = 300\,000\,\mathrm{km}$.

3. $\cos \vartheta = \dfrac{1}{n\beta}$; $\quad$ threshold at $\quad \beta > \dfrac{1}{n} \quad \Rightarrow \quad v > \dfrac{c}{n}$;

$$E_\mu = \gamma m_0 c^2 = \frac{1}{\sqrt{1 - \beta^2}} m_0 c^2 = \frac{1}{\sqrt{1 - \frac{1}{n^2}}} m_0 c^2 = \frac{n}{\sqrt{n^2 - 1}} m_0 c^2 ,$$

$$E_\mu \approx \begin{cases} 4.5\,\mathrm{GeV} \text{ in air} \\ 160.3\,\mathrm{MeV} \text{ in water} \end{cases}, \quad E_\mu^{\mathrm{kin}} \approx \begin{cases} 4.4\,\mathrm{GeV} \text{ in air} \\ 54.6\,\mathrm{MeV} \text{ in water} \end{cases} .$$

4. Number of emitted photons:

$$N_E = \frac{P}{h\nu} \approx \frac{3 \times 10^{27}\,\mathrm{W}}{10^{11}\,\mathrm{eV} \times 1.602 \times 10^{-19}\,\mathrm{J/eV}} \approx 1.873 \times 10^{35}\,\mathrm{s^{-1}} .$$

Solid angle: $\Omega = \dfrac{A}{4\pi R^2} \approx 6.16 \times 10^{-41}$.

Number of recorded photons: $N_R = \Omega N_E \approx 1.15 \times 10^{-5}/s \approx 364/a$.

Minimum flux: assumed $10/a$, $\quad P_{min} \approx P \dfrac{\Omega}{A} \dfrac{10}{364} \approx 6.35 \times 10^{-19}$ J/(cm^2 s) .

5. (a) Assume isotropic emission, which leads to a total power of $P = 4\pi r^2 P_S$, where $r = 150 \times 10^6$ km is the astronomical unit (distance Sun–Earth). One gets

$$P \approx 3.96 \times 10^{26} \text{ W} .$$

(b) In a period of 10^6 years the emitted energy is $E = P \times t \approx 1.25 \times 10^{40}$ J, which corresponds to a mass of $m = E/c^2 \approx 1.39 \times 10^{23}$ kg, which represents a relative fraction of the solar mass of

$$f = \frac{m}{M_\odot} \approx 6.9 \times 10^{-8} .$$

(c) The effective area of the Earth is $A = \pi R^2$, so that the daily energy transport to Earth is worked out to be $E = \pi R^2 P_S t$. This corresponds to a mass of

$$m = \frac{E}{c^2} \approx 1.71 \times 10^5 \text{ kg} = 171 \text{ tons} .$$

Sect. 6.4

1. The power emitted in the frequency interval $[v, v + dv]$ corresponds to the one emitted in the wavelength interval $[\lambda, \lambda + d\lambda]$, $\lambda = c/v$, i.e., $P(v)\,dv = P(\lambda)\,d\lambda$, or

$$P(\lambda) = P(v)\frac{dv}{d\lambda} \sim \frac{v^3}{e^{hv/kT} - 1}\frac{dv}{d\lambda} \sim \frac{1}{\lambda^5(e^{hc/\lambda kT} - 1)} ,$$

since $dv/d\lambda = -c/\lambda^2$.

2. The luminosity of a star is proportional to the integral over Planck's radiation formula:

$$L \sim \int_0^\infty \varrho(v, T)\,dv = \int_0^\infty \frac{8\pi h v^3}{c^3}\frac{1}{e^{hv/kT} - 1}\,dv ;$$

use $x = \frac{hv}{kT} \Rightarrow$

$$L \sim \frac{8\pi}{c^3}\left(\frac{kT}{h}\right)^3 h \int_0^\infty x^3 \frac{1}{e^x - 1}\frac{kT}{h}\,dx = \frac{8\pi}{c^3}\frac{k^4 T^4}{h^3}\int_0^\infty \frac{x^3\,dx}{e^x - 1}$$

$$= \frac{8\pi}{c^3}\frac{k^4}{h^3}\frac{\pi^4}{15}T^4 \sim T^4 .$$

In addition, the luminosity varies with the size of the surface ($\sim R^2$).

This gives the scaling law

$$\frac{L}{L_\odot} = \left(\frac{R}{R_\odot}\right)^2 \left(\frac{T}{T_\odot}\right)^4 \; ;$$

a) $R = 10\,R_\odot$, $T = T_\odot$ $\Rightarrow$ $L = 100\,L_\odot$;

b) $R = R_\odot$, $T = 10\,T_\odot$ $\Rightarrow$ $L = 10\,000\,L_\odot$.

3. The motion of an electron in a transverse magnetic field is described by

$$\frac{mv^2}{\varrho} = evB \quad \Rightarrow \quad \frac{p}{\varrho} = eB \; .$$

Since at high energies $cp \approx E$, one gets

$$\frac{cp}{\varrho} = ceB \quad \Rightarrow \quad B = \frac{E}{ce\varrho} \; ,$$

where ϱ is the bending radius. This leads to

$$P = \frac{e^2 c^3}{2\pi} C_\gamma E^2 \frac{E^2}{c^2 e^2 \varrho^2} = \frac{cC_\gamma}{2\pi} \frac{E^4}{\varrho^2} \; .$$

The energy loss for one revolution around the pulsar is

$$\Delta E = \int_0^T P\,d\tau = \frac{cC_\gamma}{2\pi} \frac{E^4}{\varrho^2} 2\pi \varrho/c = C_\gamma \frac{E^4}{\varrho}$$

$$= 8.85 \times 10^{-5} \times \frac{10^{12}\,\text{GeV}^4}{10^6\,\text{m}}\,\text{m}\,\text{GeV}^{-3} = 88.5\,\text{GeV} \; .$$

The magnetic field is

$$B = \frac{E}{ce\varrho} = \frac{10^{12}\,\text{eV} \times 1.6 \times 10^{-19}\,\text{J/eV}}{3 \times 10^8\,\text{m/s} \times 1.6 \times 10^{-19}\,\text{A s} \times 10^6\,\text{m}} = 0.0033\,\text{T} = 33\,\text{Gauss} \; .$$

4. The Lorentz force is related to the absolute value of the momentum change by $|\dot{p}| = F = evB$. The total radiated power is taken from the kinetic (or total) energy of the particle, i.e., $\dot{E}_{\text{kin}} = \dot{E} = -P$. From the centrifugal force the bending radius results to $\varrho = p/eB$, see also Problem 3 in this section. As differential equation for the energy one gets

$$\dot{E} = -P = -\frac{2}{3} \frac{e^4 B^2}{m_0^2 c^3} \gamma^2 v^2 \; .$$

a) In general $E = \gamma m_0 c^2$ holds. The ultrarelativistic limit is given by $v \to c$, hence

$$\dot{E} = -\frac{2}{3} \frac{e^4 B^2 c^3}{(m_0 c^2)^4} E^2 = -\alpha E^2 \; , \quad \alpha = \frac{2}{3} \frac{e^4 B^2 c^3}{(m_0 c^2)^4} \; .$$

The solution of this differential equation is obtained by separation of variables,

$$\int_{E_0}^{E} \frac{dE'}{E'^2} = -\alpha \int_0^t dt' \quad \Rightarrow \quad \frac{1}{E_0} - \frac{1}{E} = -\alpha t \quad \Rightarrow \quad E = \frac{E_0}{1 + \alpha E_0 t} \, .$$

It is only valid for $E \gg m_0 c^2$. (This limit is not respected for longer times.)
The bending radius follows with $E \to pc$ to

$$\varrho = \frac{E}{ceB} = \frac{1}{ceB} \frac{E_0}{1 + \alpha E_0 t} = \frac{\varrho_0}{1 + \alpha ceB\varrho_0 t} \, .$$

b) In the general (relativistic) case there is $p^2 = \gamma^2 m_0^2 v^2$, i.e., $\gamma^2 v^2 = p^2/m_0^2 = (E^2 - m_0^2 c^4)/m_0^2 c^2$, thus

$$\dot{E} = -\frac{2}{3} \frac{e^4 B^2 c^3}{(m_0 c^2)^4} (E^2 - m_0^2 c^4) = -\alpha(E^2 - m_0^2 c^4) \, , \quad \alpha \text{ as in (a)} \, .$$

The solution is calculated as in (a):

$$\int_{E_0}^{E} \frac{dE'}{E'^2 - m_0^2 c^4} = -\int_0^t dt' = -\alpha t \quad \Rightarrow$$

$$\frac{1}{2m_0 c^2} \int_{E_0}^{E} \left(\frac{1}{E' - m_0 c^2} - \frac{1}{E' + m_0 c^2} \right) dE' = \frac{1}{2m_0 c^2} \ln \frac{E' - m_0 c^2}{E' + m_0 c^2} \Big|_{E'=E_0}^{E}$$

$$= -\frac{1}{m_0 c^2} \left(\frac{1}{2} \ln \frac{1 + \frac{m_0 c^2}{E}}{1 - \frac{m_0 c^2}{E}} - \frac{1}{2} \ln \frac{1 + \frac{m_0 c^2}{E_0}}{1 - \frac{m_0 c^2}{E_0}} \right)$$

$$= -\frac{1}{m_0 c^2} \left(\text{artanh} \frac{m_0 c^2}{E} - \text{artanh} \frac{m_0 c^2}{E_0} \right) = -\alpha t \, .$$

Solving this equation for E leads to

$$E = \frac{m_0 c^2}{\tanh \left(\alpha m_0 c^2 t + \text{artanh} \frac{m_0 c^2}{E_0} \right)} = m_0 c^2 \coth \left(\alpha m_0 c^2 t + \text{artanh} \frac{m_0 c^2}{E_0} \right) \, .$$

For larger times (in the non-relativistic regime) the rest energy is appoached exponentially.
From this the bending radius results in

$$\varrho = \frac{p}{eB} = \frac{\sqrt{E^2 - m_0^2 c^4}}{ceB} = \frac{m_0 c}{eB} \sqrt{\coth^2 \left(\alpha m_0 c^2 t + \text{artanh} \frac{m_0 c^2}{E_0} \right) - 1}$$

$$= \frac{m_0 c/eB}{\sinh \left(\alpha m_0 c^2 t + \text{artanh} \frac{m_0 c^2}{E_0} \right)} = \frac{m_0 c/eB}{\sinh \left(\alpha m_0 c^2 t + \text{artanh} \frac{m_0 c}{\sqrt{\varrho_0^2 e^2 B^2 + m_0^2 c^2}} \right)} \, .$$

5. Measured flux R = source flux $R^* \times$ efficiency $\varepsilon \times$ solid angle Ω;

$$\varepsilon = 1 - e^{-\mu x} = 1 - e^{-125 \times 5.8 \times 10^{-3}} \approx 51.6\% , \quad \Omega = \frac{10^4\,\text{cm}^2}{4\pi(55\,\text{kpc})^2} \approx 2.76 \times 10^{-44} ,$$

$$R^* = \frac{R}{\varepsilon\,\Omega} \approx 1.95 \times 10^{40}/\text{s} .$$

6. In contrast to Compton scattering of photons on an electron target at rest, all three-momenta are different from zero in this case. For the four-momenta k_i, k_f (photon) and q_i, q_f (electron) one has $k_i - k_f = q_f - q_i$. Squaring this equation gives $k_i k_f = q_i q_f - m_e^2$. On the other hand, rewriting the four-momentum conservation as $q_f = q_i + k_i - k_f$ and multiplying with q_i leads to $q_i q_f - m_e^2 = q_i(k_i - k_f)$, yielding

$$k_i k_f = q_i(k_i - k_f) .$$

Let ϑ be the angle between k_i and k_f, which gives

$$\omega_i \omega_f (1 - \cos\vartheta) = E_i(\omega_i - \omega_f) - |q_i|(\omega_i \cos\varphi_i - \omega_f \cos\varphi_f) .$$

Solving for ω_f leads to

$$\omega_f = \omega_i\, \frac{1 - \sqrt{1 - (m_e/E_i)^2}\,\cos\varphi_i}{1 - \sqrt{1 - (m_e/E_i)^2}\,\cos\varphi_f + \omega_i(1 - \cos\vartheta)/E_i} .$$

This expression is still exact. If the terms ω_i/E_i and m_e/E_i are neglected, the relation quoted in the problem is obtained.

7. Starting from the derivative $dP/d\nu$,

$$dP/d\nu \sim \frac{3\nu^2\,(e^{h\nu/kT} - 1) - \nu^3\,\frac{h}{kT}\,e^{h\nu/kT}}{(e^{h\nu/kT} - 1)^2} = \nu^2\,\frac{3\,(e^x - 1) - x\,e^x}{(e^x - 1)^2} ,$$

and the condition $dP/d\nu = 0$ one obtains the equation

$$e^{-x} = 1 - \frac{x}{3} , \quad x = \frac{h\nu}{kT} .$$

An approximated solution to this transcendental equation gives $x \approx 2.8$, leading to a frequency, resp. energy of the maximum of the Planck distribution of

$$h\nu_M \approx 2.8\,kT .$$

This linear relation between the frequency in the maximum and the temperature is called *Wien's displacement law*. For $h\nu_M = 50\,\text{keV}$ a temperature of

$$T \approx 2 \times 10^8\,\text{K} .$$

is obtained.

Sect. 6.5

1. $\Delta E = \dfrac{GMm_\gamma}{R} - \dfrac{GMm_\gamma}{R+H} = GMm_\gamma \dfrac{H}{R\,(H+R)}$,

$\dfrac{\Delta E}{E} = \dfrac{GM}{c^2}\dfrac{H}{R\,(H+R)} = \dfrac{GM}{R^2c^2}\dfrac{HR}{H+R} = \dfrac{g_\odot}{c^2}\dfrac{HR}{H+R} \approx \dfrac{g_\odot}{c^2}H \approx 3 \times 10^{-12}$

for $H \ll R$.

2. a) $P \approx \dfrac{G}{c^2}\,\omega^6\,m^2\,r^4$.

For $\omega = 1/\text{year}$, $m = 5.97 \times 10^{24}\,\text{kg}$, $r = 1\,\text{AU}$ one gets

$$P \approx 1.34 \times 10^{22}\,\text{W} \ .$$

Even though this power appears rather large, it is only a fraction of about 3×10^{-5} of the solar emission in the optical range.

b) Assume $\omega = 10^3\,\text{s}^{-1}$, $m = 10\,\text{kg}$, $r = 1\,\text{m}$ which leads to

$$P \approx 7.4 \times 10^{-8}\,\text{W} \ .$$

17.7 Chapter 7

1. $p = \dfrac{F}{A} = \dfrac{mg}{A} = 1.013 \times 10^5\,\dfrac{\text{N}}{\text{m}^2}$,

$\dfrac{m}{A} = \dfrac{p}{g} = \dfrac{1.013 \times 10^5}{9.81}\,\dfrac{\text{kg}}{\text{m}^2} \approx 10\,326\,\text{kg/m}^2 \approx 1.03\,\text{kg/cm}^2$.

2. $p = p_0\,\mathrm{e}^{-20/7.99} \approx 82.9\,\text{hPa}$, $\dfrac{m}{A} \approx \dfrac{8.29 \times 10^3}{9.81}\,\dfrac{\text{kg}}{\text{m}^2} \approx 845\,\dfrac{\text{kg}}{\text{m}^2} = 84.5\,\text{g/cm}^2$.

3. The differential sea-level muon spectrum can be parameterized by

$$N(E)\,\mathrm{d}E \sim E^{-\gamma}\,\mathrm{d}E \quad \text{with} \quad \gamma = 3 \quad \Rightarrow \quad I(E) = \int N(E)\,\mathrm{d}E \sim E^{-2} \ ,$$

where $I(E) \to 0$ for $E \to \infty$. The thickness of the atmosphere varies with zenith angle like

$$d(\theta) = \dfrac{d(0)}{\cos\theta} \ .$$

For 'low' energies ($E <$ several $100\,\text{GeV}$) the muon energy loss is constant $\left(\frac{\mathrm{d}E}{\mathrm{d}x} = a\right)$

$$\Rightarrow E \sim d \ , \quad I(\theta) \sim E^{-2} \sim d^{-2} = I(0)\cos^2\theta \ .$$

4. $N(> E, R) = A(aR)^{-\gamma}$,   see (7.15) ,

$$\frac{\Delta N}{\Delta R} = -\gamma A a^{-\gamma} R^{-(\gamma+1)} , \quad \frac{\Delta N}{N} = -\gamma R^{-1} \Delta R = -\gamma \frac{\Delta R}{R} ,$$

traditionally $\Delta R = 100 \, \text{g/cm}^2$.

With $\gamma = 2$ (exponent of the integral sea-level spectrum):

$$\left| \frac{\Delta N}{N} \right| = \left| \frac{S}{N} \right| = \frac{200 \, \text{g/cm}^2}{R} = 2 \times 10^{-2} , \quad \text{e.g., for } 100 \, \text{m w.e.}$$

5. $\vartheta_{\text{r.m.s.}} = \frac{13.6 \, \text{MeV}}{\beta c p} \sqrt{\frac{x}{X_0}} \left(1 + 0.038 \ln \frac{x}{X_0} \right)$ (see [2])

$$\approx 4.87 \times 10^{-3}(1 + 0.27) \approx 6.19 \times 10^{-3} \approx 0.35° .$$

6. The lateral separation is caused by
   a) transverse momenta in the primary interaction,
   b, c) multiple scattering in the air and rock.
   a) $p_T \approx 350 \, \text{MeV}/c$,

   $$\vartheta = \frac{p_T}{p} \approx \frac{350 \, \text{MeV}/c}{100 \, \text{GeV}/c} = 3.5 \times 10^{-3} ,$$

   average displacement: $\Delta x_1 = \vartheta \, h$, where h is the production height (20 km),

   $$\Delta x_1 \approx 70 \, \text{m} .$$

   b) Multiple-scattering angle in air,

   $$\vartheta_{\text{air}} \approx 6.90 \times 10^{-4}(1 + 0.123) \approx 7.75 \times 10^{-4} \approx 0.044° ,$$

   $$\Delta x_2 = 15.5 \, \text{m} .$$

   c) Multiple scattering in rock,

   $$\Delta x_3 \approx 0.8 \, \text{m} .$$

$$\Rightarrow \Delta x = \sqrt{\Delta x_1^2 + \Delta x_2^2 + \Delta x_3^2} \approx 72 \, \text{m} .$$

7. Essentially only those geomagnetic latitudes are affected where the geomagnetic cutoff exceeds the atmospheric cutoff. This concerns latitudes between 0° and approximately 50°. The average latitude can be worked out from

$$\langle \lambda \rangle = \frac{\int_{0°}^{50°} \lambda \cos^4 \lambda \, d\lambda}{\int_{0°}^{50°} \cos^4 \lambda \, d\lambda} \approx 18.4°$$

corresponding to an average geomagnetic cutoff of $\langle E \rangle \approx 12 \, \text{GeV}$. For zero field all particles with $E \geq 2 \, \text{GeV}$ have a chance to reach sea level. Their intensity is ($\varepsilon = E/\text{GeV}$)

$$N_1(> 2\,\text{GeV}) = a \int_2^\infty \varepsilon^{-2.7}\, d\varepsilon = -(a/1.7)\,\varepsilon^{-1.7}\Big|_2^\infty = 0.181\,a \ .$$

With full field on, only particles with $E > \langle E \rangle = 12\,\text{GeV}$ will reach sea level for latitudes between $0°$ and $50°$:

$$N_2(> 12\,\text{GeV}) \approx 0.0086\,a \ .$$

These results have to be combined with the surface of the Earth that is affected. Assuming isotropic incidence and zero field, the rate of cosmic rays at sea level is proportional to the surface of the Earth,

$$\Phi_1 = A\,N_1 = 4\pi\,R^2\,N_1 \ .$$

With full field on, only the part of the surface of Earth is affected for which $0° \le \lambda \le 50°$. The relevant surfaces can be calculated from elementary geometry yielding a flux

$$\Phi_2 = A(50°–90°)\,N_1 + A(0°–50°)\,N_2 \ ,$$

where

$$A(50°–90°) = 2\pi \left[(R \cos 50°)^2 + (R(1 - \sin 50°))^2 \right]$$

and

$$A(0°–50°) = 4\pi\,R^2 - A(50°–90°) \ .$$

This crude estimate leads to $\Phi_1/\Phi_2 = 3.7$, i.e., in periods of transition when the magnetic field went through zero, the radiation load due to cosmic rays was higher by about a factor of 4.

8. In principle, neutrons are excellent candidates. Because they are neutral, they would point back to the sources of cosmic rays. Their deflection by inhomogeneous magnetic fields is negligible. The only problem is their lifetime, $\tau_0 = 885.7$ seconds. At $10^{20}\,\text{eV}$ the Lorentz factor extends this lifetime considerably to

$$\tau = \gamma\,\tau_0 = \frac{10^{20}\,\text{eV}}{m_n c^2}\,\tau_0 = 9.4 \times 10^{13}\,\text{s} \,\hat{=}\, 2.99 \times 10^6\,\text{light-years} = 0.916\,\text{Mpc} \ .$$

Still the sources would have to be very near (compare $\lambda_{\gamma p} \approx 10\,\text{Mpc}$), and there is no evidence for point sources of this energy in the close vicinity of the Milky Way.

17.8 Chapter 8

1. Due to the relative motion of the galaxy away from the observer, the energy of the photon appears decreased. The energies of emitted photons and observed photons are related by the Lorentz transformation,

$$E_{em} = \gamma\,E_{obs} + \gamma\beta c p_{\|obs} = \gamma(1 + \beta)E_{obs} \ .$$

Since $E = h\nu = hc/\lambda$, one gets

$$\frac{1}{\lambda_{em}} = \frac{1}{\lambda_{obs}} \gamma (1+\beta) \quad \Rightarrow \quad \lambda_{obs} = \lambda_{em} \gamma (1+\beta) \,,$$

$$z = \frac{\lambda_{obs} - \lambda_{em}}{\lambda_{em}} = \gamma (1+\beta) - 1 = \frac{1}{\sqrt{1-\beta^2}} (1+\beta) - 1$$

$$= \frac{1+\beta}{\sqrt{(1+\beta)(1-\beta)}} - 1 = \frac{\sqrt{1+\beta}}{\sqrt{1-\beta}} - 1 \,,$$

which reduces for $\beta \ll 1$ to

$$z = \sqrt{(1+\beta)(1+\beta)} - 1 \approx \beta.$$

2. For an orbital motion in a gravitational potential one has

$$\frac{mv^2}{R} = G\frac{mM}{R^2} \,, \quad \Rightarrow \quad \frac{1}{2}mv^2 = \frac{1}{2}G\frac{mM}{R} \,,$$

i.e., $E_{kin} = \frac{1}{2}E_{pot}$. The kinetic energy of the gas cloud is simply

$$E_{kin} = \frac{3}{2}kT\frac{M}{\mu} \,.$$

The potential energy can be obtained by integration: the mass in a spherical subvolume of radius r is

$$m = \frac{4}{3}\pi r^3 \varrho \,.$$

A spherical shell surrounding this volume contains the mass

$$dm = 4\pi r^2 \varrho \, dr$$

leading to a potential energy of gravitation of

$$dE = \frac{Gm\,dm}{r} = \frac{G}{r}\frac{4}{3}\pi r^3 \varrho \, 4\pi r^2 \varrho \, dr = \frac{(4\pi)^2}{3}\varrho^2 r^4 G \, dr \,,$$

so that the potential energy is

$$E_{pot} = \int_0^R dE = \frac{(4\pi)^2}{3}\varrho^2 G \frac{R^5}{5} \,.$$

Using $M = \frac{4}{3}\pi R^3 \varrho$, one obtains

$$E_{pot} = \frac{3}{5}G\frac{M^2}{R} \,.$$

From the relation between E_{kin} and E_{pot} above one gets

$$2E_{\text{kin}} = 3kT\frac{M}{\mu} = \frac{3}{5}G\frac{M^2}{R} \quad \text{and} \quad R = \frac{1}{5}\frac{GM\mu}{kT} \ .$$

Then

$$M = \frac{4\pi}{3}R^3\varrho = \frac{4\pi}{3}\varrho\left(\frac{1}{5}\right)^3\frac{G^3M^3\mu^3}{(kT)^3}$$

and finally

$$M \approx \left(\frac{kT}{\mu G}\right)^{3/2}\frac{1}{\sqrt{\varrho}} \times 5.46 \ .$$

If $M_{\text{cloud}} > M$, the cloud gets unstable and collapses.

3. $\dfrac{mv^2}{R} \geq \dfrac{GMm}{R^2} \quad \Rightarrow \quad v^2 \geq \dfrac{GM}{R} \ .$

Since the density is constant, $M = \frac{4}{3}\pi\varrho\,R^3$, the revolution time will be

$$T_{\text{r}} = 2\pi R/v = 2\pi\frac{R\sqrt{R}}{\sqrt{GM}} = 2\pi\frac{R^{3/2}}{\sqrt{G}\sqrt{\varrho}\,R^{3/2}\sqrt{\frac{4\pi}{3}}} = \frac{\sqrt{3\pi}}{\sqrt{G\varrho}} \ ,$$

so that

$$v = \sqrt{\frac{4\pi}{3}}R\sqrt{G\varrho} \sim R$$

if $\varrho = \text{constant}$.

This behaviour is characteristic of the orbital velocities of stars in galaxies not too far away from the galactic center (see Fig. 1.17 or Fig. 13.3).

4. $\dfrac{\dot{R}^2}{R^2} + \dfrac{k}{R^2} = \dfrac{8\pi}{3}G\varrho\ , \quad \dot{R}^2 + k = \dfrac{8\pi}{3}G\varrho R^2\ .$

Differentiating this expression with respect to time gives

$$2\dot{R}\ddot{R} = \frac{8\pi}{3}G(\dot{\varrho}R^2 + 2R\dot{R}\varrho)\ ,$$

inserting $\dot{\varrho}$ from the fluid equation leads to

$$2\dot{R}\ddot{R} = \frac{8\pi}{3}G\left(2R\dot{R}\varrho - \frac{3\dot{R}}{R}(\varrho + P)R^2\right)\ .$$

This is equivalent to

$$\ddot{R} = \frac{4\pi}{3}G(2R\varrho - 3R\varrho - 3RP) \quad \text{or} \quad \frac{\ddot{R}}{R} = -\frac{4\pi}{3}G(\varrho + 3P)\ ,$$

which is the acceleration equation.

5. Photon mass $\quad m = \dfrac{h\nu}{c^2} = \dfrac{E}{c^2}$;

energy loss of a photon in a gravitational potential ΔU: $\Delta E = m\Delta U$;

reduced photon energy: $E' = E - \Delta E = h\nu - \dfrac{h\nu}{c^2}\Delta U \;\Rightarrow$

$$\nu' = \nu\left(1 - \dfrac{\Delta U}{c^2}\right), \quad \dfrac{\Delta \nu}{\nu} = \dfrac{\Delta U}{c^2} \;;$$

gravitational potential: $\quad \Delta U = \dfrac{GM}{R} \quad \Rightarrow$

$$\dfrac{\Delta \nu}{\nu} = \dfrac{GM}{Rc^2} \quad \Rightarrow \quad \dfrac{GM}{Rc^2} \quad \text{is dimensionless;}$$

using the Schwarzschild radius $\quad R_{\mathrm{S}} = \dfrac{2GM}{c^2} \quad$ one has $\quad \dfrac{\Delta \nu}{\nu} = \dfrac{R_{\mathrm{S}}}{2R}.$

This result is, however, only valid far away from the event horizon. The exact result from general relativity reads, see, e.g., [9]: $\Delta\nu/\nu = z = 1/\sqrt{1 - R_{\mathrm{S}}/R} - 1.$

6.

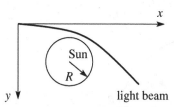

Acceleration due to gravity at the solar surface: $g = GM/R^2$.

Assumption: the deflection takes place essentially over the Sun's diameter $2R$. The photons travel on a parabola:

$$y = \dfrac{g}{2}t^2 , \; x = ct \;\Rightarrow\; y(x) = \dfrac{GM}{2R^2}\dfrac{x^2}{c^2} .$$

The deflection δ corresponds to the increase of $y(x)$ at $x = 2R$,

$$\dfrac{dy}{dx} = \dfrac{GM}{R^2c^2}x \;\Rightarrow\; y'(2R) = \dfrac{2GM}{Rc^2} = \dfrac{R_{\mathrm{S}}}{R} = \delta ,$$

$R = 6.961 \times 10^5 \,\mathrm{km}$, $M = 1.9884 \times 10^{30}\,\mathrm{kg} \;\Rightarrow\; \delta \approx 4.24 \times 10^{-6} \approx 0.87\,\mathrm{arcsec}$, 1 arcsec $= 1''$.

This is the classical result using Newton's theory. The general theory of relativity gives $\delta^* = 2\delta = 1.75\,\mathrm{arcsec}$.

The solution to this problem may alternatively be calculated following Problem 13.7, see also Problem 3.4.

7. An observer in empty space measures times with an atomic clock of frequency v_0. The signals emitted by an identical clock on the surface of the pulsar reach the observer in empty space with a frequency $v = v_0 - \Delta v$ where $\dfrac{\Delta v}{v_0} = \dfrac{\Delta U}{c^2} = \dfrac{GM}{Rc^2}$ (see Problem 5 in this chapter) $\Rightarrow$

$$\frac{f_{\text{pulsar}}}{f_{\text{empty space}}} = \frac{v_0 - \Delta v}{v_0} = 1 - \frac{GM}{Rc^2} = 1 - \varepsilon.$$

For the pulsar this gives $\varepsilon = 0.074$, i.e., the clocks in the gravitational potential on the surface of the pulsar are slow by 7.4%. For our Sun the relative slowing-down rate, e.g., with respect to Earth, is 2×10^{-6}. At the surface of the Earth clocks run slow by 1.06×10^{-8} with respect to clocks far away from any mass, from which just 7×10^{-10} results from the Earth's gravitation and the main contribution is caused by the gravitational potential of the Sun.

8. Gravitational force

$$dF = -G\frac{M(r)\, dm}{r^2} = -\frac{GM(r)}{r^2}\varrho(r)\, dr\, 4\pi r^2 , \quad \text{inward force}$$

$$dp = \frac{dF}{4\pi r^2} = -\frac{GM(r)}{r^2}\varrho(r)\, dr , \quad \frac{dp}{dr} = -\frac{GM(r)}{r^2}\varrho(r) . \tag{$*$}$$

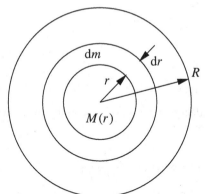

On the other hand $\dfrac{dp}{dr} \approx \dfrac{p(R) - p(r = 0)}{R} = -\dfrac{p}{R}$. Compare with $(*)$ and with the re-

placement $r \to R$ and $\varrho(r) \to$ average density ϱ one gets $\dfrac{p}{R} = \dfrac{GM}{R^2}\varrho$.

If a uniform density $\varrho(r) = \varrho$ is assumed, $(*)$ can be integrated directly using $M(r) = \dfrac{4\pi}{3}\varrho r^3$ and thus $\dfrac{dp}{dr} = -\dfrac{4\pi}{3}G\varrho^2 r$,

$$p(0) = \int_R^0 \frac{dp}{dr}\, dr = \frac{4\pi}{3}G\varrho^2 \int_0^R r\, dr = \frac{4\pi}{3}G\varrho^2 \frac{R^2}{2} = \frac{1}{2}\frac{GM}{R}\varrho .$$

This leads to $p = \dfrac{1}{2}\dfrac{GM}{R}\varrho \approx \begin{cases} 1.3 \times 10^{14}\,\text{N/m}^2 & \text{for the Sun} \\ 1.7 \times 10^{11}\,\text{N/m}^2 & \text{for the Earth} \end{cases}$.

9. Schwarzschild radius

$$\left.\begin{array}{c} R_S = \frac{2GM}{c^2} \\ M = \frac{4}{3}\pi R_S^3 \varrho \end{array}\right\} \quad \Rightarrow \quad \left(\frac{3}{4\pi\varrho}M\right)^{1/3} = \frac{2GM}{c^2} \quad \Rightarrow \quad \left(\frac{c^2}{2GM}\right)^3 \frac{3M}{4\pi} = \varrho \,.$$

$$\varrho = \frac{3c^6}{32\pi G^3 M^2} = \frac{3c^6}{32\pi G^3 M_\odot^2}\left(\frac{M_\odot}{M}\right)^2 \approx 1.8 \times 10^{19}\left(\frac{M_\odot}{M}\right)^2 \frac{\text{kg}}{\text{m}^3} \,,$$

a) $M \approx 10^{11} M_\odot \;\Rightarrow\; \varrho \approx 1.8 \times 10^{-3}\,\text{kg/m}^3$

b) $M = M_\odot \;\Rightarrow\; \varrho \approx 1.8 \times 10^{19}\,\text{kg/m}^3$

c) $M = 10^{15}\,\text{kg} \;\Rightarrow\; \varrho \approx 7.3 \times 10^{49}\,\text{kg/m}^3$

10.   a) $\dfrac{mv^2}{R} = \dfrac{GmM}{R^2} = \dfrac{Gm}{R^2}\dfrac{4}{3}\pi\varrho R^3 \quad\Rightarrow\quad v^2 = \dfrac{4}{3}\pi\varrho G R^2 \quad\Rightarrow$

$$v = R\sqrt{\frac{4}{3}\pi\varrho G} = 245\,\text{km/s} \quad \text{for } R = 20\,000 \text{ light-years} \,.$$

b) For $R > 20\,000$ light-years the orbital velocities show Keplerian characteristics,

$$\frac{mv^2}{R} \approx G\frac{mM}{R^2} \;\Rightarrow\; v^2 \approx G\frac{M}{R}\,, \quad M \approx v^2\frac{R}{G} = 1.7 \times 10^{41}\,\text{kg} = 8.6 \times 10^{10} M_\odot \,.$$

c) The energy density of photons in the universe is approximately $0.3\,\text{eV/cm}^3$. The critical density amounts to $\varrho_c = 0.945 \times 10^{-29}\,\text{g/cm}^3$, which corresponds to an energy density of $\varrho_c c^2 = 5.3 \times 10^3\,\text{eV/cm}^3$, which means $\varrho_{\text{photons}}/\varrho_c c^2 \approx 5 \times 10^{-5}$; i.e., photons contribute only a small fraction to the total Ω parameter.

17.9 Chapter 9

1. For a gas of non-relativistic particles the ideal gas law holds: $P = nT$. The density is given by $\varrho = nm$. Since $T \ll m$ for non-relativistic particles, one has $P \approx 0$. The fluid equation for $P = 0$ reads

$$\dot\varrho + \frac{3\dot R}{R}\varrho = 0 \quad\Rightarrow\quad \frac{1}{R^3}\frac{d}{dt}\left(\varrho R^3\right) = 0 \quad\Rightarrow\quad \varrho R^3 = \text{const} \quad\Rightarrow\quad \varrho \sim \frac{1}{R^3}\,.$$

2. Assume $P = 0$ which gives $\varrho \sim \frac{1}{R^3}$ (see Problem 9.1). The last relation can be parameterized by

$$\varrho = \varrho_0\left(\frac{R_0}{R}\right)^3 \,.$$

The Friedmann equation can be approximated:

$$\frac{\dot R^2}{R^2} + \frac{k}{R^2} = \frac{8\pi}{3}G\varrho \quad\Rightarrow\quad \frac{\dot R^2}{R^2} = \frac{8\pi}{3}G\varrho \,,$$

since the second term on the left-hand side in the Friedmann equation is small compared to ϱ ($\sim \frac{1}{R^3}$) for the early universe.

With the ansatz

$$R = A t^p , \quad \dot{R} = p A t^{p-1}$$

one gets

$$\frac{\dot{R}^2}{R^2} = \frac{p^2 A^2 t^{2p} t^{-2}}{A^2 t^{2p}} = \frac{p^2}{t^2} = \frac{8\pi}{3} G \varrho_0 \left(\frac{R_0}{R}\right)^3 = \frac{8\pi}{3} G \varrho_0 R_0^3 A^{-3} t^{-3p} .$$

Comparing the t dependence on the right- and left-hand sides gives

$$p = \frac{2}{3} \quad \Rightarrow \quad R = A t^{2/3} .$$

In this procedure the constant of proportionality A is automatically fixed as

$$A = \sqrt[3]{6\pi G \varrho_0} R_0 .$$

3. For the early universe one can approximate the Friedmann equation by

$$\frac{\dot{R}^2}{R^2} = \frac{8\pi}{3} G \varrho ; \quad \text{with} \quad \varrho = \frac{\pi^2}{30} g_* T^4 \quad \Rightarrow \quad \frac{\dot{R}^2}{R^2} = \frac{8\pi}{3} G \frac{\pi^2}{30} g_* T^4 = \frac{8\pi^3}{90} G g_* T^4 .$$

Since $\dfrac{\dot{R}}{R} = H$ and $G = \dfrac{1}{m_{Pl}^2}$ one has

$$H = \sqrt{\frac{8\pi^3 g_*}{90} \frac{T^2}{m_{Pl}}} .$$

4. $[G] = \text{m}^3 \, \text{s}^{-2} \, \text{kg}^{-1}$, $[c] = \text{m s}^{-1}$, $[\hbar] = \text{J s} = \text{kg m}^2 \, \text{s}^{-1}$.

$$\text{Try} \left[\frac{\hbar G}{c^3}\right] = \frac{\text{kg m}^2 \, \text{s}^{-1} \, \text{m}^3 \, \text{s}^{-2} \, \text{kg}^{-1}}{\text{m}^3 \, \text{s}^{-3}} = \text{m}^2 .$$

Therefore, $\sqrt{\dfrac{\hbar G}{c^3}}$ has the dimension of a length, and this is the Planck length.

5. The escape velocity from a massive object can be worked out from

$$\frac{1}{2} m v^2 = G \frac{m M}{R} ,$$

where m is the mass of the escaping particle, and M and R are the mass and radius of the massive object. The mass of the escaping object does not enter into the escape velocity. If the escape velocity is equal to the velocity of light, even light cannot escape from the object. If v is replaced by c this leads to

$$c^2 = \frac{2 G M}{R} \quad \text{or} \quad R = \frac{2 G M}{c^2} ,$$

which is the Schwarzschild radius. The result of this classical treatment of the problem (even though classical physics does not apply in this situation) accidentally agrees with the outcome of the correct derivation based on general relativity.

By definition the Schwarzschild radius is $R_S = \dfrac{2GM}{c^2}$ and the calculation gives

$$R_S = \begin{cases} 8.9 \text{ mm for Earth} \\ 2.95 \text{ km for the Sun} \end{cases} .$$

6. $\dfrac{GM^2}{R} > \dfrac{3}{2}kT\dfrac{M}{\mu}$, μ – mass of a hydrogen atom,

$$M > \frac{3}{2}\frac{kT}{\mu G} \qquad R = \frac{3}{2}\frac{kT}{\mu G}\left(\frac{3M}{4\pi\varrho}\right)^{1/3} ;$$

since $M = \dfrac{4}{3}\pi\varrho R^3$: $M^3 > \left(\dfrac{3}{2}\dfrac{kT}{\mu G}\right)^3\dfrac{3M}{4\pi\varrho}$,

$$\varrho > \frac{3}{4\pi}\frac{1}{M^2}\left(\frac{3}{2}\frac{kT}{\mu G}\right)^3 \approx 3.9 \times 10^{-10} \text{ kg/m}^3$$

$(k = 1.38 \times 10^{-23} \text{ J/K}, M = 2 \times 10^{30} \text{ kg}, \mu = 1.67 \times 10^{-27} \text{ kg})$.

7.   star

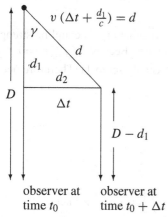

$$v\left(\Delta t + \tfrac{d_1}{c}\right) = d$$

observer at   observer at
time t_0 time $t_0 + \Delta t$

The first observation is made at time t_0. The star is at a distance D from the observer. It moves at an angle γ to the line of sight. During Δt it has moved $d_1 = d\cos\gamma$ closer to the observer. The first light beam has further to travel. The light was emitted $\Delta t + \frac{d_1}{c}$ earlier compared to the second measurement. The apparent velocity is

$$v^* = \frac{d_2}{\Delta t} = \frac{d\sin\gamma}{\Delta t} = \frac{v\left(\Delta t + \frac{d_1}{c}\right)\sin\gamma}{\Delta t} = v\left(1 + \frac{d_1}{c\Delta t}\right)\sin\gamma$$

$$= v\left(1 + \frac{d_2}{c\Delta t}\cot\gamma\right)\sin\gamma = v\left(1 + \frac{v^*}{c}\cot\gamma\right)\sin\gamma = v\sin\gamma + \frac{v}{c}v^*\cos\gamma .$$

Solve for v^*: $v^* \left(1 - \dfrac{v}{c}\cos\gamma\right) = v\sin\gamma$, $v^* = \dfrac{v\sin\gamma}{1 - \frac{v}{c}\cos\gamma}$; if v approaches c:

$$v^* = c\,\frac{\sin\gamma}{1 - \cos\gamma} = c\,\frac{2\sin\frac{\gamma}{2}\cos\frac{\gamma}{2}}{2\sin^2\frac{\gamma}{2}} = c\cot\frac{\gamma}{2} > c \quad \text{for} \quad 0 < \gamma < \frac{\pi}{2}.$$

17.10 Chapter 10

1. The number of degrees of freedom is

$$g_* = 2 + \frac{7}{8}(4 + 2N_\nu) \, ;$$

for 3 neutrino generations this gives $g_* = 10.75$. Equation (10.2) related the time to the temperature and the Planck mass,

$$t = \frac{0.301\,m_{\mathrm{Pl}}}{\sqrt{g_*}\,T^2} \quad \Rightarrow \quad tT^2 = 0.301\,\frac{m_{\mathrm{Pl}}}{\sqrt{g_*}} \, .$$

To get a number, one has to obtain a dimension time $\times$ energy^2 for tT^2. Therefore, a factor $\hbar$ is missing:

$$tT^2 = 0.301 \times 6.582 \times 10^{-22}\,\mathrm{MeV\,s} \times 1.22 \times 10^{22}\,\mathrm{MeV} \times \frac{1}{\sqrt{g_*}} \approx 0.74\,\mathrm{s\,MeV^2} \, .$$

2. The process $\nu_e n \to e^- p$ is governed by the weak interaction. Its coupling strength is given by $G_F = 10^{-5}/m_p^2$, where m_p is the proton mass. Because of $\lambda/2\pi = \hbar/p$, the length can be measured in units of energy^{-1} ($\hbar$ is usually set to 1). Therefore one gets from

$$\sigma(\nu_e n \to e^- p) \sim G_F^2\,f(s) \, ,$$

knowing that $G_F^2 \sim \text{energy}^{-4}$:

$$f(s) \sim s \, ,$$

$$\sigma(\nu_e n \to e^- p)\left\{\text{length}^2 \,\widehat{=}\, \text{energy}^{-2}\right\} = \left(G_F^2\,s\right)\left\{\text{energy}^{-4} \times \text{energy}^2\right\} \, .$$

Since for relativistic particles $v \approx c$ and $s = (\frac{3}{2}kT)^2 \sim T^2$, one gets

$$\sigma \sim G_F^2 T^2 \, .$$

3. The hadronic cross section $\sigma_{\mathrm{hadr}}(e^+ e^- \to Z \to \text{hadrons})$ at the Z resonance can be described by a Breit–Wigner distribution

$$\sigma_{\mathrm{hadr}} = \sigma_{\mathrm{hadr}}^0\,\frac{s\Gamma_Z^2}{(s - M_Z^2)^2 + s^2\Gamma_Z^2/M_Z^2}\,(1 + \delta_{\mathrm{rad}}(s)) \, ,$$

where

σ^0_{hadr}	–	peak cross section,
$\sqrt{s}$	–	center-of-mass energy,
Γ_Z	–	total width of the Z,
M_Z	–	mass of the Z,
δ_{rad}	–	radiative correction.

The peak cross section σ^0_{hadr} for the process $e^+e^- \to Z \to$ hadrons is given by

$$\sigma^0_{hadr} = \frac{12\pi}{M_Z^2} \frac{\Gamma_{e^+e^-} \Gamma_{hadr}}{\Gamma_Z^2} ,$$

where $\Gamma_{e^+e^-}$ is the partial e^+e^- width of the Z and Γ_{hadr} the partial width for $Z \to$ hadrons.

The total Z width can be written as

$$\Gamma_Z = \Gamma_{hadr} + \Gamma_{e^+e^-} + \Gamma_{\mu^+\mu^-} + \Gamma_{\tau^+\tau^-} + \Gamma_{inv} ,$$

where Γ_{inv} describes the invisible decay of the Z into neutrino pairs ($\nu_e \bar{\nu}_e$, $\nu_\mu \bar{\nu}_\mu$, $\nu_\tau \bar{\nu}_\tau$, and possibly other neutrino pairs from a suspected fourth neutrino family).

Lepton universality is well established. Therefore,

$$\Gamma_Z = \Gamma_{hadr} + 3\Gamma_{\ell\bar{\ell}} + \Gamma_{inv} .$$

$$\frac{\Gamma_{inv}}{\Gamma_{\ell\bar{\ell}}} = \frac{\Gamma_Z}{\Gamma_{\ell\bar{\ell}}} - 3 - \frac{\Gamma_{hadr}}{\Gamma_{\ell\bar{\ell}}} = \frac{\Gamma_Z}{\Gamma_{\ell\bar{\ell}}} - 3 - R , \qquad (*)$$

where R is the usual ratio

$$R = \frac{\sigma(e^+e^- \to \text{hadrons})}{\sigma(e^+e^- \to \ell^+\ell^-)} .$$

The experiment provides σ^0_{hadr}, Γ_Z, and M_Z.

The measurement of the peak cross section σ^0_{hadr} and Γ_Z together with M_Z fixes the product $\Gamma_{\ell\bar{\ell}} \Gamma_{hadr}$.

The measurement of the peak cross section for the process $e^+e^- \to Z \to \mu^+\mu^-$,

$$\sigma^0_{\mu\mu} = \frac{12\pi}{M_Z^2} \frac{\Gamma_{e^+e^-} \Gamma_{\mu^+\mu^-}}{\Gamma_Z^2} ,$$

determines $\Gamma_{\ell\bar{\ell}}$, if lepton universality is assumed. Therefore, also Γ_{hadr} is now known, so that Γ_{inv} can be obtained from $(*)$.

In the framework of the Standard Model the width of

$$Z \to \nu_x \bar{\nu}_x$$

is calculated to be $\approx 170\,\text{MeV}$. The LEP experiments resulted in

$$\Gamma_{inv} \approx 500\,\text{MeV}$$

indicating very clearly that there are only $500/170 \approx 3$ neutrino generations.

17.11 Chapter 11

1. This probability can be worked out from

$$\phi = \sigma_{Th}\, N\, d\,,$$

where σ_{Th} is the Thomson cross section (665 mb), N the number of target atoms per cm^3, and d the distance traveled.

Since the universe is flat, one has $\varrho = \varrho_{crit}$. Because of $\varrho_{crit} = 9.47 \times 10^{-30}\, g/cm^3$ and $m_H = 1.67 \times 10^{-24}\, g$, and the fact that only 4% of the total matter density is in the form of baryonic matter, one has $N = 2.27 \times 10^{-7}\, cm^{-3}$. The distance travelled from the surface of last scattering corresponds to the age of the universe (14 billion years),

$$d = 14 \times 10^9 \times 3.156 \times 10^7\, s \times 2.998 \times 10^{10}\, cm/s = 1.32 \times 10^{28}\, cm$$

resulting in

$$\phi = 1.99 \times 10^{-3} \approx 0.2\%\,.$$

2. The critical density $\varrho_c = 3H^2/8\pi G$ as obtained from the Friedmann equation has to be modified if the effect of the cosmological constant is taken into account:

$$\varrho_{c,\Lambda} = \frac{3H^2 - \Lambda c^2}{8\pi G}\,.$$

Since $\varrho_{c,\Lambda}$ cannot be negative, one can derive a limit from this equation:

$$3H^2 - \Lambda c^2 > 0\,,\quad \Lambda < \frac{3H^2}{c^2} = 1.766 \times 10^{-56}\, cm^{-2}\,.$$

This leads to an energy density

$$\varrho \leq \frac{c^4}{8\pi G}\Lambda = 8.51 \times 10^{-10}\, J/m^3 = 5.3\, GeV/m^3 = 5.3\, keV/cm^3\,.$$

3. $p_1 = -\dfrac{dE_{class}}{dV} = -\dfrac{d}{dV}\left(-\dfrac{GmM}{R}\right) = GmM\dfrac{d}{dV}\left(\dfrac{1}{V^{1/3}}\right)\left(\dfrac{4\pi}{3}\right)^{1/3}$

$\sim \dfrac{d}{dV}\left(\dfrac{1}{V^{1/3}}\right) = -\dfrac{1}{3}V^{-4/3} \sim -\dfrac{1}{R^4}\,,$

$p_2 = -\dfrac{dE_\Lambda}{dV} \sim -\dfrac{d}{dV}(-\Lambda R^2) \sim \dfrac{d}{dV}V^{2/3} = \dfrac{2}{3}V^{-1/3} \sim +\dfrac{1}{R}\,.$

p_1 is an inward pressure due to the normal gravitational pull, whereas p_2 is an outward pressure representing a repulsive gravity.

4. Planck distribution

$$\varrho(v, T) = \frac{8\pi h}{c^3} v^3 \frac{1}{e^{hv/kT} - 1} \,, \quad \langle v \rangle = \frac{\int_0^\infty v \, \varrho(v, T) \, dv}{\int_0^\infty \varrho(v, T) \, dv} \,;$$

substitution $\frac{hv}{kT} = x$:

$$\int_0^\infty \varrho(v, T) \, dv = \frac{8\pi h}{c^3} \left(\frac{kT}{h}\right)^3 \int_0^\infty \frac{x^3}{e^x - 1} \frac{kT}{h} \, dx$$

$$= \frac{8\pi h}{c^3} \left(\frac{kT}{h}\right)^4 \int_0^\infty \frac{x^3}{e^x - 1} \, dx \,,$$

$$\int_0^\infty v \, \varrho(v, T) \, dv = \frac{8\pi h}{c^3} \left(\frac{kT}{h}\right)^5 \int_0^\infty \frac{x^4}{e^x - 1} \, dx \,;$$

$$\int_0^\infty \frac{x^3}{e^x - 1} \, dx = 3! \, \zeta(4) \,, \quad \zeta(4) = \pi^4/90 \,, \quad \zeta - \text{Riemann's zeta function,}$$

$$\int_0^\infty \frac{x^4}{e^x - 1} \, dx = 4! \, \zeta(5) \,, \quad \zeta(5) = 1.036\,927\,7551\ldots \,;$$

$$\langle hv \rangle = \frac{h \frac{8\pi h}{c^3} \left(\frac{kT}{h}\right)^5 \times 4! \, \zeta(5)}{\frac{8\pi h}{c^3} \left(\frac{kT}{h}\right)^4 \times 3! \, \zeta(4)} = h \frac{kT}{h} \times 4 \times \frac{\zeta(5)}{\zeta(4)}$$

$$= kT \times 4 \times \frac{\zeta(5)}{\pi^4} \times 90 = \frac{360}{\pi^4} kT \, \zeta(5) \approx 900\,\mu\text{eV} \,.$$

5. a) At present: number density of bb photons: $410/\text{cm}^3$, average energy $\langle E \rangle = 900\,\mu\text{eV}$ (from the previous problem). This leads to a first estimate of the present energy density of

$$\varrho_0 \approx 0.37\,\text{eV}/\text{cm}^3 \,.$$

Since, however, the temperature enters with the fourth power into the energy density one has to use

$$\varrho_0 = \frac{\pi^2}{15} T^4$$

for photons (number of degrees of freedom $g = 2$, see (9.10)). One has to include the adequate factors of k and $\hbar c$ to get the correct numerical result:

$$\varrho_0 = \frac{\pi^2}{15} T^4 \frac{k^4}{(\hbar c)^3} \approx 0.26\,\text{eV}/\text{cm}^3$$

$(k = 8.617 \times 10^{-5}\,\text{eV}\,\text{K}^{-1}, \hbar c = 0.197\,\text{GeV}\,\text{fm}).$

b) At last scattering ($t = 380\,000$ a, temperature at the time of last scattering: $T_{\text{dec}} = 0.3$ eV, see Chap. 11):

$$T_{\text{dec}} = 0.3\,\text{eV}/k \approx 3500\,\text{K} ,$$

$$\varrho_{\text{dec}} = \varrho_0 \left(\frac{T_{\text{dec}}}{T_0}\right)^4 \approx 0.26\,\text{eV/cm}^3 \times \left(\frac{3500}{2.725}\right)^4 \approx 0.7\,\text{TeV/cm}^3 .$$

6. Naïvely one would expect the fraction of neutral hydrogen to become significant when the temperature drops below 13.6 eV. But this happens only at much lower temperatures because there are so many more photons than baryons, and the photon energy distribution, i.e., the Planck distribution, has a long tail towards high energies. The baryon-to-photon ratio, $\eta \approx 5 \times 10^{-10}$, is extremely small. Therefore, the temperature must be significantly lower than this before the number of photons with $E > 13.6\,\text{eV}$ is comparable to the number of baryons. Furthermore, interaction or ionization can take place in several steps via excited states of the hydrogen atom, the H_2 molecule, or the H_2^+ ion. One finds that the numbers of neutral and ionized atoms become equal at a *recombination temperature* of $T_{\text{rec}} \approx 0.3\,\text{eV}$ (3500 K). At this point the universe transforms from an ionized plasma to an essentially neutral gas of hydrogen and helium.

17.12 Chapter 12

1. The Friedmann equation for $k = 0$, corresponding to the dominance of Λ, reads, see also Problem 11.2,

$$H^2 = \frac{8\pi G}{3}(\varrho + \varrho_{\text{v}}) .$$

With

$$\Lambda = \frac{8\pi G}{c^2}\varrho_{\text{v}} \quad \Rightarrow \quad H^2 - \frac{1}{3}\Lambda c^2 = \frac{8\pi G}{3}\varrho .$$

Since $\varrho > 0$ one has the inequality

$$H^2 - \frac{1}{3}\Lambda c^2 \geq 0 \quad \text{or} \quad \Lambda \leq \frac{3H^2}{c^2} \approx 2 \times 10^{-56}\,\text{cm}^{-2} .$$

This is just a reflection of the fact that in the visible universe there is no obvious effect of the curvature of space. The size of the visible flat universe being 10^{28} cm can be converted into

$$\Lambda_1 \leq 10^{-56}\,\text{cm}^{-2} .$$

If one assumes on the other hand that Einstein's theory of relativity is valid down to the Planck scale, then one would expect

$$\Lambda_2 \approx (\ell_{\text{Pl}}^2)^{-1} \approx 10^{66}\,\text{cm}^{-2} .$$

The difference between the two estimates is 122 orders of magnitude.

The related mass densities for the two scenarios are estimated to be

$$\varrho_v^1 = \frac{\Lambda_1 c^2}{8\pi G} \approx \frac{10^{-52}\,\mathrm{m}^{-2} \times 9 \times 10^{16}\,\mathrm{m}^2/\mathrm{s}^2\,\mathrm{kg}\,\mathrm{s}^2}{8\pi \times 6.67 \times 10^{-11}\,\mathrm{m}^3}$$

$$\approx 5.37 \times 10^{-27}\,\mathrm{kg/m}^3 = 5.37 \times 10^{-30}\,\mathrm{g/cm}^3\ ,$$

$$\varrho_v^2 = \frac{\Lambda_2 c^2}{8\pi G} \approx 5.37 \times 10^{95}\,\mathrm{kg/m}^3 = 5.37 \times 10^{92}\,\mathrm{g/cm}^3\ .$$

2. Starting from

$$H^2 = \frac{8\pi G}{3}(\varrho + \varrho_v) \quad \Rightarrow \quad H^2 - \frac{1}{3}\Lambda c^2 = \frac{8\pi G}{3}\varrho$$

and since $H = \dot{R}/R$ one obtains (for $\varrho = \mathrm{const}$)

$$\frac{\dot{R}}{R} = \sqrt{\frac{1}{3}\Lambda c^2 + \frac{8\pi G}{3}\varrho} \quad \Rightarrow \quad R = R_i \exp\left(\sqrt{\frac{1}{3}\Lambda c^2 + \frac{8\pi G}{3}\varrho}\ t\right)\ ,$$

which represents an expanding universe.

3. The Friedmann equation extended by the Λ term for a flat universe is

$$\frac{\dot{R}^2}{R^2} - \frac{1}{3}\Lambda c^2 = \frac{8\pi G}{3}\varrho\ .$$

For $\frac{1}{3}\Lambda c^2 \gg \frac{8\pi G}{3}\varrho$ this equation simplifies to

$$\frac{\dot{R}}{R} = \sqrt{\frac{1}{3}\Lambda c^2} = \sqrt{\frac{1}{3}\Lambda_0 c^2(1 + \alpha t)} = a\sqrt{1 + \alpha t} \quad \text{with} \quad a = \sqrt{\frac{1}{3}\Lambda_0 c^2}\ ,$$

$$\ln R = \int a\sqrt{1 + \alpha t}\,dt + \mathrm{const} = a\frac{2}{3\,\alpha}(1 + \alpha t)^{3/2} + \mathrm{const}\ ,$$

$$R = \exp\left(\frac{2a}{3\alpha}(1 + \alpha t)^{3/2} + \mathrm{const}\right)\ ;$$

boundary condition $R(t = 0) = R_i$

$$R(t = 0) = \exp\left(\frac{2a}{3\alpha} + \mathrm{const}\right) = R_i \quad \Rightarrow$$

$$\frac{2a}{3\alpha} + \mathrm{const} = \ln R_i \quad \Rightarrow \quad \mathrm{const} = \ln R_i - \frac{2a}{3\alpha}\ ,$$

$$R = R_i \exp\left(\frac{2a}{3\alpha}(1 + \alpha t)^{3/2} - \frac{2a}{3\alpha}\right) = R_i \exp\left(\frac{2a}{3\alpha}[(1 + \alpha t)^{3/2} - 1]\right)$$

$$\text{with} \quad a = \sqrt{\frac{1}{3}\Lambda_0 c^2}\ .$$

For large t this result shows a dependence like

$$R \sim \exp\left(\beta t^{3/2}\right) \quad \text{with} \quad \beta = \frac{2a\sqrt{\alpha}}{3}\ .$$

4. $\dfrac{R_0}{R_{mr}} = \left(\dfrac{t_0}{t_{mr}}\right)^{\frac{2}{3}}$,

where R_{mr} is the size of the universe at the time of matter–radiation equality $t_{mr} = 50\,000$ a. With $t_0 = 14$ billion years, one has

$$\frac{R_0}{R_{mr}} \approx 4280 \ .$$

The size of the universe at that time was

$$R_{mr} \approx \frac{R_0}{4280} \approx 3.27 \times 10^6 \ \text{light-years} \approx 3.09 \times 10^{22}\, \text{m} \ .$$

Extrapolating to earlier times requires to assume $R \sim \sqrt{t}$, since for times $t < t_{mr}$ the universe was radiation dominated:

$$\frac{R_{mr}}{R_{infl}} = \left(\frac{t_{mr}}{t_{infl}}\right)^{\frac{1}{2}} \approx \left(\frac{50\,000\, \text{a}}{10^{-36}\, \text{s}}\right)^{\frac{1}{2}} \approx 1.26 \times 10^{24} \ .$$

Correspondingly, the size of the universe at the end of inflation was

$$R_{infl} \approx \frac{R_{mr}}{1.26 \times 10^{24}} \approx 2.45\, \text{cm} \ .$$

If inflation had ended at 10^{-32} s, one would have obtained $R_{infl} \approx 2.45$ m.

17.13 Chapter 13

1. $\dfrac{mv^2}{r} = G\dfrac{mM}{r^2} \quad \Rightarrow \quad M = \dfrac{v^2 r}{G}$,

$$M \approx \frac{(1.10 \times 10^5\, \text{m/s})^2 \times 2.5 \times 3.086 \times 10^{16}\, \text{m}}{6.67 \times 10^{-11}\, \text{m}^3\, \text{s}^{-2}}\ \text{kg} \approx 1.4 \times 10^{37}\, \text{kg} \approx 7 \times 10^6\, M_\odot \ .$$

2. Non-relativistic calculation: $p_W - p'_W = p_e$;

$$\frac{p_W^2}{2m_W} = \frac{p_W'^2}{2m_W} + \frac{(p_W - p'_W)^2}{2m_e} = \frac{p_W'^2}{2m_W} + \frac{p_W^2}{2m_e} + \frac{p_W'^2}{2m_e} - \frac{p_W \cdot p'_W}{m_e} \ ,$$

$$p_W'^2 \left(\frac{1}{m_W} + \frac{1}{m_e}\right) - \frac{2p_W \cdot p'_W}{m_e} = p_W^2 \left(\frac{1}{m_W} - \frac{1}{m_e}\right) \ ,$$

$$p_W'^2 - 2\frac{m_W}{m_W + m_e} p_W \cdot p'_W = p_W^2 \frac{m_e - m_W}{m_W + m_e} \ ,$$

$$p_W'^2 - 2\frac{m_W}{m_W + m_e} p_W \cdot p'_W + \left(\frac{m_W}{m_W + m_e}\right)^2 p_W^2 = p_W^2 \left(\frac{m_e - m_W}{m_W + m_e} + \frac{m_W^2}{(m_W + m_e)^2}\right) \ ,$$

$$\left(p'_W - \frac{m_W}{m_W + m_e} p_W\right)^2 = p_W^2 \frac{m_e^2}{(m_W + m_e)^2} \ .$$

Central collision

$$p'_W - \frac{m_W}{m_W + m_e} p_W = \pm p_W \frac{m_e}{m_W + m_e} \quad \Rightarrow \quad p'_W = \frac{m_W - m_e}{m_W + m_e} p_W \ ,$$

since only the negative sign is physically meaningful.

$$\Delta E = \frac{1}{2} m_W (v_1^2 - v_1'^2) = E_W^{kin}\left(1 - \frac{v_1'^2}{v_1^2}\right) = E_W^{kin}\left(1 - \left(\frac{m_W - m_e}{m_W + m_e}\right)^2\right)$$

$$\approx 1\,\mathrm{GeV} \times 2 \times 10^{-5} = 20\,\mathrm{keV} \ .$$

3. Classical Fermi gas of neutrinos:

$$E_F = \frac{p^2}{2m_\nu} = \frac{\hbar^2 k^2}{2m_\nu} \ , \quad k - \text{wave vector} \ .$$

In a quantized Fermi gas there is one k vector per $2\pi/L$ if one assumes that the neutrino gas is contained in a cube of side L. Number of states (at $T = 0$) for 2 spin states:

$$N = 2 \frac{\frac{4}{3}\pi k_F^3}{(2\pi/L)^3} = V \frac{1}{3\pi^2} k_F^3 \quad \Rightarrow \quad k_F = (3\pi^2 n)^{1/3}$$

with $n = N/V$ particle density,

$$\Rightarrow E_F = \frac{\hbar^2}{2m}(3\pi^2 n)^{2/3} \ .$$

Relativistic Fermi gas:

$$E_F = p_F c = \hbar k_F c = \hbar c (3\pi^2 n)^{1/3}$$

assuming relativistic neutrinos one would get, e.g., for 300 neutrinos per cm^3.
$\hbar c = 197.3\,\mathrm{MeV\,fm}$,

$$E_F = \hbar c \,(3\pi^2 \times 300)^{1/3}\,\mathrm{cm}^{-1} = 197.3 \times 10^6\,\mathrm{eV} \times 10^{-13}\,\mathrm{cm} \times 20.71\,\mathrm{cm}^{-1}$$
$$\approx 409\,\mu\mathrm{eV} \ .$$

4. Since $v \ll c$ is expected, one can use classical kinematics:

$$\frac{1}{2}mv^2 = G\frac{mM}{R} \quad \Rightarrow \quad v = \sqrt{\frac{2GM}{R}} \quad \Rightarrow$$

$$v \approx \sqrt{\frac{2 \times 6.67 \times 10^{-11}\,\mathrm{m^3\,kg^{-1}\,s^{-2}} \times 10^{10} \times 2 \times 10^{30}\,\mathrm{kg}}{3 \times 10^3 \times 3.086 \times 10^{16}\,\mathrm{m}}} \approx 1.7 \times 10^5\,\mathrm{m/s} \ ,$$

$$\beta \approx 5.66 \times 10^{-4} \ .$$

The axion mass does not enter the calculation, i.e., the escape velocity does not depend on the mass.

A relativistic treatment, $(\gamma - 1)m_0 c^2 = G\gamma m_0 M/R$, leads to a similar result:

$$v = \beta c = \sqrt{\frac{2GM}{R}}\sqrt{1 - \frac{GM}{2Rc^2}} \ .$$

Since this expression is not based on general relativity, it is of limited use.

5. Spherically symmetric mass distribution.

  a) Mass inside a sphere: $M(r) = \int_0^r \varrho(r')\,dV' = 4\pi \int_0^r \varrho(r')r'^2\,dr'$;

  potential energy of mass m: $E_{\mathrm{pot}}^{(m)}(r) = Gm \int_\infty^r \frac{M(r')}{r'^2}\,dr'$,

  verification: $-\dfrac{\partial E_{\mathrm{pot}}^{(m)}(r)}{\partial r} = F = -G\dfrac{mM(r)}{r^2}$.

  b) Potential energy of mass shell $dM = M'(r)\,dr$: $dE_{\mathrm{pot}} = GM'(r)\,dr \int_\infty^r \frac{M(r')}{r'^2}\,dr'$;

  total potential energy: $E_{\mathrm{pot}} = G \int_0^\infty M'(r)\,dr \int_\infty^r \frac{M(r')}{r'^2}\,dr'$;

  integration by parts: $E_{\mathrm{pot}} = \left. GM(r) \int_\infty^r \frac{M(r')}{r'^2}\,dr' \right|_{r=0}^\infty - G \int_0^\infty \frac{M^2(r)}{r^2}\,dr$;

  margin term vanishes for finite masses: $E_{\mathrm{pot}} = -G \int_0^\infty \frac{M^2(r)}{r^2}\,dr$.

  c) Mass M on a shell, i.e., $M(r < R) = 0$ and $M(r > R) = M$:

$$E_{\mathrm{pot}}^{(\mathrm{shell})} = -GM^2 \int_R^\infty \frac{1}{r^2}\,dr = \left. G\frac{M^2}{r} \right|_{r=R}^\infty = -G\frac{M^2}{R} \ .$$

  d) Homogeneous mass distribution: $M(r < R) = Mr^3/R^3$ and $M(r > R) = M$:

$$E_{\mathrm{pot}}^{(\mathrm{hom})} = -G\frac{M^2}{R^6} \int_0^R r^4\,dr + E_{\mathrm{pot}}^{(\mathrm{shell})} = \left. -G\frac{M^2}{R^6}\frac{r^5}{5} \right|_{r=0}^R - G\frac{M^2}{R} = -\frac{6}{5}G\frac{M^2}{R} \ .$$

6. $n_{\max} m_\nu = \dfrac{m_\nu^4 v_{\mathrm{f}}^3}{3\pi^2 \hbar^3}$.   (*)

If $n_{\max} m_\nu$ is assumed to be on the order of a typical dark-matter density ($\Omega = 0.3$) one can solve (*) for m_ν,

$$m_\nu = \left(\frac{3\pi^2 \hbar^3\, n_{\max} m_\nu}{v_{\mathrm{f}}^3} \right)^{1/4} \ .$$

The mass of a typical galaxy can be estimated as $M = 10^{11}\,M_{\odot} = 2 \times 10^{41}$ kg. With an assumed radius of $r = 20$ kpc one gets

$$v_f = 2.11 \times 10^5\,\text{m/s}\ .$$

With $n_{\max} m_\nu = 0.3\,\varrho_c \approx 3 \times 10^{-30}\,\text{g/cm}^3$ one obtains

$$m_\nu \geq 1\,\text{eV}\ .$$

7.

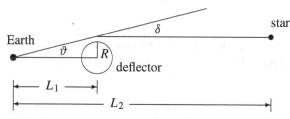

$\delta = \dfrac{2R_S}{R}$ as mentioned earlier (see Problem 8.6); $R_S = \dfrac{2GM_d}{c^2}$;

for $L_2 \gg L_1$: $\delta \approx \vartheta = \dfrac{R}{L_1} \Rightarrow R = L_1\delta$,

$$\delta = \frac{2R_S}{R} = \frac{2R_S}{L_1\delta} \Rightarrow \delta = \sqrt{\frac{2R_S}{L_1}} = \sqrt{\frac{4GM_d}{L_1c^2}} \sim \sqrt{M_d}\ .$$

Ring radius $R_E = L_1\delta = \sqrt{2L_1R_S}$, $L_1 = 10$ kpc $\approx 10 \times 10^3 \times 3.26 \times 3.15 \times 10^7\,\text{s} \times 3 \times 10^8\,\text{m}\,\text{s}^{-1} \approx 3.08 \times 10^{20}$ m: $R_E \approx \sqrt{2 \times 3.08 \times 10^{20}\,\text{m} \times 3000\,\text{m}} \approx 1.36 \times 10^{12}\,\text{m}$;

opening angle for the Einstein ring: $\gamma = 2\delta = \dfrac{2R_E}{L_1} = 8.8 \times 10^{-9} \cong 0.0018$ arcsec

$\Rightarrow$ too small to be observable $\Rightarrow$ only brightness excursion visible.

17.14 Chapter 14

1. The weight of a human is proportional to its volume, which in turn is proportional to the cube of its size;

$$W = W_0\,R^2\,20\,R\ ,$$

if one assumes that the 'height' of a human is 20 times its 'radius'. The strength of a human is proportional to its cross section

$$S = S_0\,R^2\ .$$

For a mass of 100 kg, an assumed radius of 10 cm, a human has to carry its own weight plus, maybe, an additional 100 kg,

$$W_0 = \frac{100\,\text{kg}}{20\,R^3} = 5000\,\frac{\text{kg}}{\text{m}^3} \ , \quad S_0 = \frac{200\,\text{kg}}{R^2} = 20\,000\,\frac{\text{kg}}{\text{m}^2} \ .$$

From this the stability limit (weight $\widehat{=}$ strength) can be derived,

$$5000\,\frac{\text{kg}}{\text{m}^3} \times 20\,R^3 = 20\,000\,\frac{\text{kg}}{\text{m}^2} \times R^2 \ ,$$

which leads to

$$R_{\text{max}} = \frac{1}{5}\,\text{m} \quad \Rightarrow \quad H_{\text{max}} = 4\,\text{m} \ .$$

If the gravity were to double, the strength would be composed of 100 kg and a body mass of 50 kg (corresponding to 100 kg 'weight', since $g^* = 2g$).
Since

$$V = \frac{1}{2} V_0 \quad \Rightarrow \quad R_{\text{max}}(g^* = 2g) = \frac{1}{\sqrt[3]{2}}\,R^0_{\text{max}} \ ,$$

$$R_{\text{max}}(g^* = 2g) = 0.1587\,\text{m} \quad \Rightarrow \quad H_{\text{max}}(g^* = 2g) = 3.17\,\text{m} \ .$$

2. Carbon is produced in the so-called triple-alpha process. In step one ^{8}Be is produced in $\alpha\alpha$ collisions

$$^4\text{He} + {}^4\text{He} \ \rightarrow \ {}^8\text{Be} + \gamma - 91.78\,\text{keV} \ .$$

The ^{8}Be produced in this step is unstable and decays back into helium nuclei in 2.6 µs. It therefore requires a high helium density to induce a reaction with ^{8}Be before it has decayed,

$$^8\text{Be} + {}^4\text{He} \ \rightarrow \ {}^{12}\text{C}^* + 7.367\,\text{MeV} \ .$$

The excited $^{12}\text{C}^*$ state is unstable, but a few of these excited carbon nuclei emit a γ ray quickly enough to become stable before they disintegrate. The net energy release of the triple-alpha process is 7.275 MeV.
Only at extremely high temperatures ($\approx 10^8$ K) the bottleneck of carbon production from helium in a highly improbable reaction can be accomplished because of a lucky mood of nature: high temperatures are required to overcome the Coulomb barrier for helium fusion. Also high densities of helium nuclei are needed to make the triple-alpha process possible.
The reaction rate for the triple-alpha process depends on the number density N_α of α particles and the temperature T of the α plasma:

$$\sigma(3\alpha \rightarrow {}^{12}\text{C}) \sim N_\alpha^3\,\frac{1}{T^3}\exp\left(-\frac{\varepsilon}{kT}\right)\Gamma_\gamma \ ,$$

where ε is a parameter (≈ 0.4 MeV). For high temperatures up to $T \approx 1.5 \times 10^9$ K the exponential wins over the power T^3. The maximum cross section is derived to be in the following way:

$$0 = \frac{d}{dT}\left[\frac{1}{T^3}\exp\left(-\frac{\varepsilon}{kT}\right)\right] = \left(-\frac{3}{T^4} + \frac{\varepsilon}{kT^5}\right)\exp\left(-\frac{\varepsilon}{kT}\right) \quad \Rightarrow$$

$$T_{max} = \frac{\varepsilon}{3k} \approx 1.5 \times 10^9\,\text{K} \quad \Rightarrow \quad \sigma_{max}(3\alpha \to {}^{12}\text{C}) \sim N_\alpha^3 \left(\frac{3k}{\varepsilon}\right)^3 e^{-3}\,\Gamma_\gamma\,.$$

Γ_γ is the electromagnetic decay width to the first excited state of ^{12}C.
The conditions for the triple-alpha process can be summarized as follows:

- high plasma temperatures ($\approx 10^8$ K),

- high α-particle density,

- large, resonance-like cross section $\sigma(3\alpha)$.

3. Gravity sets the escape velocity (v_{esc}) and the temperature determines the mean speed with which the molecules are moving (v_{mol}). The condition for a planet to retain an atmosphere is

$$v_{esc} \gg v_{mol}\,.$$

v_{esc} can be obtained from

$$\frac{1}{2}mv^2 = G\frac{m\,M_{planet}}{R_{planet}} \quad \Rightarrow \quad v_{esc} = \sqrt{2GM_{planet}/R_{planet}} \quad (= 11.2\,\text{km/s for Earth})\,.$$

v_{mol} can be calculated from

$$\frac{1}{2}mv^2 = \frac{3}{2}kT \quad \Rightarrow \quad v_{mol} = \sqrt{3kT/m} \quad (= 517\,\text{m/s for N}_2\text{ for Earth})\,.$$

For $v_{esc} = v_{mol}$ one obtains

$$2G\,\frac{M_{planet}}{R_{planet}} = \frac{3kT}{m}\,.$$

Let us assume earthlike conditions ($T = 300\,\text{K}$, $m = m(^{14}\text{N}_2) = 4.65 \times 10^{-23}$ g $= 4.65 \times 10^{-26}$ kg, $k = 1.38 \times 10^{-23}\,\text{J K}^{-1}$):

$$2G\,\frac{M_{planet}}{R_{planet}} = 267\,097\,\text{J/kg} \quad \Rightarrow \quad \frac{M_{planet}}{R_{planet}} = 2 \times 10^{15}\,\text{kg/m}\,.$$

With $M = \frac{4}{3}\pi R^3 \varrho$, and assuming an earthlike density $\varrho = 5.5\,\text{g/cm}^3 = 5500\,\text{kg/m}^3$ one gets

$$R = \sqrt[3]{\frac{3M}{4\pi\varrho}} \approx 0.035\,\text{m} \times \sqrt[3]{M/\text{kg}}\,,$$

$$\frac{M_{planet}}{\sqrt[3]{M_{planet}}} \approx 2 \times 10^{15}\,\frac{\text{kg}}{\text{m}} \times 0.035\,\frac{\text{m}}{\text{kg}^{1/3}} = 7.0 \times 10^{13}\,\text{kg}^{2/3}\,.$$

This results in

$$M_{\text{planet}} \approx 5.9 \times 10^{20} \text{ kg} .$$

The result of this estimate depends very much on earthlike properties and the assumption that other effects, like rotational velocity ($\rightarrow$ centrifugal force) are negligible.

4. a) The long-range forces in three dimensions are the gravitational force and the electromagnetic force, the latter of which reduces in the non-relativistic (static) limit to the Coulomb force. The force is proportional to the masses (Newton's gravitation law) or to the charges (Coulomb's law), respectively. Both forces scale with $1/r^2$, where r is the distance between the two bodies. The surface of a three-dimensional sphere is $4\pi r^2$, and the forces are, roughly speaking, proportional to the solid angle under which the bodies see each other. The surface of an n-dimensional sphere is (already from dimensional considerations) given by

$$s_n(r) = s_n \, r^{n-1} , \quad \text{i.e.,} \quad g_n = n - 1 ,$$

where the constant s_n is the surface of the corresponding unit sphere. Isotropic long-range two-body forces in n dimensions should therefore scale as

$$F(r) \sim \frac{1}{s_n(r)} \sim \frac{1}{r^{n-1}} .$$

b) The potential of a radial force is given by ($F \cdot r < 0$ for attractive forces, $n > 2$)

$$V(r) = \int_{r_0}^{r} F(r') \, dr' \sim \int_{r_0}^{r} \frac{dr'}{r'^{n-1}} = -\frac{1}{n-2} \left(\frac{1}{r^{n-2}} - \frac{1}{r_0^{n-2}} \right)$$

$$\sim -\frac{1}{r^{n-2}} + \text{const} .$$

A particle of mass m, at distance r from a center, and having velocity v on a circular orbit has an angular momentum of $L = rp = mvr$. (In general, the velocity component perpendicular to r is considered.) The centrifugal force is then given by

$$F_c(r) = \frac{mv^2}{r} = \frac{L^2}{mr^3} ,$$

where the last expression is also valid for general motion and L is a constant for central forces. The corresponding centrifugal potential therefore reads

$$V_c(r) = -\int_{r_0}^{r} F_c(r') \, dr' = \frac{L^2}{2mr^2} + \text{const} .$$

The effective potential is the sum of V and V_c,

$$V_{\text{eff}}(r) = V(r) + V_c(r) = -\frac{C_n}{r^{n-2}} + \frac{L^2}{2mr^2} + \text{const} .$$

Circular orbits take place for a vanishing effective radial force,

$$0 = -F_{\text{eff}}(r) = \frac{dV_{\text{eff}}(r)}{dr} = \frac{(n-2)C_n}{r^{n-1}} - \frac{L^2}{mr^3} \quad \Rightarrow \quad r_{\text{orb}} = \left(\frac{m\tilde{C}_n}{L^2}\right)^{\frac{1}{n-4}}$$

with $\tilde{C}_n = (n-2)C_n$, $n \neq 4$, also applicable for $n = 2$. Stable orbits are given for potential minima:

$$0 < \left.\frac{d^2 V_{\text{eff}}(r)}{dr^2}\right|_{r_{\text{orb}}} = \left.\left(-\frac{(n-1)\tilde{C}_n}{r^n} + \frac{3L^2}{mr^4}\right)\right|_{r_{\text{orb}}} = (4-n)\frac{L^2}{m}\left(\frac{L^2}{m\tilde{C}_n}\right)^{\frac{4}{n-4}}.$$

Thus, stable motion is only possible for $n < 4$. For $n \geq 4$ the motion is either unbounded in space or eventually leads to a collision of the bodies after finite time. As discussed in books of classical mechanics, also the spatial direction of non-circular motion is conserved for Newton's law, described by the Runge–Lenz vector, leading to 'true' ellipses. This is a special feature of $V \sim 1/r$, so for $n = 2$ with a logarithmic potential the motion in contrast is ergodic, i.e., every energetically reachable space point eventually lies in the vicinity of a trajectory point.

c) The field energy of the force in the radial interval $[\lambda, \Lambda]$ reads

$$W = \int_{\lambda \leq r \leq \Lambda} w(r)\, dV_n = \int_\lambda^\Lambda w(r)s_n(r)\, dr \sim s_n \int_\lambda^\Lambda F^2(r)r^{n-1}\, dr .$$

With the expression $F(r) \sim 1/r^{n-1}$ one yields

$$W \sim \int_\lambda^\Lambda \frac{dr}{r^{n-1}} = \begin{cases} \ln(\Lambda/\lambda) , & n = 2 \\ (\lambda^{-(n-2)} - \Lambda^{-(n-2)})/(n-2) , & n > 2 \end{cases},$$

which is similar to the expression of the potential energy. For $n > 2$ the limit $\Lambda \to \infty$ leads to a finite result, whereas W diverges for $\lambda \to 0$. In the case $n = 2$ both limits are divergent. Quantum corrections, here for the so-called self-energy corrections, are supposed to become significant for small distances. Therefore, the limit $\lambda \to 0$ may diverge, whereas the limit $\Lambda \to \infty$ should exist. Thus, $n > 2$ dimensions are considered valid from this aspect. The degree of divergence is then smallest for $n = 3$.

17.15 Chapter 15

1. There is not yet a solution. If you have solved the problem successfully, you should book a flight to Stockholm because you will be the next laureate for the Nobel Prize in physics.

A Mathematical Appendix

A.1 Selected Formulae

> *"Don't worry about your difficulties in mathematics; I can assure you that mine are still greater."*
>
> Albert Einstein

The solution of physics problems often involves mathematics. In most cases nature is not so kind as to allow a precise mathematical treatment. Many times approximations are not only rather convenient but also necessary, because the general solution of specific problems can be very demanding and sometimes even impossible.

In addition to these approximations, which often involve power series, where only the leading terms are relevant, basic knowledge of calculus and statistics is required. In the following the most frequently used mathematical aids shall be briefly presented.

1. Power Series

Binomial expansion: **binomial expansion**

$$(1 \pm x)^m =$$
$$1 \pm mx + \frac{m(m-1)}{2!}x^2 \pm \frac{m(m-1)(m-2)}{3!}x^3 + \cdots$$
$$+ (\pm 1)^n \frac{m(m-1)\cdots(m-n+1)}{n!}x^n + \cdots .$$

For integer positive m this series is finite. The coefficients are **binomial coefficients**

$$\frac{m(m-1)\cdots(m-n+1)}{n!} = \binom{m}{n} .$$

If m is not a positive integer, the series is infinite and convergent for $|x| < 1$. This expansion often provides a simplification of otherwise complicated expressions.

A few examples for most commonly used binomial expansions:

examples for binomial expansions

$$(1 \pm x)^{1/2} = 1 \pm \frac{1}{2}x - \frac{1}{8}x^2 \pm \frac{1}{16}x^3 - \frac{5}{128}x^4 \pm \cdots ,$$

$$(1 \pm x)^{-1/2} = 1 \mp \frac{1}{2}x + \frac{3}{8}x^2 \mp \frac{5}{16}x^3 + \frac{35}{128}x^4 \mp \cdots ,$$

$$(1 \pm x)^{-1} = 1 \mp x + x^2 \mp x^3 + x^4 \mp \cdots ,$$

$$(1 \pm x)^4 = 1 \pm 4x + 6x^2 \pm 4x^3 + x^4 \quad \text{finite} .$$

trigonometric functions   Trigonometric functions:

$$\sin x = x - \frac{x^3}{3!} + \frac{x^5}{5!} - \cdots + (-1)^n \frac{x^{2n+1}}{(2n+1)!} \pm \cdots ,$$

$$\cos x = 1 - \frac{x^2}{2!} + \frac{x^4}{4!} - \cdots + (-1)^n \frac{x^{2n}}{(2n)!} \pm \cdots ,$$

$$\tan x = x + \frac{1}{3}x^3 + \frac{2}{15}x^5 + \frac{17}{315}x^7 + \cdots , \quad |x| < \frac{\pi}{2} ,$$

$$\cot x = \frac{1}{x} - \frac{x}{3} - \frac{x^3}{45} - \frac{2x^5}{945} - \cdots , \quad 0 < |x| < \pi .$$

exponential function   Exponential function:

$$e^x = 1 + \frac{x}{1!} + \frac{x^2}{2!} + \frac{x^3}{3!} + \cdots + \frac{x^n}{n!} + \cdots .$$

natural logarithm   Logarithmic function:

$$\ln(1 + x) = x - \frac{x^2}{2} + \frac{x^3}{3} - \frac{x^4}{4} + \cdots + (-1)^{n+1}\frac{x^n}{n} .$$

2. Indefinite Integrals

powers

$$\int x^n \, dx = \frac{x^{n+1}}{n+1} , \quad (n \neq -1) ,$$

$$\int \frac{dx}{x} = \ln x ,$$

powers of linear functions

$$\int (ax + b)^n \, dx = \frac{1}{a(n+1)}(ax + b)^{n+1} , \quad (n \neq -1) ,$$

$$\int \frac{dx}{ax + b} = \frac{1}{a}\ln(ax + b) ,$$

$$\int e^x \, dx = e^x \,,$$ exponential function

$$\int x\, e^{ax} \, dx = \frac{e^{ax}}{a^2}(ax - 1) \,,$$

$$\int \frac{dx}{1 + e^{ax}} = \frac{1}{a} \ln \frac{e^{ax}}{1 + e^{ax}} \,,$$

$$\int \sin x \, dx = -\cos x \,,$$ trigonometric functions

$$\int \cos x \, dx = \sin x \,,$$

$$\int \tan x \, dx = -\ln \cos x \,,$$

$$\int \ln x \, dx = x \ln x - x \,.$$ natural logarithm

3. Specific Integrals

$$\int_0^{\pi/2} \cos x \sin x \, dx = \frac{1}{2} \,,$$ triginometric

$$\int_0^{\infty} e^{-ax^2} \, dx = \frac{1}{2}\sqrt{\frac{\pi}{a}} \,,$$ Gaussian

$$\int_0^{\infty} \frac{x \, dx}{e^x - 1} = \frac{\pi^2}{6} \,,$$ exponentials

$$\int_0^{\infty} \frac{x \, dx}{e^x + 1} = \frac{\pi^2}{12} \,,$$

$$\int_0^{\infty} \frac{\sin ax}{x} \, dx = \begin{cases} \frac{\pi}{2} & \text{for } a > 0 \\ -\frac{\pi}{2} & \text{for } a < 0 \end{cases} \,,$$ $\sin ax/x$

$$\int_0^1 \frac{\ln x}{x - 1} \, dx = \frac{\pi^2}{6} \,.$$ logarithm

4. Probability Distributions

Binomial: binomial distribution

$$f(r, n, p) = \frac{n!}{r!(n-r)!} p^r q^{n-r} \,,$$

$$r = 0, 1, 2, \ldots, n \,, \quad 0 \le p \le 1 \,, \quad q = 1 - p \,;$$

mean: $\langle r \rangle = np \,,$ variance: $\sigma^2 = npq \,.$

Poisson distribution   Poisson:

$$f(r, \mu) = \frac{\mu^r \, e^{-\mu}}{r!} \, , \quad r = 0, 1, 2, \dots , \quad \mu > 0 \, ;$$

mean: $\langle r \rangle = \mu$,   variance: $\sigma^2 = \mu$.

Gaussian distribution   Gaussian:

$$f(x, \mu, \sigma^2) = \frac{1}{\sigma \sqrt{2\pi}} \exp \left\{ -\frac{(x - \mu)^2}{2\sigma^2} \right\} \, , \quad \sigma > 0 \, ;$$

mean: $\langle x \rangle = \mu$,   variance: σ^2 .

Landau distribution   Approximation for the Landau distribution:

$$L(\lambda) = \frac{1}{\sqrt{2\pi}} \exp \left\{ -\frac{1}{2} (\lambda + e^{-\lambda}) \right\} \, ,$$

where λ is the deviation from the most probable value.

5. Errors and Error Propagation

mean value   Mean value of n independent measurements:

$$\langle x \rangle = \frac{1}{n} \sum_{i=1}^{n} x_i \, ;$$

variance   variance of n independent measurements:

$$s^2 = \frac{1}{n} \sum_{i=1}^{n} (x_i - \langle x_i \rangle)^2 = \frac{1}{n} \sum_{i=1}^{n} x_i^2 - \langle x \rangle^2 \, ,$$

standard deviation   where s is called the standard deviation. A best estimate for the standard estimation of the mean is

$$\sigma = \frac{s}{\sqrt{n - 1}} \, .$$

If $f(x, y, z)$ and $\sigma_x, \sigma_y, \sigma_z$ are the function and standard
independent, uncorrelated   deviations of the independent, uncorrelated variables, then
variables

$$\sigma_f^2 = \left(\frac{\partial f}{\partial x} \right)^2 \sigma_x^2 + \left(\frac{\partial f}{\partial y} \right)^2 \sigma_y^2 + \left(\frac{\partial f}{\partial z} \right)^2 \sigma_z^2 \, .$$

If $D(z)$ is the distribution function of the variable z around the true value z_0 with expectation value $\langle z \rangle$ and standard deviation σ_z, the quantity

$$1 - \alpha = \int_{\langle z \rangle - \delta}^{\langle z \rangle + \delta} D(z)\,\mathrm{d}z$$

represents the probability that the true value z_0 lies in the interval $\pm\delta$ around the measured value $\langle z \rangle$; i.e., $100(1 - \alpha)\%$ measured values are within $\pm\delta$. If $D(z)$ is a Gaussian distribution one has

confidence interval

δ	$1 - \alpha$
1σ	68.27%
2σ	95.45%
3σ	99.73%

In experimental distributions frequently the full width at half maximum, Δz, is easily measured. For Gaussian distributions Δz is related to the standard deviation by

full width at half maximum

$$\Delta z(\text{fwhm}) = 2\sqrt{2\ln 2}\,\sigma_z \approx 2.355\,\sigma_z \;.$$

A.2 Mathematics for Angular Variations of the CMB

> *"As physics advances farther and farther every day and develops new axioms, it will require fresh assistance from mathematics."*
>
> *Francis Bacon*

In this appendix the mathematics needed to describe the variations in the CMB temperature as a function of direction is reviewed. In particular, some of the important properties of the spherical harmonic functions $Y_{lm}(\theta, \phi)$ will be collected. More information can be found in standard texts on mathematical methods of physics such as [46].

spherical harmonics

First it will be recollected what these functions are needed for in astroparticles physics. Suppose a quantity (here the temperature) as a function of *direction* has been measured, which one can take as being specified by the standard polar coordinate angles θ and ϕ. This applies, e.g., for the directional measurements of the blackbody radiation. But one is not able – or at least it is highly impractical – to try to understand individually every measurement for every direction. Rather, it is preferable to parameterize the data with some function and see if one can understand the most important characteristics of this function.

dependence on the direction

'periodic functions'   When, however, a function to describe the measured temperature as a function of direction is chosen, one cannot take a simple polynomial in θ and ϕ, because this would not satisfy the obvious continuity requirements, e.g., that the function at $\phi = 0$ matches that at $\phi = 2\pi$. By using spherical harmonics as the basis functions for the expansion, these requirements are automatically taken into account.

important differential   Now one has to remember how the spherical harmon-
equations   ics are defined. Several important differential equations of mathematical physics (Schrödinger, Helmholtz, Laplace) can be written in the form

$$\left(\nabla^2 + v(r)\right)\psi = 0 , \tag{A.1}$$

nabla operator   where ∇ is the usual nabla operator, as defined by

$$\nabla = e_x \frac{\partial}{\partial x} + e_y \frac{\partial}{\partial y} + e_z \frac{\partial}{\partial z} . \tag{A.2}$$

Here $v(r)$ is an arbitrary function depending only on the radial coordinate r. In separation of variables in spherical co-
separation of variables   ordinates, a solution of the form

$$\psi(r, \theta, \phi) = R(r)\,\Theta(\theta)\,\Phi(\phi) \tag{A.3}$$

is tried. Substituting this back into (A.1) gives for the angu-
angular parts   lar parts

$$\frac{d^2\Phi}{d\phi^2} = -m^2\Phi ,$$
$$\frac{d^2\Theta}{d\theta^2} + \frac{\cos\theta}{\sin\theta}\frac{d\Theta}{d\theta} + \left[l(l+1) - \frac{m^2}{\sin^2\theta}\right]\Theta = 0 ,$$
$$\tag{A.4}$$

where $l = 0, 1, \ldots$ and $m = -l, \ldots, l$ are separation con-
azimuthal solution   stants. The solution for Φ is

$$\Phi(\phi) = \frac{1}{\sqrt{2\pi}}\,e^{im\phi} . \tag{A.5}$$

polar solution   The solution for Θ is proportional to the associated Legendre function $P_l^m(\cos\theta)$. The product of the two angular
spherical harmonic function   parts is called the spherical harmonic function $Y_{lm}(\theta, \phi)$,

$$Y_{lm}(\theta, \phi) = \Theta(\theta)\Phi(\phi)$$
$$= \sqrt{\frac{2l+1}{4\pi}\frac{(l-m)!}{(l+m)!}}\,P_l^m(\cos\theta)\,e^{im\phi} . \tag{A.6}$$

Some of the spherical harmonics are given below:

$$Y_{00}(\theta, \phi) = \frac{1}{\sqrt{4\pi}} , \tag{A.7}$$

$$Y_{11}(\theta, \phi) = -\sqrt{\frac{3}{8\pi}} \sin\theta \, e^{i\phi} , \tag{A.8}$$

$$Y_{10}(\theta, \phi) = \sqrt{\frac{3}{4\pi}} \cos\theta , \tag{A.9}$$

$$Y_{22}(\theta, \phi) = \sqrt{\frac{15}{32\pi}} \sin^2\theta \, e^{2i\phi} , \tag{A.10}$$

$$Y_{21}(\theta, \phi) = -\sqrt{\frac{15}{8\pi}} \sin\theta \cos\theta \, e^{i\phi} , \tag{A.11}$$

$$Y_{20}(\theta, \phi) = \sqrt{\frac{5}{16\pi}} (3\cos^2\theta - 1) . \tag{A.12}$$

The importance of spherical harmonics for this investigation is that they form a complete orthogonal set of functions. That is, any arbitrary function $f(\theta, \phi)$ can be expanded in a *Laplace series* as

complete orthogonal set of functions

Laplace series

$$f(\theta, \phi) = \sum_{l=0}^{\infty} \sum_{m=-l}^{l} a_{lm} Y_{lm}(\theta, \phi) . \tag{A.13}$$

To determine the coefficients a_{lm}, one uses the orthogonality relation

orthogonality relation

$$\iint \sin\theta \, d\theta \, d\phi \, Y_{lm}(\theta, \phi) Y^*_{l'm'}(\theta, \phi) = \delta_{l'l} \delta_{m'm} . \tag{A.14}$$

If both sides of (A.13) are multiplied by $Y_{l'm'}$, integration over θ and ϕ leads to

calculation of coefficients

$$a_{lm} = \iint \sin\theta \, d\theta \, d\phi \, Y^*_{lm}(\theta, \phi) f(\theta, \phi) . \tag{A.15}$$

So, in principle, once a function $f(\theta, \phi)$ is specified, the coefficients of its Laplace series can be found.

The same spherical harmonics are found in the *multipole expansion* of the potential from an electric charge distribution. The terminology is usually borrowed from this example and the terms in the series are referred to as multipole moments. The $l = 0$ term is the monopole, $l = 1$ the dipole, etc.

multipole expansion

Laplace series for the temperature variations of the CMB

To quantify the temperature variations of the CMB, the Laplace series can be used to describe

$$\Delta T(\theta, \phi) = T(\theta, \phi) - \langle T \rangle$$

$$= \sum_{l \geq 1} \sum_{m=-l}^{l} a_{lm} Y_{lm}(\theta, \phi), \qquad (A.16)$$

where $\langle T \rangle$ is the temperature averaged over all directions. Here the sum starts at $l = 1$, not $l = 0$, since by construction the $l = 0$ term gives the average temperature, which has been subtracted off. In some references one expands $\Delta T / \langle T \rangle$ rather than ΔT. This gives the equivalent information but with the coefficients simply differing from those in (A.16) by a factor of $\langle T \rangle$.

finite series: practical limit

In practice one determines the coefficients a_{lm} up to some $l_{\max}$ by means of a statistical parameter estimation technique such as the method of maximum likelihood. This procedure will use as input the measured temperatures and information about their accuracy to determine estimates for the coefficients a_{lm} and their uncertainties.

Once one has estimates for the coefficients a_{lm}, one can summarize the amplitude of regular variation with angle by defining

$$C_l = \frac{1}{2l+1} \sum_{m=-l}^{l} |a_{lm}|^2. \qquad (A.17)$$

angular power spectrum

The set of numbers C_l is called the *angular power spectrum*. The value of C_l represents the level of structure found at an angular separation

$$\Delta\theta = \frac{180°}{l}. \qquad (A.18)$$

The measuring device will in general only be able to resolve angles down to some minimum value; this determines the maximum measurable l.

B Results from Statistical Physics: Thermodynamics of the Early Universe

In this appendix some results from statistical and thermal physics will be recalled that will be needed to describe the early universe. To start with, the Fermi–Dirac and Bose–Einstein distributions for the number of particles per unit volume of momentum space will be derived:

Fermi–Dirac, Bose–Einstein distribution

$$f(\boldsymbol{p}) = g \frac{V}{(2\pi)^3} \frac{1}{e^{(E-\mu)/T} \pm 1} \, , \tag{B.1}$$

where $E = \sqrt{p^2 + m^2}$ is the energy and g is the number of internal degrees of freedom for the particle, V is the volume of the system, T is the temperature, and μ is the chemical potential. (The Boltzmann constant k is set to unity as usual.) The minus sign in (B.1) is used for bosons and the plus for fermions. These distributions will be required to derive the number density n of particles, the energy density ϱ, and the pressure P.

chemical potential

Some of the relations may differ from those covered in a typical course in statistical mechanics. This is for two main reasons. First, the particles in the very hot early universe typically have velocities comparable to the speed of light, therefore the relativistic equation $E^2 = p^2 + m^2$ will be needed to relate energy and momentum. Second, the temperatures will be so high that particles are continually being created and destroyed, e.g., through reactions such as $\gamma\gamma \leftrightarrow e^+e^-$. This is in contrast to the physics of low-temperature systems, where the number of particles in a system is usually constrained to be constant. The familiar exception is blackbody radiation, since massless photons can be created and destroyed at any non-zero temperature. For a gas of relativistic particles it will be found that the expressions for n, ϱ, and P are similar to those for blackbody radiation.

relativistic treatment

variable particle numbers

blackbody radiation

B.1 Statistical Mechanics Review

"The general connection between energy and temperature may only be established by probability considerations. Two systems are in statistical equilibrium when a transfer of energy does not increase the probability."

Max Planck

Consider a system with volume $V = L^3$ and energy U, which could be a cube of the very early universe. The number of particles will not be fixed since the temperatures considered here will be so high that particles can be continually created and destroyed. For the moment only a single particle type will be considered but eventually the situation will be generalized to include all possible types.

The system can be in any one of a very large number of possible microstates. The fundamental postulate of statistical mechanics is that all microstates consistent with the imposed constraints (volume, energy) are equally likely. A given microstate, e.g., an N-particle wave function $\psi(x_1, \ldots, x_N)$ specifies everything about the system, but this is far more than one wants to know. To reduce the information to a more digestible level, one can determine from the microstate the momentum distribution of the particles, i.e., the expected number of particles in each cell $d^3 p$ of momentum space.

fundamental postulate: equipartition of energy

There will be many microstates that lead to the same distribution, but one distribution in particular will have overwhelmingly more possible microstates than the others. To good approximation all the others can be ignored and this *equilibrium distribution* can be regarded as the most likely. Once it has been found, one can determine from it the other quantities needed, such as the energy density and pressure.

equilibrium distribution

So, to find the equilibrium distribution one needs to determine the number of possible microstates consistent with a distribution and then find the one for which this is a maximum. This is treated in standard books on statistical mechanics, e.g., [47]. Here only the main steps will be reviewed.

N-particle wave function

It is assumed that the system's N-particle wave function can be expressed as a sum of N terms, each of which is the product of N one-particle wave functions of the form

$$\psi_A(x) \sim e^{i p_A \cdot x} .$$

(B.2)

The total wave function is thus

$$\psi(\pmb{x}_1, \ldots, \pmb{x}_N) =$$
$$= \frac{1}{\sqrt{N!}} \sum P(i, j, \ldots) \psi_A(\pmb{x}_i) \psi_B(\pmb{x}_j) \cdots , \qquad \text{(B.3)}$$

where the sum includes all possible permutations of the co-ordinates $\pmb{x}_i$. For a system of identical bosons, the factor P is equal to one, whereas for identical fermions it is plus or minus one depending on whether the permutation is obtained from an even or odd number of exchanges of particle coordinates. This results in a wave function that is symmetric for bosons and antisymmetric for fermions upon interchange of any pair of coordinates. As a consequence, the total wave function for a system of fermions is zero if the same one-particle wave function appears more than once in the product of terms; this is the Pauli exclusion principle. **symmetrization for bosons** **antisymmetrization for fermions** **Pauli exclusion principle**

Although the most general solution to the N-particle Schrödinger equation does not factorize in the way ψ has been written in (B.3), this form will be valid to good approximation for systems of weakly interacting particles. For high-temperature systems such as the early universe, (B.3) is assumed to hold.

Further, one assumes that the one-particle wave functions should obey periodic boundary conditions in the volume $V = L^3$. The plane-wave form for the one-particle wave functions in (B.2) then implies that the momentum vectors $\pmb{p}$ cannot take on arbitrary values but that they must satisfy **periodic boundary conditions: discretizing momentum**

$$\pmb{p} = \frac{2\pi}{L}(n_x, n_y, n_z) , \qquad \text{(B.4)}$$

where n_x, n_y, and n_z are integers. Thus, the possible momenta for the one-particle states are given by a cubic lattice of points in momentum space with separation $2\pi/L$.

For a given N-particle wave function, where N will in general be very large, the possible momentum vectors for the one-particle states will follow some distribution in momentum space. That is, one will find a certain number dN of one-particle states for each element d^3p in momentum space, and **momentum distribution**

$$f(\pmb{p}) = \frac{d^3 N}{d^3 p} \qquad \text{(B.5)}$$

will be called the momentum distribution.

A given distribution $f(\boldsymbol{p})$ could result from a number of distinct N-particle wave functions, i.e., from a number of different microstates. All available microstates are equally likely, but the overwhelming majority of them will correspond to a single specific $f(\boldsymbol{p})$, the equilibrium distribution. This is what one needs to find.

To find this equilibrium momentum distribution, one **number of microstates** must determine the number of microstates t for a given $f(\boldsymbol{p})$. To do this, one considers the momentum space to be divided into cells of size $\delta^3 p$. The number of particles in the ith cell is

$$v_i = f(\boldsymbol{p_i})\,\delta^3 p \;. \tag{B.6}$$

The number of possible one-particle momentum states in the cell is $\delta^3 p$ divided by the number of states per unit volume **total number** in momentum space, $(2\pi/L)^3$. The total number of one- **of one-particle states** particle states in $\delta^3 p$ is therefore[1]

$$\gamma_i = g\,\frac{\delta^3 p}{(2\pi/L)^3} \;, \tag{B.7}$$

number of degrees where g represents the number of internal (e.g., spin) de- **of freedom** grees of freedom for the particle. For an electron with spin $1/2$, for example, one has $g = 2$.

It is assumed that the element $\delta^3 p$ is large compared to the volume of momentum space per available state, which is $(2\pi/L)^3$, but small compared to the typical momenta of the particles. Within this approximation, the set of numbers v_i for all i contains the same information as $f(\boldsymbol{p})$.

system of bosons For a system of bosons, there is no restriction on the number of particles that can have the same momentum. Therefore, each of the γ_i states can have from zero up to v_i particles. The number of ways of distributing the v_i particles among the γ_i states is a standard problem of combinatorics (see, e.g., [47]). One obtains

$$\frac{(v_i + \gamma_i - 1)!}{v_i!\,(\gamma_i - 1)!} \tag{B.8}$$

total number of microstates possible arrangements. The total number of microstates for the distribution is therefore

---

[1] In many references the number of particles is called n_i and the number of states g_i. Unfortunately, these letters need to be used with different meanings later in this appendix, so here v_i and γ_i will be used instead.

$$t_{\text{BE}}[f(\boldsymbol{p})] = \prod_i \frac{(\nu_i + \gamma_i - 1)!}{\nu_i!(\gamma_i - 1)!} \approx \prod_i \frac{(\nu_i + \gamma_i)!}{\nu_i!\gamma_i!} ,$$

$$(\text{B.9})$$

where the product extends over all cells in momentum space. The subscript BE in (B.9) stands for Bose–Einstein since this relation holds for a collection of identical bosons.

For fermions, the antisymmetric nature of the total wave function implies that it can contain a given one-particle state at most only once. Therefore, each of the γ_i states in the ith cell in momentum space can be occupied either once or not at all. This implies $\gamma_i \geq \nu_i$. The number of possible arrangements of ν_i particles in the γ_i states where each state is occupied zero or one time is another standard problem of combinatorics, for which one finds

system of fermions

$$\frac{\gamma_i!}{\nu_i!(\gamma_i - \nu_i)!} .$$

$$(\text{B.10})$$

The total number of combinations for all cells is thus

total number of microstates

$$t_{\text{FD}}[f(\boldsymbol{p})] = \prod_i \frac{\gamma_i!}{\nu_i!(\gamma_i - \nu_i)!} ,$$

$$(\text{B.11})$$

where FD stands for Fermi–Dirac.

As the number of microstates $t[f(\boldsymbol{p})]$ is astronomically large, it is more convenient to work with its logarithm, and furthermore one can use Stirling's approximation,

logarithm of the number of microstates

$$\ln N! \approx N \ln N - N ,$$

$$(\text{B.12})$$

valid for large N. This gives

$$\ln t_{\text{BE}}[f(\boldsymbol{p})] \approx \qquad (\text{B.13})$$
$$\approx \sum_i [(\nu_i + \gamma_i) \ln(\nu_i + \gamma_i) - \nu_i \ln \nu_i - \gamma_i \ln \gamma_i]$$

and

$$\ln t_{\text{FD}}[f(\boldsymbol{p})] \approx \qquad (\text{B.14})$$
$$\approx \sum_i [\gamma_i \ln \gamma_i - \nu_i \ln \nu_i - (\gamma_i - \nu_i) \ln(\gamma_i - \nu_i)]$$

for bosons and fermions, respectively.

The next step is to find the distribution $f(\boldsymbol{p})$ which maximizes $\ln t[f(\boldsymbol{p})]$. Before doing this, however, the problem should be generalized to allow for more than one type of particle. As long as a particle's mass is small compared to

more than one type of particle

the temperature, it will be continually created and destroyed, and for sufficiently early times this will be true for all particle types. So one will have a set of distribution functions $f_a(\boldsymbol{p})$, where the index $a = e, \mu, \tau, u, d, s, \ldots$ is a label for the particle type. One can write this set of functions as a vector, $\boldsymbol{f} \equiv (f_e, f_\mu, f_\tau, f_u, f_d, \ldots)$.

maximization

energy conservation

In order to find the set of distributions $\boldsymbol{f}(\boldsymbol{p})$, one needs to maximize the total number of microstates $t[\boldsymbol{f}(\boldsymbol{p})]$ subject to two types of constraints. First, it is required that the sum of the energies of all of the particles be equal to the total energy U, i.e.,

$$
\begin{aligned}
U &= \sum_a \int E_a f_a(\boldsymbol{p}) \, \mathrm{d}^3 p \\
&= \sum_a \int \sqrt{p^2 + m_a^2} \, f_a(\boldsymbol{p}) \, \mathrm{d}^3 p \; .
\end{aligned}
\tag{B.15}
$$

conserved quantities

Second, although particles may be created and destroyed, certain quantities are conserved. Suppose that the system has a total conserved charge Q (e.g., zero), baryon number B, and lepton number L. Suppose further that in thermal equilibrium the system has N_a particles of type a. These requirements can be written as

$$
Q = \sum_a Q_a N_a \; ,
\tag{B.16}
$$

$$
B = \sum_a B_a N_a \; ,
\tag{B.17}
$$

$$
L = \sum_a L_a N_a \; .
\tag{B.18}
$$

Note that the values N_a are not explicitly constrained, but rather only the total Q, B, and L. Thus, the quantity that one wants to maximize can be expressed as

$$
\begin{aligned}
\phi(\boldsymbol{f}(\boldsymbol{p}), \alpha_Q, \alpha_B, \alpha_L, \beta) &= \\
&= \sum_a \ln t_a[f_a(\boldsymbol{p})] + \beta \left(U - \sum_a \int E_a f_a(\boldsymbol{p}) \, \mathrm{d}^3 p \right) \\
&\quad + \alpha_Q \left(Q - \sum_a Q_a N_a \right) + \alpha_B \left(B - \sum_a B_a N_a \right) \\
&\quad + \alpha_L \left(L - \sum_a L_a N_a \right) ,
\end{aligned}
\tag{B.19}
$$

Lagrange multipliers

where β, α_Q, α_B, and α_L are Lagrange multipliers. Setting the derivatives of ϕ with respect to the Lagrange multipliers to zero ensures that the corresponding constraints are fulfilled. To find the set of distributions f which maximize (B.19), one substitutes $f_a(p_i) = v_{ai}/\delta^3 p$. Then the integrals are converted to sums, and further one has $\sum_i v_{ai} = N_a$. In addition, the number of microstates can be obtained from (B.13) for bosons and (B.14) for fermions. The derivative of ϕ with respect to v_{ai} is

discretizing momentum integrals

$$\frac{\partial \phi}{\partial v_{ai}} = \ln(\gamma_{ai} \pm v_{ai}) - \ln v_{ai} - \beta E_{ai}$$

$$- \alpha_Q Q_a - \alpha_B B_a - \alpha_L L_a \,, \tag{B.20}$$

where $\gamma_{ai} = g_a \delta^3 p/(2\pi/L)^3$ is the number of states available to a particle of type a in the cell i, and for the derivation in this section the upper sign refers to bosons and the lower sign to fermions. Setting (B.20) equal to zero and solving for v_{ai} gives

solution for number of particles

$$v_{ai} = \frac{\gamma_{ai}}{\exp\left[\alpha_Q Q_a + \alpha_B B_a + \alpha_L L_a + \beta E_{ai}\right] \mp 1} \,. \tag{B.21}$$

Re-expressing this in terms of the functions $f_a(p_i) = v_{ai}/\delta^3 p$ gives

$$f_a(\boldsymbol{p}) = \frac{g_a(L/2\pi)^3}{\exp\left[\alpha_Q Q_a + \alpha_B B_a + \alpha_L L_a + \beta E_a\right] \mp 1} \,. \tag{B.22}$$

The temperature can be defined as

definition of temperature

$$T = 1/\beta \,, \tag{B.23}$$

and it can be shown that this has all of the desired properties of the usual thermodynamic temperature. Furthermore, the chemical potential for particle type a can be defined as

chemical potential

$$\mu_a = -T(\alpha_Q Q_a + \alpha_B B_a + \alpha_L L_a) \,, \tag{B.24}$$

which can be modified in the obvious way to include a different set of conserved quantities. Note that, although the Lagrange multipliers are specific to the system, i.e., are the same for all particle types, the chemical potential depends on the charge, baryon number, and lepton number of the particle. In a reaction where, say, $a + b \leftrightarrow c + d$, (B.24) implies $\mu_a + \mu_b = \mu_c + \mu_d$.

resulting momentum
distribution

Using these modified names for the Lagrange multipliers gives the desired result for the momentum distribution,

$$f_a(\boldsymbol{p}) = \frac{g_a (L/2\pi)^3}{e^{(E_a - \mu_a)/T} \mp 1} \, , \tag{B.25}$$

internal degrees of freedom

where one uses the minus sign if particle type a is a boson and plus if it is a fermion. The number of internal degrees of freedom, g_a, is usually $2J + 1$ for a particle of spin J, but it could include other degrees of freedom besides spin such as colour.

B.2 Number and Energy Densities

"There are 10^{11} stars in the galaxy. That used to be a huge number. But it's only a hundred billion. It's less than the national deficit! We used to call them astronomical numbers. Now we should call them economical numbers."

Richard P. Feynman

From the Planck distribution given by (6.81), (B.1) one can proceed to determine the number and energy per unit volume for all of the particle types. The function (B.25) gives the number of particles of type a in a momentum-space volume $d^3 p$. The number density n is obtained by integrating this over all of momentum space and dividing by the volume $V = L^3$, i.e.,

number density n

$$n = \frac{1}{V} \int f(\boldsymbol{p}) \, d^3 p = \frac{g}{(2\pi)^3} \int \frac{d^3 p}{e^{(E-\mu)/T} \pm 1} \, , \tag{B.26}$$

where for clarity the index indicating the particle type has been dropped. Since the integrand only depends on the magnitude of the momentum through $E = \sqrt{p^2 + m^2}$, one can take the element $d^3 p$ to be a spherical shell with radius p and thickness dp, so that $d^3 p \, \hat{=} \, 4\pi p^2 \, dp$. From $E^2 = p^2 + m^2$ one gets $2E \, dE = 2p \, dp$ and therefore

n: energy integral

$$n = \frac{g}{2\pi^2} \int_m^\infty \frac{\sqrt{E^2 - m^2} \, E \, dE}{e^{(E-\mu)/T} \pm 1} \, . \tag{B.27}$$

n: relativistic limit

The integral (B.27) can be carried out in closed form only for certain limiting cases. In the limit where the particles are relativistic, i.e., $T \gg m$, and also if $T \gg \mu$, one finds

$$
n = \begin{cases} \dfrac{\zeta(3)}{\pi^2} g T^3 & \text{for bosons,} \\[2mm] \dfrac{3}{4} \dfrac{\zeta(3)}{\pi^2} g T^3 & \text{for fermions.} \end{cases} \tag{B.28}
$$

Here ζ is the Riemann zeta function and $\zeta(3) \approx 1.20\,206\ldots$. Notice that in particle physics units the number density has dimension of energy cubed. To convert this to a normal number per unit volume, one has to divide by $(\hbar c)^3 \approx (0.2\,\text{GeV fm})^3$.

In the non-relativistic limit ($T \ll m$), the integral (B.27) becomes

n: non-relativistic limit

$$
n = g \left(\frac{mT}{2\pi} \right)^{3/2} e^{-(m-\mu)/T} , \tag{B.29}
$$

where the same result is obtained for both the Fermi–Dirac and Bose–Einstein distributions. One sees that for a non-relativistic particle species, the number density is exponentially suppressed by the factor $e^{-m/T}$, the so-called Boltzmann factor. This may seem counter intuitive, since the density of air molecules in a room is certainly not suppressed by this factor, although they are non-relativistic. One must take into account the fact that the chemical potentials depend in general on the temperature, and this dependence is exactly such that all relevant quantities are conserved. In the case of the air molecules, μ varies with temperature so as to exactly compensate the factor $T^{3/2} e^{-m/T}$.

For very high temperatures, to good approximation all of the chemical potentials can be set to zero. The total number of particles will be large compared to the net values of any of the conserved quantum numbers, and the constraints effectively play no rôle.

very high temperatures

To find the energy density ϱ one multiplies the number of particles in $\text{d}^3 p$ by the energy and integrate over all momenta,

energy density ϱ

$$
\begin{aligned}
\varrho &= \frac{g}{(2\pi)^3} \int \frac{E \, \text{d}^3 p}{e^{(E-\mu)/T} \pm 1} \\
&= \frac{g}{2\pi^2} \int_m^\infty \frac{\sqrt{E^2 - m^2} \, E^2 \, \text{d}E}{e^{(E-\mu)/T} \pm 1} .
\end{aligned} \tag{B.30}
$$

As with n, the integral can only be carried out in closed form for certain limiting cases. In the relativistic limit, $T \gg m$, one finds

ϱ: relativistic limit

$$\varrho = \begin{cases} \dfrac{\pi^2}{30} g T^4 & \text{for bosons,} \\[2ex] \dfrac{7}{8}\dfrac{\pi^2}{30} g T^4 & \text{for fermions.} \end{cases} \tag{B.31}$$

ϱ: non-relativistic limit   In the non-relativistic limit one has

$$\varrho = mn , \tag{B.32}$$

with the number density n given by (B.29).

From the number and energy densities one can obtain **average energy per particle** the average energy per particle, $\langle E \rangle = \varrho/n$. For $T \gg m$ one finds

$$\langle E \rangle = \begin{cases} \dfrac{\pi^4}{30\,\zeta(3)} T \approx 2.701\,T & \text{for bosons,} \\[2ex] \dfrac{7\pi^4}{180\,\zeta(3)} T \approx 3.151\,T & \text{for fermions.} \end{cases} \tag{B.33}$$

In the non-relativistic limit, the average energy, written as the sum of mass and kinetic terms, reads

$$\langle E \rangle = m + \frac{3}{2} T , \tag{B.34}$$

which is dominated by the mass m for low T.

B.3 Equations of State

> "If your theory is found to be against the second law of thermodynamics, I give you no hope; there is nothing for it but to collapse in deepest humiliation."
>
> Arthur Eddington

Finally in this appendix an equation of state will be derived, that is, a relation between energy density ϱ and pressure P. This will be needed in conjunction with the acceleration and fluid equations in order to solve the Friedmann equation for $R(t)$.

There are several routes to the desired relation. The ap- **first law of thermodynamics** proach that starts from the first law of thermodynamics is the most obvious one,

$$dU = T\,dS - P\,dV , \tag{B.35}$$

which relates the total energy U, temperature T, entropy S, pressure P, and volume V of the system. The differential dU can also be written as

$$dU = \left(\frac{\partial U}{\partial S}\right)_V dS + \left(\frac{\partial U}{\partial V}\right)_S dV , \qquad \text{(B.36)}$$

where the subscripts indicate what is kept constant when computing the partial derivatives. Equating the coefficients of dV in (B.35) and (B.36) gives the pressure,

pressure

$$P = -\left(\frac{\partial U}{\partial V}\right)_S . \qquad \text{(B.37)}$$

Recall that the entropy is simply the logarithm of the total number of microstates Ω, and that to good approximation this is given by the number of microstates of the equilibrium distribution $t[f(\boldsymbol{p})]$. That is,

entropy and number of microstates

$$S = \ln \Omega \approx \ln t[f(\boldsymbol{p})] . \qquad \text{(B.38)}$$

The important thing to notice here is that the entropy is entirely determined by the distribution $f(\boldsymbol{p})$. Therefore, to keep the entropy constant when computing $(\partial U/\partial V)_S$, one simply needs to regard the distribution $f(\boldsymbol{p})$ as remaining constant when V is changed.

The total energy U is

total energy

$$U = \int E f(\boldsymbol{p}) d^3 p , \qquad \text{(B.39)}$$

and the pressure is therefore

$$P = -\left(\frac{\partial U}{\partial V}\right)_S = -\int \frac{\partial E}{\partial V} f(\boldsymbol{p}) d^3 p . \qquad \text{(B.40)}$$

The derivative of E with respect to the volume $V = L^3$ is

$$\frac{\partial E}{\partial V} = \frac{\partial E}{\partial p} \frac{\partial p}{\partial V} = \frac{\partial E}{\partial p} \frac{\partial p}{\partial L} \Big/ \frac{\partial V}{\partial L} . \qquad \text{(B.41)}$$

One has $\partial V/\partial L = 3L^2$ and furthermore $E = \sqrt{p^2 + m^2}$, so

$$\frac{\partial E}{\partial p} = \frac{1}{2} \left(p^2 + m^2\right)^{-1/2} 2p = \frac{p}{E} . \qquad \text{(B.42)}$$

From (B.4) one gets that $p \sim L^{-1}$, and therefore $\partial p/\partial L = -p/L$. Substituting these into (B.41) gives

$$\frac{\partial E}{\partial V} = \frac{p}{E}\left(\frac{-p}{L}\right)\frac{1}{3L^2} = \frac{-p^2}{3EV} \; . \tag{B.43}$$

general expression for the pressure Putting this into (B.40) provides the general expression for the pressure,

$$P = \frac{1}{3V}\int \frac{p^2}{E} f(p)\, d^3p \; . \tag{B.44}$$

In the relativistic limit the particle's rest mass can be neglected, so, $E = \sqrt{p^2 + m^2} \approx p$. Equation (B.44) then becomes

$$P = \frac{1}{3V}\int E\, f(p)\, d^3p \; . \tag{B.45}$$

But the total energy U is (B.39)

$$U = \int E\, f(p)\, d^3p \tag{B.46}$$

pressure in the relativistic limit and $\varrho = U/V$, so the final result for the pressure for a gas of relativistic particles is simply

$$P = \frac{\varrho}{3} \; . \tag{B.47}$$

This is the well-known result from blackbody radiation, but one realizes here that it applies for any particle type in the relativistic limit $T \gg m$.

In the non-relativistic limit, the pressure is given by the **non-relativistic limit:** ideal gas law,
ideal gas law

$$P = nT \; . \tag{B.48}$$

In this case, however, the energy density is simply $\varrho = mn$, so for $T \ll m$ one has $P \ll \varrho$ and in the acceleration and fluid equations one can approximate $P \approx 0$.

Finally, the case of vacuum energy density from a cos-
vacuum energy density mological constant can be treated,
from a cosmological constant

$$\varrho_v = \frac{\Lambda}{8\pi G} \; . \tag{B.49}$$

If one takes $U/V = \varrho_v$ as constant, then the pressure is

$$P = -\left(\frac{\partial U}{\partial V}\right)_S = -\frac{U}{V} = -\varrho_v \; . \tag{B.50}$$

negative pressure Thus, a vacuum energy density leads to a negative pressure.

C Definition of Equatorial and Galactic Coordinates

> *"In the technical language of astronomy, the richness of the star field depends mainly on the galactic latitude, just as the Earth's climate depends mainly on the geographic latitude, and not to any great extent on the longitude."*
>
> *Sir James Jeans*

Optical astronomers mostly prefer equatorial coordinates, like right ascension and declination, while astrophysicists often use galactic coordinates.

For *equatorial coordinates*, which are centered on the Earth, the plane of the Earth's equator is chosen as the plane of reference. This plane is called the celestial equator. There is another plane which is defined by the motion of the Earth around the Sun. This plane has an inclination with respect to the celestial equator of 23.5 degrees. The plane of the Earth's orbit is called the ecliptic. At the periphery these two planes intersect in two points. The one where the Sun crosses the celestial equator from the south is the vernal equinox.

The coordinate measured along the celestial equator eastward from the vernal equinox is called *right ascension*, usually named α. The distance perpendicular to the celestial equator is named *declination*, named δ. Right ascension varies from 0 to 360 degrees, or sometimes – for convenience – from 0 to 24 hours. Declination varies from -90 to $+90$ degrees (see Fig. C.1).

equatorial coordinates

ecliptic

right ascension

declination

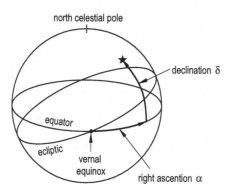

Fig. C.1
Definition of the equatorial coordinates right ascension and declination

galactic coordinates   In contrast, *galactic coordinates* are fixed on our galaxy.
                                  The spherical coordinates (r, l, b) are centered on the Sun,
longitude   with l being the galactic *longitude* and b the *latitude*. r is the
latitude   *distance* of the celestial object from the Sun. The galactic
distance from the Sun   longitude l is the angle between the direction to the galactic
                                  center and the projection of the direction to the star onto the
                                  galactic plane. It counts from the direction to the galactic
                                  center ($l = 0$ degrees) via the galactic anticenter (180 de-
                                  grees, away from the Sun) back to the galactic center. The
                                  latitude b varies from the $+90$ degrees (perpendicular above
                                  the galactic plane) to -90 degrees (Fig. C.2).

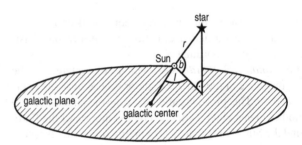

Fig. C.2
Definition of the galactic
coordinates latitude and longitude

D Important Constants for Astroparticle Physics[1]

"Men who wish to know about the world must learn about it in its particular details."

Heraclitus

		relative uncertainty
velocity of light c	$299\,792\,458$ m/s	exact
gravitational constant G	$6.6742 \times 10^{-11}\,\mathrm{m^3\,kg^{-1}\,s^{-2}}$	1.5×10^{-4}
Planck's constant h	$6.626\,0693 \times 10^{-34}\,\mathrm{J\,s}$	1.7×10^{-7}
	$= 4.135\,6675 \times 10^{-15}\,\mathrm{eV\,s}$	
$\hbar = h/2\pi$	$1.054\,571\,68 \times 10^{-34}\,\mathrm{J\,s}$	1.7×10^{-7}
	$= 6.582\,119\,15 \times 10^{-16}\,\mathrm{eV\,s}$	8.5×10^{-8}
Planck mass $\sqrt{\hbar c/G}$	$2.176\,45 \times 10^{-8}\,\mathrm{kg}$	7.4×10^{-5}
	$= 1.220\,90 \times 10^{19}\,\mathrm{GeV}/c^2$	
Planck length $\sqrt{G\hbar/c^3}$	$1.616\,24 \times 10^{-35}\,\mathrm{m}$	7.4×10^{-5}
Planck time $\sqrt{G\hbar/c^5}$	$5.391\,19 \times 10^{-44}\,\mathrm{s}$	7.4×10^{-5}
elementary charge e	$1.602\,176\,53 \times 10^{-19}\,\mathrm{C}$	8.5×10^{-8}
fine-structure constant $\alpha = \frac{e^2}{4\pi\varepsilon_0\hbar c}$	$1/137.035\,999\,11$	3.3×10^{-9}
Rydberg energy Ry	$13.605\,6923\,\mathrm{eV}$	8.5×10^{-8}
electron mass m_e	$9.109\,3826 \times 10^{-31}\,\mathrm{kg}$	1.7×10^{-7}
	$0.510\,998\,918\,\mathrm{MeV}/c^2$	8.6×10^{-8}
proton mass m_p	$1.672\,621\,71 \times 10^{-27}\,\mathrm{kg}$	1.7×10^{-7}
	$= 938.272\,029\,\mathrm{MeV}/c^2$	8.6×10^{-8}
neutron mass m_n	$1.674\,9287 \times 10^{-27}\,\mathrm{kg}$	6×10^{-7}
	$= 939.565\,36\,\mathrm{MeV}/c^2$	3×10^{-7}
$m_n - m_p$	$= 1.293\,3317\,\mathrm{MeV}/c^2$	6×10^{-6}
Avogadro constant N_A	$6.022\,1415 \times 10^{23}\,\mathrm{mol^{-1}}$	1.7×10^{-7}
Boltzmann constant k	$1.380\,6505 \times 10^{-23}\,\mathrm{J\,K^{-1}}$	1.8×10^{-6}
	$= 8.617\,343 \times 10^{-5}\,\mathrm{eV\,K^{-1}}$	
Stefan–Boltzmann constant σ	$5.670\,400 \times 10^{-8}\,\mathrm{W\,m^{-2}\,K^{-4}}$	7×10^{-6}
electron volt, eV	$1.602\,176\,53 \times 10^{-19}\,\mathrm{J}$	8.5×10^{-8}

---

[1] see also Eidelman et al., Phys. Letters B592, 1 (2004)

		relative uncertainty
standard atmosphere, atm	101 325 Pa	exact
acceleration due to gravity g	$9.806\,65\ \text{m s}^{-2}$	exact
Hubble constant H_0	$71\ \text{km s}^{-1}\ \text{Mpc}^{-1}$	5%
age of the universe t_0	$13.7 \times 10^9\ \text{a}$	1.5%
Hubble distance c/H_0	4 225 Mpc	5%
critical density $\varrho_c = 3H_0^2/8\pi G$	$9.469 \times 10^{-30}\ \text{g/cm}^3$	$\approx 10\%$
density of galaxies	$\approx 0.02\ \text{Mpc}^{-3}$	
temperature of the blackbody radiation	2.725 K	4×10^{-4}
number density of blackbody photons	$410.4\ \text{cm}^{-3}$	1.2×10^{-3}
astronomical unit, AU	$1.495\,978\,706\,60 \times 10^{11}\ \text{m}$	1.3×10^{-10}
parsec, pc (1 AE/1 arcsec)	$3.085\,677\,5807 \times 10^{16}\ \text{m}$	1.3×10^{-10}
light-year, LY	$0.3066\,\text{pc} = 0.9461 \times 10^{16}\ \text{m}$	
Schwarzschild radius of the Sun: $2GM_\odot/c^2$	2.953 250 08 km	3.6×10^{-4}
mass of the Sun $M_\odot$	$1.988\,44 \times 10^{30}\ \text{kg}$	1.5×10^{-4}
solar constant	$1\,360\ \text{W/m}^2$	0.2%
solar luminosity $L_\odot$	$3.846 \times 10^{26}\ \text{W}$	0.2%
mass of the Earth $M_\oplus$	$5.9723 \times 10^{24}\ \text{kg}$	1.5×10^{-4}
radius of the Earth $R_\oplus$	$6.378\,140 \times 10^6\ \text{m}$	
Schwarzschild radius of the Earth: $2GM_\oplus/c^2$	0.887 056 22 cm	1.5×10^{-4}
velocity of the solar system about the galactic center	220 km/s	9%
distance of the Sun from the galactic center	8.0 kpc	6%
matter density of the universe Ω_m	0.27	18%
baryon density of the universe Ω_b	0.044	9%
dark-matter density of the universe Ω_dm	0.22	18%
radiation density of the universe Ω_γ	4.9×10^{-5}	10%
neutrino density of the universe Ω_ν	≤ 0.015	95% C.L.
dark-energy density of the universe Λ	0.73	5%
total energy density of the universe Ω_tot ($\Omega_\text{tot} \equiv \Omega$)	1.02	2%
number density of baryons n_b	$2.5 \times 10^{-7}/\text{cm}^3$	4%
baryon-to-photon ratio η	6.1×10^{-10}	3%
0°C	273.15 K	

References

"References are like signposts to other data."

Steve Cook

[1] CERN Courier, May 2004, p. 13.

[2] Particle Data Group, S. Eidelman et al., Phys. Lett. **B592** (2004) 1.

[3] Claus Grupen, *Particle Detectors*, Cambridge University Press, Cambridge, 1996.

[4] M. Kleifges and H. Gemmeke, Auger Collaboration, IEEE Conference Rome, October 2004, to be published in the Proceedings.

[5] Y. Totsuka, IEEE conference, Portland, 2003.

[6] G. R. Farrar, T. Piran, Phys. Rev. Lett. **84** (April 2000) 3527.

[7] Michael S. Turner and J. Anthony Tyson, Rev. Mod. Phys. **71** (1999) 145.

[8] Andrew Liddle, *An Introduction to Modern Cosmology*, Wiley, Chichester, 1999.

[9] Torsten Fließbach, *Allgemeine Relativitätstheorie*, Spektrum, Heidelberg, 1998.

[10] Bradley W. Carroll and Dale A. Ostlie, *An Introduction to Modern Astrophysics*, Addison-Wesley, Reading, Mass., 1996.

[11] S. Perlmutter et al., web site of the Supernova Cosmology Project, `supernova.lbl.gov`.

[12] See the web site of the Relativistic Heavy Ion Collider, `www.bnl.gov/RHIC/`.

[13] See the web site of the Alpha Magnetic Spectrometer, `ams.cern.ch`.

[14] A. G. Cohen, *CP Violation and the Origins of Matter*, Proceedings of the 29th SLAC Summer Institute, 2001.

[15] A. G. Cohen, A. De Rujula and S. L. Glashow, Astrophys. J. **495** (1998) 539–549.

[16] A. D. Sakharov, JETP Letters, **5** (1967) 24.

[17] Edward W. Kolb and Michael S. Turner, *The Early Universe*, Addison-Wesley, Reading, Mass., 1990.

[18] BBN code can be downloaded from `www-thphys.physics.ox.ac.uk/users/SubirSarkar/bbn.html`

[19] S. Burles and D. Tytler, Astrophys. J. **499** (1998) 699; **507** (1998) 732.

[20] The LEP and SLD experiments, *A Combination of Preliminary Electroweak Measurements and Constraints on the Standard Model*, CERN EP/2000-016 (2000).

[21] D. Denegri, B. Sadoulet and M. Spiro, Rev. Mod. Phys. **62** (1990) 1.

[22] G. Gamow, Phys. Rev. **74** (1948) 505.

[23] Ralph A. Alpher and Robert C. Herman, Phys. Rev. **75** (1949) 1089.

[24] A. A. Penzias and R. W. Wilson, Astrophys. J. **142** (1965) 419.

[25] R. H. Dicke, P. J. E. Peebles, P. G. Roll and D. T. Wilkinson, Astrophy. J. **142** (1965) 414.

[26] D. J. Fixsen et al., Astrophys. J. **473** (1996) 576;
    `http://lambda.gsfc.nasa.gov/product/cobe/`
       `firas_overview.cfm.`

[27] C. L. Bennett et al., Astrophy. J. **464** (1996) L1–L4; web site of the COBE satellite,
    `http://lambda.gsfc.nasa.gov/product/cobe/.`

[28] Web site of the Wilkinson Microwave Anisotropy Probe, `map.gsfc.nasa.gov.`

[29] G. Hinshaw et al., Astrophys. J. Suppl. **148** (2003) 135; `map.gsfc.nasa.gov.`

[30] Barbara Ryden, *Introduction to Cosmology*, Addison-Wesley, San Francisco, 2003.

[31] Wayne Hu, `background.uchicago.edu/~whu`; see also W. Hu and S. Dodelson, Ann. Rev. Astron. Astrophys. **40** (2002) 171, astro-ph/0110414.

[32] U. Seljak and M. Zaldarriaga, Astrophy. J. **469** (1996) 437–444; code for the program `CMBFAST` is available from `www.cmbfast.org.`

[33] CERN Courier July/August 2002; A. C. S. Readhead et al., astro-ph/0402359 (2004).

[34] Web site of the PLANCK experiment, `astro.estec.esa.nl/Planck.`

[35] Milton Abramowitz ans Irene Stegun, *Handbook of Mathematical Functions*, 9th printig, Dover Publ., New York, 1970.

[36] I. S. Gradstein and I. M. Ryshik, *Tables of Series, Products, and Integrals*, Harri Deutsch, Thun, 1981.

[37] B. Cabrera, Phys. Rev. Lett. **48** (1982) 1378.

[38] John A. Peacock, *Cosmological Physics*, Cambridge University Press, p. 329, Cambridge 2003.

[39] A. H. Guth, Phys. Rev. D **23** (1981) 347.

[40] A. D. Linde, Phys. Lett. **108B** (1982) 389.

[41] A. Albrecht and P. J. Steinhardt, Phys. Rev. Lett. **48** (1982) 1220.

[42] M. Tegmark et al., Astrophys. J. **606** (2004) 702–740.

[43] Alan H. Guth, *Inflation* in Carnegie Observatories Astrophysics Series, Vol. 2, *Measuring and Modeling the Universe*, ed. W. L. Freedman, Cambridge University Press, 2004; astro-ph/0404546.

[44] F. Zwicky, Astrophys. J. **86** (1937) 217.

[45] `www.hep.physik.uni-siegen.de/~grupen/astro/pot2drot.pdf`

[46] George B. Arfken and Hans-Jürgen Weber, *Mathematical Methods for Physicists*, 4th edition, Academic Press, New York (1995).

[47] Tony Guénault, *Statistical Physics*, 2nd edition, Kluwer Academic Publishers, Secaucus, NJ, 1995.

Further Reading

"Education is the best provision of old age."

Aristotle

Claus Grupen, *Astroteilchenphysik: das Universum im Licht der kosmischen Strahlung*, Vieweg, Braunschweig, 2000.

Lars Bergström and Ariel Goobar, *Cosmology and Particle Astrophysics*, Wiley, Chichester, 1999.

H. V. Klapdor-Kleingrothaus and K. Zuber, *Particle Astrophysics*, IOP, 1997.

G. Börner, *The Early Universe: Facts and Fiction*, Springer, 1988.

S. Weinberg, *The First Three Minutes*, BasicBooks, New York, 1993.

Alan Guth, *The Inflationary Universe*, Vintage, Vancouver, USA, 1998.

G. Cowan, *Lecture Notes on Particle Physics*, RHUL Physics Dept. course notes for PH3520, 2002.

Web site of the European Organisation for Nuclear Research (CERN): www.cern.ch.

H. Georgi and S. L. Glashow, Phys. Rev. Lett. **32** (1974) 438.

Web site of the Super-Kamiokande experiment:
www-sk.icrr.u-tokyo.ac.jp/doc/sk.

M. Shiozawa et al., (Super-Kamiokande collaboration), Phys. Rev. Lett. **81** (1998) 3319.

G. Cowan, *Statistical Data Analysis*, Oxford University Press, 1998.

A. V. Filippenko and A. G. Riess, astro-ph/9807008 (1998).

T. D. Lee and C. N. Yang, Phys. Rev. **104** (1956) 254.

C. S. Wu et al., Phys. Rev. **105** (1957) 1413.

J. H. Christenson, J. W. Cronin, V. L. Fitch and R. Turlay, Phys. Rev. Lett. **13** (1964) 138.

Harold P. Furth, *Perils of Modern Living*, originally published in the New Yorker Magazine, 1956.

Ken Nollett, www.phys.washington.edu/~nollett.

David N. Schramm and Michael S. Turner, Rev. Mod. Phys. **70** (1998) 303.

Yuri I. Izotov and Trinh X. Thuan, *Heavy-element abundances in blue compact galaxies*, Astrophys. J. **511** (1999) 639–659.

Scott Burles, Kenneth M. Nollet, James N. Truran and Michael S. Turner, Phys. Rev. Lett. **82** (1999) 4176; astro-ph/9901157.

Web site of the ALEPH collaboration: `alephwww.cern.ch`, and the ALEPH public pages, `alephwww.cern.ch/Public.html`.

C. L. Bennett et al., Astrophys. J. Suppl. **148** (2003) 1; `map.gsfc.nasa.gov`.

Keith A. Olive, Primordial Big Bang Nucleosynthesis, astro-ph/9901231 (1999).

R. Hagedorn, *Relativistic Kinematics*, W. A. Benjamin Inc. Reading, Mass., 1963.

Otto C. Allkofer, Fortschr. d. Physik **15**, 113–196, 1967.

S. Hayakawa, *Cosmic Ray Physics*, Wiley-Interscience, New York, 1969.

Albrecht Unsöld, *Der Neue Kosmos*, Springer, Heidelberg, 1974.

Otto C. Allkofer, *Introduction to Cosmic Radiation*, Thiemig, München, 1975.

D. J. Adams, *Cosmic X-ray Astronomy*, Adam Hilger Ltd., Bristol, 1980.

Joseph Silk, *The Big Bang – The Creation and Evolution of the Universe*, Freeman, New York, 1980.

Frank H. Shu, *The Physical Universe: An Introduction to Astronomy*, Univ. Science Books, Mill Valley, California, 1982.

Rodney Hillier, *Gamma-Ray Astronomy*, Oxford Studies in Physics, Clarendon Press, Oxford 1984.

O. C. Allkofer, P. K. F. Grieder, *Cosmic Rays on Earth*, Fachinformationszentrum Karlsruhe, 1984.

Hans Schäfer, *Elektromagnetische Strahlung – Informationen aus dem Weltall*, Vieweg, Wiesbaden, 1985.

W. D. Arnett et al., *Supernova 1987A*, Ann. Rev. Astron. Astrophysics **27** (1989) 629–700.

Thomas K. Gaisser, *Cosmic Rays and Particle Physics*, Cambridge University Press, Cambridge, 1990.

John Gribbin, *Auf der Suche nach dem Omega-Punkt*, Piper, München, 1990.

Martin Pohl, Reinhard Schlickeiser, *Exoten im Gamma-Licht*, Astronomie heute, Bild der Wissenschaft **3**, S. 92/93, 1993.

Neil Gehrels et al., *The Compton Gamma-Ray Observatory*, Scientific American, 38–45, Dec. 1993.

Charles D. Dermer, Reinhard Schlickeiser, Astrophys. J. **416** (1993) 458–484.

J. W. Rohlf, *Modern Physics from α to Z^0*, J. Wiley & Sons, New York, 1994.

T. Kifune, *The Prospects for Very High Energy Gamma-Ray Astronomy*, Inst. f. Cosmic Ray Research, Univ. Tokyo, ICRR-326-94-21, 1994.

H. V. Klapdor-Kleingrothaus, A. Staudt, *Teilchenphysik ohne Beschleuniger*, Teubner, Stuttgart, 1995.

Kitty Ferguson, *Das Universum des Stephen Hawking*, Econ, Düsseldorf, 1995.

Ulf Borgeest, Karl-Jochen Schramm, *Bilder von Gravitationslinsen*, Sterne und Weltraum, Nr. 1, 1995, S. 24–31.

Peter L. Biermann, *The Origin of Cosmic Rays*, MPI Radioastronomie Bonn, MPIfR 605, astro-ph/9501003, Phys. Rev. **D51** (1995) 3450.

Daniel Vignaud, *Solar and Supernovae Neutrinos*, 4th School on Non-Accelerator Particle Astrophysics, Trieste 1995, DAPNIA/SPP 95-26, Saclay.

Thomas K. Gaisser, Francis Halzen, Todor Stanev, Phys. Rep. **258**, No. 3, 1995.

Lexikon der Astronomie, Spektrum-Verlag, Heidelberg, 1995.

S. P. Plunkett et al., Astrophys. Space Sci. **231** (1995) 271; astro-ph/9508083, 1995.

Mordehai Milgrom, Vladimir Usov, Astrophys. J. **449** (1995) 37; astro-ph/9505009, 1995.

V. Berezinsky, *High Energy Neutrino Astronomy*, Gran Sasso Lab. INFN, LNGS-95/04, 1995.

John V. Jelley, Trevor C. Weekes, *Ground-Based Gamma-Ray Astronomy*, Sky & Telescope, Sept. 95, 20–24, 1995.

Volker Schönfelder, *Exotische Astronomie mit dem Compton-Observatorium*, Physik in unserer Zeit **26**, 262–271, 1995.

A. C. Melissinos, *Lecture Notes on Particle Astrophysics*, Univ. of Rochester 1995, UR-1841, Sept. 1996.

Kitty Ferguson, *Gottes Freiheit und die Gesetze der Schöpfung*, Econ, Düsseldorf, 1996.

Kitty Ferguson, *Prisons of Light – Black Holes*, Cambridge University Press, Cambridge, 1996.

Claus Grupen, *Particle Detectors*, Cambridge University Press, Cambridge, 1996.

Michel Spiro, Eric Aubourg, *Experimental Particle Astrophysics*, Int. Conf. on High Energy Physics, Varsovie (Pologne), 1996.

Karl Mannheim, Dieter Hartmann, Burkhardt Funk, Astrophys. J. **467** (1996) 532–536; astro-ph/9605108, 1996.

G. Battistoni, O. Palamara, *Physics and Astrophysics with Multiple Muons*, Gran Sasso Lab. INFN/AE 96/19, 1996.

John Ellis, Nucl. Phys. Proc. Suppl. **48** (1996) 522–544, CERN-TH/96-10, astro-ph/9602077, 1996.

H. V. Klapdor-Kleingrothaus, K. Zuber, *Teilchenastrophysik*, Teubner, Stuttgart, 1997, and *Particle Astrophysics*, Inst. of Physics Publ., 2000.

Georg G. Raffelt, *Astro-Particle Physics*, Europhysics Conference on High-Energy Physics, August 1997, Jerusalem, hep-ph/9712548, 1997.

Francis Halzen, *The Search for the Source of the Highest Energy Cosmic Rays*, Univ. of Madison Preprint MAD-PH 97-990, astro-ph/9704020, 1997.

G. Burdman, F. Halzen, R. Gandhi, Phys. Lett. **B417** (1998) 107-113, Univ. of Madison Preprint MAD-PH 97-1014, hep-ph/9709399, 1997.

Paolo Lipari, *Cosmology, Particle Physics and High Energy Astrophysics*, Conf. Proc. Frontier Objects in Astrophysics and Particle Physics, F. Giovannelli and G. Mannocchi (Eds.), Vol 57, p. 595, 1997.

Georg G. Raffelt, *Dark Matter: Motivation, Candidates and Searches*, Proc. 1997 European School of High Energy Physics, Menstrup, Denmark, hep-ph/9712538, 1997.

C. N. de Marzo, Nucl. Phys. Proc. Suppl. **70** (1999) 515–517, physics/9712039, 1997.

Hinrich Meyer, *Photons from the Universe: New Frontiers in Astronomy*, Univ. Wuppertal, Germany, hep-ph/9710362, 1997.

Particle Data Group (S. Eidelman et al.), *Review of Particle Physics*, Phys. Lett. **B592** (2004) 1–1109.

R. K. Bock, A. Vasilescu, *The Particle Detector Briefbook*, Springer, Berlin, Heidelberg, 1998.

Jonathan Allday, *Quarks, Leptons and the Big Bang*, Inst. of Physics Publ., Bristol, 1998.

Craig J. Hogan, *The Little Book of the Big Bang*, Copernicus, Springer, New York, 1998.

Lawrence M. Krauss, *A New Cosmological Paradigm: The Cosmological Constant and Dark Matter*, hep-ph/9807376, Case Western Reserve University, Cleveland, 1998.

Francis Halzen, *Ice Fishing for Neutrinos*, AMANDA Homepage,
    http://amanda.berkeley.edu/www/ice-fishing.html.

Jochen Greiner, *Gamma-Ray Bursts: Old and New*, Astroph. Inst. Potsdam, astro-ph/9802222, 1998.

Volker Schönfelder, *Gammastrahlung aus dem Kosmos*, Phys. Bl. **54**, 325–330, 1998.

K. S. Capelle et al., Astropart. Phys. **8** (1998) 321–328, astro-ph/9801313, 1998.

Laura Whitlock, *Gamma-Ray Bursts*,
    http://imagine.gsfc.nasa.gov/docs/science/know_l1/
    burst.html.

Glennys R. Farrar, *Can Ultra High Energy Cosmic Rays be Evidence for New Particle Physics?*, Rutgers Univ., USA, astro-ph/9801020, 1998.

Super-Kamiokande Collaboration, Phys. Lett. **B433** (1998) 9–18, hep-ex/9803006, 1998.

M. S. Turner, Publ. Astron. Soc. Pac. **111** (1999) 264–273, astro-ph/9811364, 1998.

J. P. Henry, U. G. Briel, H. Böhringer, *Die Entwicklung von Galaxienhaufen*, Spektrum der Wissenschaft, 2/1999, S. 64–69.

Guido Drexlin, *Neutrino-Oszillationen*, Phys. Bl., Heft 2, S. 25–31, 1999.

G. Veneziano, *Challenging the Big Bang: A Longer History of Time*, CERN-Courier, March 1999, p. 18.

M. A. Bucher, D. N. Spergel, *Was vor dem Urknall geschah*, Spektrum der Wissenschaft, 3/1999, S. 55.

C. J. Hogan, R. P. Kirshner, N. B. Suntzeff, *Die Vermessung der Raumzeit mit Supernovae*, Spektrum der Wissenschaft, 3/1999, S. 40.

L. M. Krauss, *Neuer Auftrieb für ein beschleunigtes Universum*, Spektrum der Wissenschaft, 3/1999, S. 47.

J. Trümper, *ROSAT und seine Nachfolger*, Phys. Bl. 55/9, S. 45, 1999.

Craig J. Hogan, Rev. Mod. Phys. **72** (2000) 1149–1161.

H. Blümer, *Die höchsten Energien im Universum*, Physik in unserer Zeit, S. 234–239, Nov. 1999.

Brian Greene, *"The Elegant Universe"*, Vintage 2000, London, 1999.

H. Völk, H. Blümer & K.-H. Kampert, H. Krawczynski et. al., Ch. Spiering, M. Simon, J. Jochum & F. von Feilitzsch: Phys. Bl., Schwerpunkt Astroteilchenphysik, März 2000, S. 35–68.

Paul J. Steinhardt, *"Quintessential Cosmology and Cosmic Acceleration"*, Selection of review papers under the heading *"Dark Energy and the Cosmological Constant"* http://pancake.uchicago.edu/~carroll/reviewarticles.html.

Peter K. F. Grieder, *Cosmic Rays at Earth*, Elsevier Science, 2001.

S. Hayakawa, *Cosmic Ray Physics*, Wiley Interscience, N.Y., 1969.

K. C. Cole, *The Hole in the Universe*, The Harvest Book/Harcourt Inc., San Diego, 2001.

Brian Greene, *The Fabric of the Cosmos: Space, Time and the Texture of Reality*, Alfred A. Knopf, N.Y., 2004.

John A. Bahcall, *Neutrino Astronomy*, Cambridge University Press, 1990.

Rodney Hillier, *Gamma Ray Astronomy*, Oxford Studies in Physics, 1984.

John A. Peacock, *Cosmological Physics*, Cambridge University Press, 2003.

Martin Rees, *Just Six Numbers: The Deep Forces That Shape the Universe*, Basic Books, New York, 2001.

Martin Rees, *New Perspectives in Astrophysical Cosmology*, Cambridge University Press, 2000.

S. M. Bilenky, *The History of Neutrino Oscillations*, hep-ph/0410090, 2004.

A. V. Filippenko and A. G. Riess, Phys. Rep. **307** (1999) 31.

Photo Credits

"Equipped with his five senses, man explores the universe around him and calls the adventure science."

Edwin Powell Hubble

{1} T. Credner and S. Kohle, University of Bonn, Calar Alto Observatory

{2} Prof. Dr. D. Kuhn, University of Innsbruck, Austria

{3} Courtesy of The Archives, California Institute of Technology, Photo ID 1.22-5

{4} C. Butler and G. Rochester, Manchester University

{5} Prof. Dr. A. A. Penzias, Bell Labs, USA

{6} D. Malin, Anglo-Australian Observatory

{7} Particle Data Group, European Physical Journal C3, (1998), 144;
http://pdg.lbl.gov

{8} Prof. Dr. Y. Totsuka, Institute for Cosmic Ray Research, Tokyo, Japan

{9} T. Kajita, Y. Totsuka, Rev. Mod. Phys. 73 (2001) p. 85

{10} Y. Fukuda et al., Phys. Rev. Lett. 81 (1998) 1562–1567

{11} Prof. Dr. R. Davis, Brookhaven National Laboratory, USA

{12} R. Svoboda and K. Gordan (Louisiana State University),
http://antwrp.gsfc.nasa.gov/apod/ap980605.html.
Equivalent figure in: A. B. McDonald et al., Rev. Sci. Instrum. 75 (2004) 293–316,
astro-ph/0311343

{13} AMANDA Collaboration, Christian Spiering, private communication 2004

{14} Dr. D. Heck, Dr. J. Knapp, Forschungszentrum Karlsruhe, Germany

{15} Cangaroo Collaboration, Woomera, Australia

{16} CGRO Project Scientist Dr. Neil Gehrels, Goddard Space Flight Center, USA (1999)

{17} BATSE-Team, NASA, M. S. Briggs,
(http://www.batse.msfc.nasa.gov/batse/grb/skymap/)

{18} Prof. Dr. J. Trümper, Max-Planck Institute for Extraterrestrial Physics, München, Garching

{19} ESA/XMM-Newton,
http://xmm.vilspa.esa.es/external/xmm_science/
gallery/public
courtesy by L. Strüder

{20} ESA/XMM-Newton/Patrick Henry et al., courtesy by L. Strüder

{21} Prof. Dr. M. Simon, University of Siegen, Germany

{22} R. Tcaciuc, University of Siegen, 2004

{23} ALEPH Collaboration, CERN, courtesy by P. Dornan, Imperial College

{24} Frejus-Experiment, France/Italy, courtesy by Chr. Berger, RWTH Aachen

{25} Prof. Dr. P. Sokolsky, University of Utah, Salt Lake City, USA

{26} LBNL-53543, R. A. Knop et al., Astrophys. J. **598** (2003) 102–137;
      `http://supernova.lbl.gov/`, courtesy by Saul Perlmutter

{27} Courtesy of the COBE Science Working Group

{28} Courtesy of the WMAP Science Team

{29} Dr. L. J. King, University of Manchester, UK; NASA; King et al. 1998, MNRAS 295, L41

{30} Prof. Dr. G. G. Raffelt, Max-Planck Institute for Physics (Werner-Heisenberg Institute), München, Germany

Index*

"A document without an index is like a country without a map."

Mike Unwalla

---

* Underlined page numbers refer to main entries. Page numbers in italics apply to the glossary.